Natural Products
from
PLANTS

Natural Products
from
PLANTS

Peter B. Kaufman • Leland J. Cseke
Sara Warber • James A. Duke
Harry L. Brielmann

CRC

CRC Press

Boca Raton   Boston   London   New York   Washington, D.C.

Acquiring Editor:	Harvey Kane
Project Editor:	Maggie Mogck
Marketing Manager:	Becky McEldowney
Cover design:	Denise Craig

Library of Congress Cataloging-in-Publication Data

Natural products from plants / Peter B. Kaufman . . . [et al.].
     p. cm.
    Includes bibliographical references and indexes.
    ISBN 0-8493-3134-X (alk. paper)
    1. Botanical chemistry. 2. Plant products. I. Kaufman, Peter B.
QK861.N38 1998
581.6'3—dc21

97-49038
CIP

No claim to original U.S. Government works
International Standard Book Number 0-8493-3134-X
Library of Congress Card Number 97-49038
Printed in the United States of America 1 2 3 4 5 6 7 8 9 0
Printed on acid-free paper

Preface

We have taught many undergraduate and graduate courses about natural products in plants, i.e., Plant Biotechnology, Practical Botany, and Plants, People, and Environment. As a result, it became apparent there was a need for a comprehensive, thorough collection of information regarding exactly what kinds of natural products plants produce and why they produce them. Currently, such information is contained within thousands of somewhat disjointed reports about the helpful qualities and toxic effects of different plant species throughout the world. The aim of this book is to help bring more unity and understanding to this complicated and often contradictory jumble of information. Such a book is a necessity for many, including: biochemists, natural product chemists, and molecular biologists; research investigators in industry, federal labs, and universities; physicians, nurses, and nurse practitioners; pre-medical and medical students; ethnobotanists, ecologists, and conservationists; nutritionists; organic gardeners and farmers; and those interested in herbs and herbal medicine. With the growing interest in this field by professionals and the general public alike, it was important for us to produce a book that encompasses as much information as possible on the natural products produced by plants as well as their importance in today's world. We hope that this book helps to meet this need.

Our most compelling reasons for writing a book on natural products in plants include the following:

- There is a general lack of knowledge and much misinformation about natural products in plants and their uses by people.
- Many of the natural products in plants of medicinal value offer us new sources of drugs which have been used effectively for centuries in traditional medicine. There are many compounds used in medicine today whose original derivatives were of plant origin.
- Plants are sources of poisons, addictive drugs, and hallucinogens. These have importance in human medicine and in human social action and behavior.
- Many people are interested in using natural products from plants for preventive medicine, but these people must be made aware of potential harmful effects of such compounds.
- Plants provide us with thousands of novel compounds which give us fragrances, flavorings, dyes, fibers, foods, beverages, building materials, heavy metal chelators (important in bioremediation), biocides, and plant growth regulators.

- Knowledge of how and why plants produce such a vast array of metabolites gives us new insights into how plants use these compounds to deter predators and pathogens, attract and deter pollinators, prevent other plants from competing with themselves for the same resources, and defend themselves against environmental stress.

This book has been organized in order to provide relevant and practical information on each of the above topics. Chapter 1 provides a discussion of the various types of compounds found in plants. We then, in Chapter 2, discuss how and why these compounds are made by plants. In Chapter 3, we consider how the synthesis of these compounds is regulated by environmental stresses, biotic factors, biochemical regulators, and gene expression to provide a better understanding of how these compounds benefit the plants themselves. Chapter 4 provides information about the good and bad uses by humans of natural products from plants. In Chapter 5, using classic examples from medicine and cell biology, we discuss the modes of action of natural products at target sites. The principle of synergy between separate kinds of compounds from a single plant source and from more than one plant source is discussed in Chapter 6. Chapter 7 is about the ways natural products are extracted from plants, how they are analyzed quantitatively, as well as how they are identified by physical chemical methods. Classical cases of traditional uses of natural products from plants by humans are presented in Chapter 8. Finally, Chapter 9 takes a global view of various strategies that are used to conserve plants that produce natural products of value to humans.

The organization of individual chapters of this book are according to a format we hope will aid the reader in comprehension of the material and stimulate one to probe the chapter topics further. The Appendix to this book helps one to embark on the latter endeavor.

Regarding terminology pertaining to plant metabolites, we shall refrain from using the terms "primary and secondary metabolites". These labels have caused a lot of confusion in the literature and their continued use certainly cannot be defended on chemical grounds. So-called "primary metabolites" have referred to those compounds that produce energy, such as adenosine triphosphate. So-called "secondary metabolites" have referred to those compounds synthesized by plants that do not produce energy. These ideas, in our view, are obsolete and not useful. Why? Because many of the compounds/metabolites considered to be "secondary" are really essential for carbon fixation and reduction through photosynthesis, glycolysis, fermentation, and the tricarboxylic acid cycle. Also, many of the metabolites that have been classified as "secondary" are really essential to the survival of the plant at particular times in its developmental life cycle. So, we shall abandon the older terminology of "primary and secondary metabolites" and simply use the terms metabolite and product for any of the compounds that plants synthesize because, as far as we can ascertain, all have some survival value to the plant in both time and space.

Authors

Peter B. Kaufman, Ph.D., is a professor of Biology in the Plant Molecular, Cellular, and Developmental Biology Group of the Biology Department and a member of the faculty of the Biomedical Engineering Program at the University of Michigan, Ann Arbor.

He received his B.Sc. in Plant Science from Cornell University in Ithaca, NY in 1949 and his Ph.D. in Plant Biology from the University of California, Davis, in 1954.

Dr. Kaufman is a Fellow of the American Association for the Advancement of Science and received the Distinguished Service Award from the American Society for Gravitational and Space Biology in 1995. He served on the Editorial Board of Plant Physiology for 10 years and is the author of more than 192 research papers. He has published eight professional books to date and teaches a popular course on Plants, People, and the Environment in the Residential College at the University of Michigan. He has received research grants from the National Science Foundation, the National Aeronautics and Space Administration, the U.S. Department of Agriculture BARD Program with Israel, and Parke-Davis/Warner Lambert Pharmaceutical Research Laboratories in Ann Arbor.

Leland J. Cseke, M.Sc., is a Ph.D. candidate in the Plant Molecular, Cellular, and Developmental Biology Program at the University of Michigan's Biology Department. He received his B.Sc. degree from Michigan Technological University specializing in biochemistry, and his M.Sc. degree in cellular and molecular biology from the University of Michigan in 1994.

Cseke has presented his research findings on the physiological effects of stress on secondary metabolite production at several scientific meetings. He has been a graduate student instructor in many university courses including several focusing on plant biotechnology, and he is a co-author of the *Handbook of Molecular and Cellular Methods in Biology and Medicine* from CRC Press. Cseke is currently finishing his dissertation on the molecular evolution of monoterpene sythases in plants with special focus on enzymes that produce floral scent and their impact on pollinator attraction.

Sara Warber, M.D., is a family physician with a long-standing interest in botanical medicine which predates her entrance into medical school. She completed a combined residency and fellowship in family medicine at the University of Michigan, and is currently a Robert Woods Johnson Clinical Scholars Program Fellow at the University. Her interests include research into the safe and efficacious use of herbal medicines. She is collaborating on research and education related to the use of other complimentary and alternative modalities in the optimization of health.

In addition, Dr. Warber is designing community-oriented research to facilitate improved Native American health through better understanding of cultural dimensions and traditional ways of healing. She lives with her husband and two sons in Ann Arbor, MI, and enjoys spending time in the many remaining wild habitats surrounding the Great Lakes.

James A. Duke, Ph.D., recently retired from the USDA where he served as an economic botanist for 30 years. In retirement he is serving as adviser with the American Botanical Council, Herbalife, Microbotanica, Nature's Herbs, and conducting ecotours in Belize, Costa Rica, Kenya, Peru, and Tanzania, where he teaches continuing education courses in medical botany. He is the author of over 20 books, the most recent being the best seller, *The Green Pharmacy.*

Dr. Duke graduated Phi Beta Kappa from the University of North Carolina in 1961. Before joining the USDA, he spent several years in Central and South America studying neotropical ethnobotany and living with various ethnic groups while closely observing their deep dependence on forest products. He is very interested in natural foods and nutritional approaches to preventive medicine and spent two years advising the Designer Food Program at the National Institutes of Health. He is a popular lecturer on the subjects of ethnobotany, herbs, medicinal plants, and new crops and their ecology, and has taped dozens of TV and radio shows.

Dr. Duke belongs to a number of organizations including the American Botanical Council, the International Society for Tropical Root Crops, the Society for Conservation Biology, and the Washington Academy of Sciences.

Harry L. Brielmann, Jr., Ph.D., obtained his B.Sc. degree in chemistry from the University of Connecticut in 1981. From 1981 to 1983 he was a research chemist for The American Health Foundation in Valhalla, NY. He held a similar position with Ensign-Bickford Industries, Inc. in Simsbury, CT between 1983 and 1987. Dr. Brielmann obtained his Master's degree in chemistry from the University of Massachusetts in Boston in 1989 and his Ph.D. in chemistry from Wesleyan University in Connecticut in 1994.

He has had three postdoctoral research appointments — one at the University of Hawaii in marine natural products chemistry; one at the University of Michigan in collaboration with Parke-Davis/Warner Lambert Pharmaceutical Research Laboratory on discovery of natural products of medicinal value from plants from the Great Lakes region; and one at Wayne State University's Department of Chemistry in organic synthesis chemistry. He currently holds a research position involving combinatorial chemistry with a biotechnology company in Connecticut.

Contributors

Mary Jo Bogenschutz-Godwin
Tropical Reigns
Division of Bogenschutz-Godwin, Inc.
987 St. Margarets Drive
Annapolis, MD 21401

James E. Hoyt, M.Sc.
Computer Systems Specialist,
  Technical Services
University Library, University of
  Michigan
Ann Arbor, MI 48109

Keewaydinoquay (Kee) Peschel
Ojibwa Medicine Woman
Leland, MI

Barbara J. Madsen, Ph.D.
Department of Biology
University of Michigan
Ann Arbor, MI 48109-1048

Maureen McKenzie, Ph.D.
CEO Naniquah Corporation
PO Box 4
Girdwood, AK 99587

Akiro Okubo, Ph.D.
Department of Applied Biological
  Chemistry
Division of Agriculture and Agricultural
  Life Sciences
University of Toyko
Bunkyo-ku, Tokyo 113, Japan

Undergraduate students from
the Independent Studies in
  Biology and Undergradate Research
  Opportunity Program (UROP),
  Department of Biology, University of
  Michigan, Ann Arbor:
Stephanie Bergman,
Donald DeSander,
Ashish Goyal,
Stacie Klein,
Lynn Pennacchini,
Matthew Perry,
Atul Rustgi, and
Kathryn Timberlake

Acknowledgments

We thank the following individuals who have assisted us or provided us with information in the preparation of this book.

- UROP (Undergraduate Research Opportunity Program) students at the University of Michigan who worked on the case studies with individual plants and the important natural products that they produce.
- Mark Plummer, Ph.D., research chemist at Parke-Davis Pharmaceutical Research Laboratories in Ann Arbor, MI, who provided us with information on the bioassays used to test plant extracts for medicinal properties.
- Keewaydinoquay Peschel, Ojibwa Medicine woman, from whom we have learned a great deal about the plants she uses to treat people in her Ojibwa band who have medical problems (see essay about her life story in Chapter 8).
- Tracy Moore, President of Xylomed Research, Inc., who has provided us with a great deal of information about the tree of joy, *Camptotheca accuminata*, and its use as a source of camptothecin used to treat patients with prostate cancer.
- Maureen McKenzie, Ph.D., CEO of Naniquah Corporation, Girdwood, AK and Michael Parks, President of Naniquah Corp., who work with the Aleutian Amerinds in Alaska in connection with natural products of medicinal value used by these peoples in their medicine. Dr. McKenzie also contributed the excellent essay on Naniquah Corporation in Chapter 9 and both she and Michael Parks contributed the figures representating present-day Amerinds of Alaska, some of the diverse ecosystems of Alaska, and plants of medicinal importance to the Amerinds. Dr. McKenzie provided the resource material on the use of aconite poison in whaling in Alaska and Asia, as documented in Chapter 4.
- David Bay, photographer in the Department of Biology at the University of Michigan, who did much of the photographic work needed for the illustrations used in this book.
- Laura Olsen, Ph.D., Assistant Professor in the Department of Biology at the University of Michigan, who provided us with helpful information on cellular organelles and membrane systems and on plant peroxysomes and glyoxysomes.
- James Hoyt, M.Sc., Computer Systems Specialist, University Library, University of Michigan, who provided invaluable information on computer databases for the Appendix of this book, made excellent editorial suggestions for Chapters 2, 3, and 4, and contributed a case study on the neem tree and its natural products to Chapter 8.

- Casey Lu, Ph.D., Assistant Professor of Biology, and Mike Messler, Ph.D., and Professor of Biology at Humboldt State University, Arcata, CA, who provided important information and photographs on collecting medicinal plants in the field and for the excellent photograph of a cut redwood tree to illustrate lignin (see Figure 2.3).
- Eldon Newcomb, Ph.D., Professor of Botany at the University of Wisconsin, Madison, WI, who provided the high-quality transmission electron micrographs of plant cells and their organelles in Chapter 2.
- Arlene O'Sullivan, Word Processing Operator III, Department of Biology, University of Michigan, Ann Arbor, MI, who prepared the computer versions of the diagrams in Figures 2.1 and 3.1.

Table of Contents

Dedication

to
Keewaydinoquay and all the unnamed others who have preserved
the knowledge of plants for future generations

1 Phytochemicals: The Chemical Components of Plants

Harry L. Brielmann, Jr.

CONTENTS

0-8493-3134-X/99/$0.00+$.50
© 1999 by CRC Press LLC

1

1.1   INTRODUCTION

Phytochemicals, as the word implies, are the individual chemicals from which plants are made. In this chapter we will look at these materials, specifically the organic components of higher plants. Numerous journals, individual books, and encyclopedic series of books have been written on this subject. The goal here is to review this area in a concise form that is easily understandable and is written with the practicing natural-products chemist in mind. The reader not familar with chemistry may be somewhat intimidated by the material presented here, but we believe that understanding the chemical composition of plants is a prequisite to the remaining topics of this book. For those interested in reviewing a specific area in greater detail, the bibliography section provides numerous references for each organic group covered.

During the course of this survey, several themes will be emphasized. In order to demonstrate the rich diversity of structures which exist, numerous examples of each type of plant natural product will be provided. Often these will be derived from common plants with which most of us are familiar. Special attention will be paid to the herbal drugs, from which numerous natural products of all types can be found. Finally, some mention will be made of marine plants, from which many truly unique individual bioactive components have been isolated.

The general categories of plant natural products are organized very broadly in terms of increasing oxidation state. This begins with the lipids, including the simple and functionalized hydrocarbons, as well as the terpenes, which are treated separately. Following this are the unsaturated natural products, including the polyacetylene and aromatic compounds. We then cross over into the realm of the primarily hydrophilic molecules including the sugars, and continue with those which can form salts, including the alkaloids, the amino acids, and the nucleosides. Overall, this scheme provides a simple organizational pattern for the phytochemicals. It is consistent with the way chemists often categorize organic chemicals in general, and roughly is equivalent to a normal-phase chromatographic analysis of a given plant species. Like any organizational scheme for this subject, be it taxonomic, phylogenetic, or biochemical, it should serve only as a rough guide.

1.2 LIPIDS

Lipids are water-insoluble biomolecules that are soluble in nonpolar solvents. This very broad definition allows for the inclusion of a variety of structural types, all of which contain a large hydrocarbon region in their structure. Biologically, lipids provide (1) the main structural components of membranes, (2) sources of fuel for storage and transport, and (3) protective surface coatings. The lipid-based cell surface components also may be involved in cell recognition, species specificity, and tissue immunity.

1.2.1 HYDROCARBONS

Comprising a relatively small group of compounds, the least polar organic natural products are the *hydrocarbons*. These aliphatic hydrocarbons usually have an odd number of carbon atoms, resulting from the decarboxylation of their fatty acid counterparts. Several representative structures are shown in Figure 1.1. Devoid of any heteroatoms, these compounds have relatively simple structures. Hydrocarbons may be either *saturated* or *unsaturated* — the latter contain multiple bonds. Each double bond results in two less hydrogen atoms relative to the saturated counterpart, and is thus in a higher oxidation state. Note that those highly branched hydrocarbons derived from isoprene can exist as hydrocarbons; however, these materials (terpenes) will be considered separately in Section 1.2.3.

Hydrocarbons

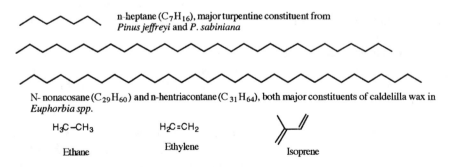

n-heptane (C_7H_{16}), major turpentine constituent from *Pinus jeffreyi* and *P. sabiniana*

N-nonacosane ($C_{29}H_{60}$) and n-hentriacontane ($C_{31}H_{64}$), both major constituents of caldelilla wax in *Euphorbia spp.*

H_3C-CH_3 $H_2C=CH_2$

Ethane   Ethylene   Isoprene

FIGURE 1.1   Hydrocarbon natural products.

1.2.1.1   Saturated Hydrocarbons

Saturated hydrocarbons are the simplest and least polar organic natural products. Common examples such as hexane $CH_3(CH_2)_4CH_3$ are not generally found in plants, but rather are derived from fossilized plant and animal matter. Turpentines, commonly used as paint removers, consist of simple hydrocarbons, particularly *n*-heptane, as found in conifers, including *Pinus jeffreyi* and *P. sabiniana*. In living plants, saturated hydrocarbons are universally distributed as the waxy coatings on leaves, and as cuticle waxes on the surfaces of fruits. Typical examples include *n*-nonacosane and hentriacontane (Figure 1.1). Several plants are rich in aliphatic hydrocarbons used in vegetable oils. For example, olive oil contains hydrocarbons ranging from C_{13} to C_{28}. Branched simple alkanes (again excluding terpenes) rarely occur in significant quantity in plants.

1.2.1.2   Unsaturated Hydrocarbons

The simplest **unsaturated hydrocarbon** is ethlyene, an important plant hormone. Larger unsaturated hydrocarbons are also common as plant waxes. Exceptionally high amounts of alkenes have been detected in rye pollen, rose petals, and sugar cane. As the chain length and degree of unsaturation increases, the hydrocarbons become waxy and then solid at room temperature. Waxes may be either long chain hydrocarbons or esters of fatty acids.

1.2.1.2.1   Polyacetylenes

Unsaturated natural products can, of course, not only contain double bonds, but triple bonds, either in the form of acetylenes or nitriles. The **polyacetylenes** are a unique group of naturally occurring hydrocarbon derivatives characterized by one or more acetylenic groups in their structures. The *sp* hybridization of the triple bond results in a linear shape for this region of the molecule. A listing of some typical polyacetylenes (Figure 1.2) shows that these molecules often contain a wide variety of additional functional groups. The domestic carrot, for example, contains four polyacetylenes, the major one being falcarinol, which is a mild neurotoxin found only to be present in 2 mg·kg^{-1} (dry weight) of carrot roots. Other plants such as the water dropwort, *Oenanthe crocata*, are commonly found near streams in the Northern Hemisphere and contain several toxic polyacetylenes and should not be consumed.

Polyacetylenes have been found to have a fairly specific distribution in plant families, existing regularly only in the Campanulacae, Asteraceae, Araliaceae, Pittosporacae, and Apiaceae. Polyacetylenes are also found in the higher fungi, where their typical chain length is from C_8 to C_{14}, whereas the polyacetylenes from higher plants are typically from 14 to 18 carbons in length.

Biosynthetically, the polyacetylenes are likely to be derived by enzymatic dehydrogenation from the corresponding olefins. The toxicity of many of the polyacetylenes including those in the aforementioned water dropwort (*Oenanthe crocata*) as well as fools parsley (*Aethusa cynapium*) may form the basis for their role in some plants. Similarly, both wyerone acid in the broad bean (*Vicia faba*) and safynol in safflower oil from *Carthamus tinctorius* have been shown to act as

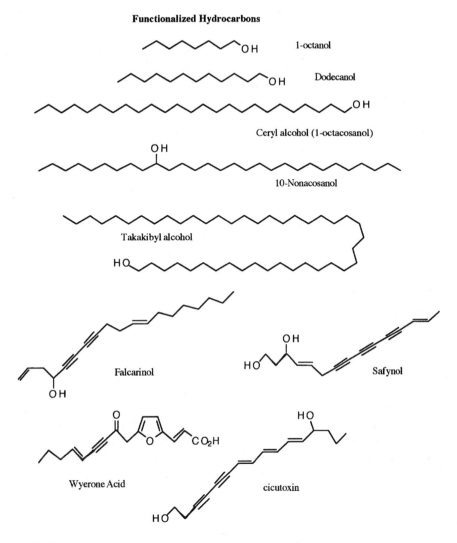

FIGURE 1.2 Some common functionalized hydrocarbon plant natural products, including polyacetylenes.

natural ***phytoalexins***, helping to deter the microorganisms which attack these plants.

1.2.2 FUNCTIONALIZED HYDROCARBONS

Excluding the lipids and the terpenes, simple ***functionalized hydrocarbons*** are less abundant but not uncommon in plants. Here, we consider these in ascending order from halide to alcohol and sulfur-containing hydrocarbons, then to aldehydes and ketones, stopping just before the hydrocarbon acids (lipids). Some typical function-alized hydrocarbons and polyacetylenes are shown in Figure 1.2.

1.2.2.1 Halogenated Hydrocarbons

Although virtually unknown among terrestrial plant natural products, the marine environment has long been recognized as a source for hydrocarbons that contain both chlorine and bromine. Hundreds of different halogenated natural products have been isolated, particularly from the red algae and the animals that feed upon them. Various species of red algae from the genus *Laurencia*, for example, have been found to contain numerous halogenated natural products, including laurinterol, spirolaurenone, and laurencin (Figure 1.3)

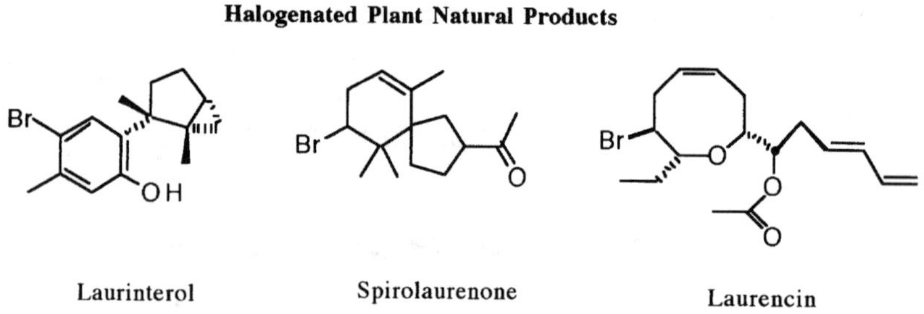

Halogenated Plant Natural Products

Laurinterol   Spirolaurenone   Laurencin

FIGURE 1.3 Halogenated plant natural products.

1.2.2.2 Alcohols

A large variety of volatile *alcohols*, e.g., aldehydes, ketones, and esters, occur in small concentrations in plants and have been classically referred to as essential oils. Their role may be related to their often strong odors, attracting them to insect pollinators and animal seed disseminators (see Chapter 2). All of the straight-chain alcohols from C_1 (wood alcohol) to C_{10} have been found in plants in either free or esterified form. Several, including ceryl alcohol, $CH_3(CH_2)_{24}CH_2OH$, a regular constituent of cuticular waxes, are shown in Figure 1.2.

1.2.2.3 Sulfides

Hydrocarbon *sulfides* are found in relatively few plants. Those that contain them, such as skunk cabbage, are readily recognizable by their obnoxious odors. Sulfides, including the simple hydrocarbon sulfides, are common in the species of *Allium*, many of which are lachrymators (substances which make eyes water) and have pungent odors. Thiophenes are limited primarily to the Asteraceae family, and are found in association with the polyacetylenes. The glucosinolates or mustard oil glucosides are also readily detected and help to create the flavors of the mustard, radish, onions, and garlic. Finally, perhaps the most common sulfur containing natural products are the amino acids cysteine and methionine.

1.2.2.4 Aldehydes and Ketones

Simple *aldehyde* and *ketone* plant natural products are uncommon. Rare examples include hentriacontan-14-16-dione ($C_{31}H_{60}O_2$), a major wax constituent of cereals and other grasses.

1.2.2.5 Esters

Esters are the condensation products of alcohols and acids. They tend to have strong and often pleasant odors. A listing of volatile ester constituents of various fruits is shown in Table 1.1.

TABLE 1.1
Volatile Ester Components of Strawberries, Apples, and Pineapples

Strawberries	Apples	Pineapples
Ethyl butyrate	Ethyl acetate	Ethyl acetate
Ethyl isovalerate	Ethyl butyrate	Methyl isocaproate
Isoamyl acetate	Ethyl valerate	Methyl isovalerate
Ethyl caproate	Propyl butyrate	Methyl caprylate
2-Hexenyl acetate		Ethyl acrylate

1.2.2.6 Fatty Acids

Fatty acids are the simplest lipids. They are compounds that are usually characterized by a polar hydrophilic head region connected to a long hydrophobic hydrocarbon tail. Some lipids, including the fats, are used for energy storage but most are used to form lipid constituents of membranes, i.e., membranes that surround cellular organelles as well as protoplasts (the plasma membrane).

There are well over a hundred different types of fatty acids, though the most common in plants are oleic and palmitic acid. The hydrocarbon chain may be saturated, as in palmitic acid, or unsaturated, as in oleic acid. Fatty acids differ from each other primarily in chain length and the locations of multiple bonds. Thus, palmitic acid (16 carbons, saturated) is symbolized 16:0 and oleic acid, which has 18 carbons with one *cis* double bond at carbon 9 is symbolized $18:1^{\Delta 9}$. Double bonds are assumed to be *cis* unless otherwise indicated. Several common fatty acids are shown in Figure 1.4.

Although fatty acids are utilized as building-block components of the saponifiable lipids, only traces occur in the free-acid form in cells and tissues. Normally these exist in various bound forms and may comprise up to 7% of the weight of dried leaves. They include long chain esters (waxes), triacylglycerols (fats), as well as glycerophospholipids and sphingolipids (membrane lipids) (Figure 1.5).

Fatty Acids

Name	Structure
Lauric	$CH_3(CH_2)_{10}CO_2H$
Myristic	$CH_3(CH_2)_{12}CO_2H$
Palmitic	$CH_3(CH_2)_{14}CO_2H$
Stearic	$CH_3(CH_2)_{16}CO_2H$
Arachidic	$CH_3(CH_2)_{18}CO_2H$
Behenic	$CH_3(CH_2)_{20}CO_2H$
Oleic	$CH_3(CH_2)_5CH=CH(CH_2)_7CO_2H$
Linoleic	$CH_3(CH_2)_4CH=CHCH_2CH=CHCH_2CH=CH(CH_2)_7CO_2H$
Arachidonic	$CH_3(CH_2)_7CH=CH(CH_2)_{11}CO_2H$

FIGURE 1.4 Some common fatty acids.

Common Lipids

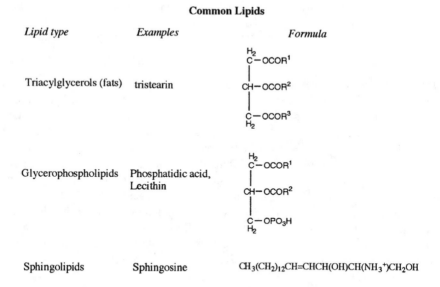

Lipid type	*Examples*	*Formula*
Triacylglycerols (fats)	tristearin	
Glycerophospholipids	Phosphatidic acid, Lecithin	
Sphingolipids	Sphingosine	$CH_3(CH_2)_{12}CH=CHCH(OH)CH(NH_3^+)CH_2OH$

FIGURE 1.5 Some common fatty acid esters (lipids).

Some generalizations can be made concerning the various fatty acids of higher plants. The most abundant have an even number of carbons ranging from C_{14} to C_{22}. Unsaturated fatty acids predominate in higher plants, with oleic acid (C_{18}) being one of the most common. Unsaturated fatty acids have lower melting points than saturated fatty acids of the same chain length.

Waxes containing polymeric esters formed by the linking of several ω-hydroxy-acids are especially prominent in the waxy coatings of conifer needles. The two most common acids in such waxes are sabinic [$HOCH_2(CH_2)_{10}CO_2H$] and juniperic acid [$HOCH_2(CH_2)_{14}CO_2H$]. The lipid constituents of cork and cuticle are known as suberin and cutin, respectively, and are composed of high molecular weight fatty acid esters.

1.2.3 TERPENES

The *terpenes* are among the most widespread and chemically diverse groups of natural products. Fortunately, despite their structural diversity, they have a simple unifying feature by which they are defined and by which they may be easily classified. Terpenes are a unique group of hydrocarbon-based natural products whose structure may be derived from isoprene, giving rise to structures which may be divided into isopentane (2-methylbutane) units (Figure 1.6).

isoprene
(2-methyl-1,3-butadiene)

isopentane
(2-methylbutane)
(isopentane)

FIGURE 1.6 The terpenes are comprised of isoprene units.

Terpenes are thus classified by the number of 5-carbon units they contain:

Hemiterpenes	C_5
Monoterpene	C_{10}
Sesquiterpene	C_{15}
Diterpene	C_{20}
Sesterterpene	C_{25} (very rare)
Triterpene	C_{30}
Tetraterpenes	C_{40}

Like all natural products, within this simple classification lies an enormous amount of structural diversity which leads to a wide variety of terpene-like (or *terpenoid*) compounds. Note that the simplest examples of the terpenes are technically hydrocarbons, though they are considered separately here because of their common structural features. Not surprisingly, the terpenes are of a similar biogenetic origin, in which isopentenyl pyrophosphate and dimethylallyl pyrophosphate combine to yield geranyl pyrophosphate, leading to monoterpenes. Similarly, compounds derived from farnesyl pyrophosphate lead to sesquiterpenes, and triterpenes are formed from two equivalents of farnesyl pyrophosphate. These various combinations and oxidations give rise to a large variety of terpenes, which will be surveyed briefly here.

The function of terpenes in plants (see Chapter 2) is generally considered to be both ecological and physiological. Many of them inhibit the growth of competing plants (allelopathy). Some are known to be insecticidal; others are found to attract insect pollinators. The plant hormone, abscissic acid, is one of the sesquiterpenes. One, gibberellic acid, is another one of the major plant hormones (over 90 gibberellins have been identified). The variety of structures that the terpenes possess is vast.

1.2.3.1   Hemiterpenes: C_5

Isoprene itself does not occur free in nature but several five-carbon compounds are known which contain the isopentane skeleton, including isoamyl alcohol, isovaleraldehyde, tiglic acid, angelic acid, and β-furoic acid. Several common plant hemiterpenes are shown in Figure 1.7.

Hemiterpenes

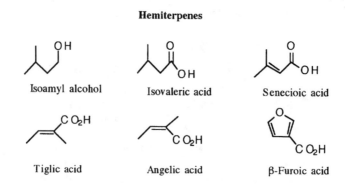

FIGURE 1.7   Some common plant hemiterpenes.

1.2.3.2   Monoterpenes: C_{10}

Nearly all possible decane arrangements appear to exist in nature. This gives the term *terpenoid* a particularly elastic meaning and is remiscent of some of the current combinatorial efforts employed in the pharmaceutical industry. The ***monoterpenoids*** are the major component of many essential oils and, as such, have economic importance as flavors and perfumes. Common aliphatic examples include myrcene, geraniol, and linalool. Open chain structures include many well-known compounds, including menthol, camphor, pinene, and limonene. A variety of common monoterpenes are shown in Figure 1.8.

Most of the monoterpenes listed in Figure 1.8 come from common sources with which most of us are familiar. ***Myrcene*** is found in the essential oil of bay leaves as well as hops. It is used as an intermediate in the manufacture of perfumes. ***Geraniol***, which is isomeric with linalool, constitutes the major part of the oil of roses and is also found in essential oils of citronella, lemon grass, and others. ***Menthol*** is a well-known monoterpene which is found in the essential oil of peppermint and other members of the mint family. ***Carvone*** is a common monoterpene which is one of the main odoriferous components of caraway seed (*Carum carvi*). ***Linalool*** is one of the principle constituents of coriander (*Coriandrum sativum*), a

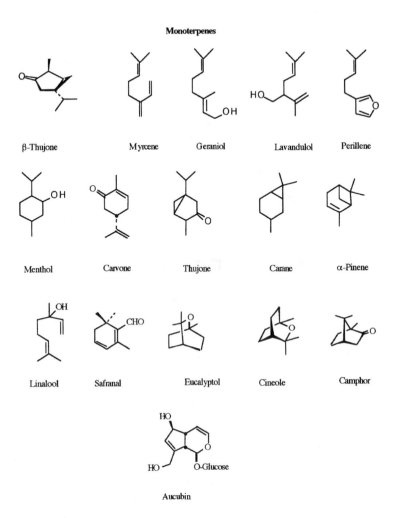

FIGURE 1.8   Some common plant monoterpenes.

common spice. *Safranal* is chiefly responsible for the characteristic odor of saffron (*Crocus sativus*). *Eucalyptol*, also known as cineole, is the main component of the essential oil of eucalyptus leaf (*Eucalyptus* spp.). Eucalyptol, along with *camphor*, form the major constituents of rosemary oil. Mullein, a common tomentose biennial, produces a number of iridoid glycosides, including *aucubin*.

1.2.3.3   Sesquiterpenes: C_{15}

Derived from three isoprene units, the C_{15} sesquiterpenes exist in a wide variety of forms, including linear, bicyclic, and tricyclic frameworks. Like the monoterpenes, most of the sesquiterpenes are considered to be essential oils because they belong to the steam distillable fraction often containing the characteristic odoriferous components of the plant. An important member of this series is *farnesol* whose pyrophosphate

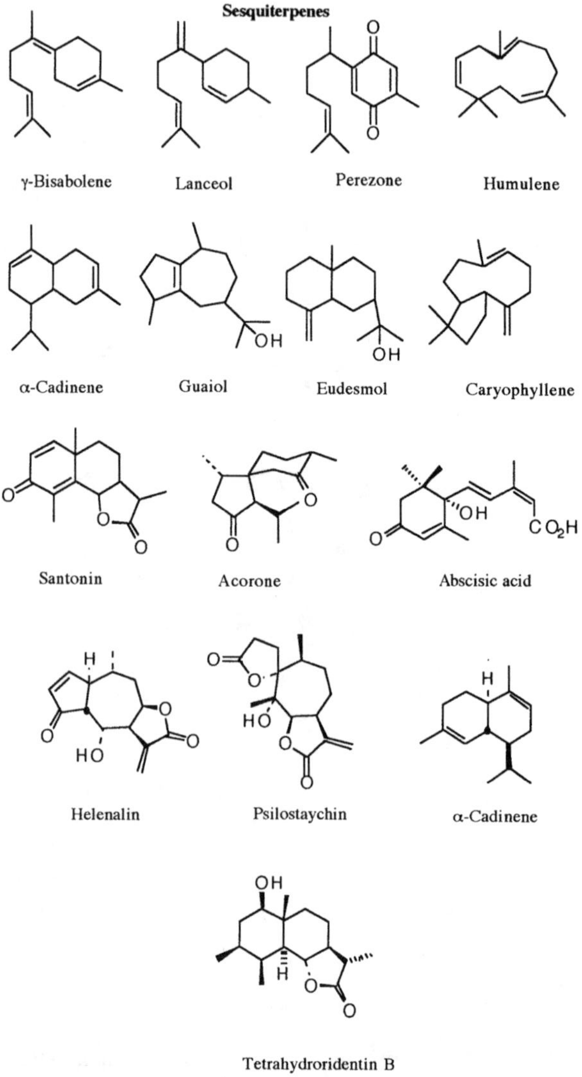

Sesquiterpenes

γ-Bisabolene   Lanceol   Perezone   Humulene

α-Cadinene   Guaiol   Eudesmol   Caryophyllene

Santonin   Acorone   Abscisic acid

Helenalin   Psilostaychin   α-Cadinene

Tetrahydroridentin B

FIGURE 1.9   Some common plant sesquiterpenes.

serves as a key intermediate in terpenoid biosynthesis (see Chapter 2). Some common sesquiterpenes are shown in Figure 1.9.

The **cadinenes** occur as essential oils from juniper and cedar trees and santonin is an antihelmintic that is isolated from wormwood (*Artemisia maritima*). **Caryophyllene**, first synthesized in 1964, is one of the principal components of oil of cloves. **Helenalin** is one of numerous pseudoguaianolide sesquiterpene lactones isolated from arnica oil (*Arnica montana*). **Psilostaychin**, an eudesmane-type sesquiterpene lactone, is one of over 100 identified constituents of mugwort (*Artemisia vulgaris*). **Acorone** is a sesquiterpene diketone present in the essential oil of sweet flag (*Acorus calamus*). The sesquiterpene α-**cadinene** is one of the more than 70 isolated components from the essential oil of juniper berries. It has been used as a diuretic and antiseptic. Finally, **tetrahydroridentin B** is one of the bitter eudes-molides unique to the common dandelion (*Taraxacum officinale*).

1.2.3.4 Diterpenes: C$_{20}$

The *diterpenes* are a widely varied group of compounds based on four isoprene groups, most of which are of limited distribution in the plant kingdom. Because of their higher boiling points, they are not considered to be essential oils. Instead, they are classically considered to be resins, the material that remains after steam distil-lation of a plant extract. The diterpenes exist in a variety of structural types (a selection is shown in Figure 1.10).

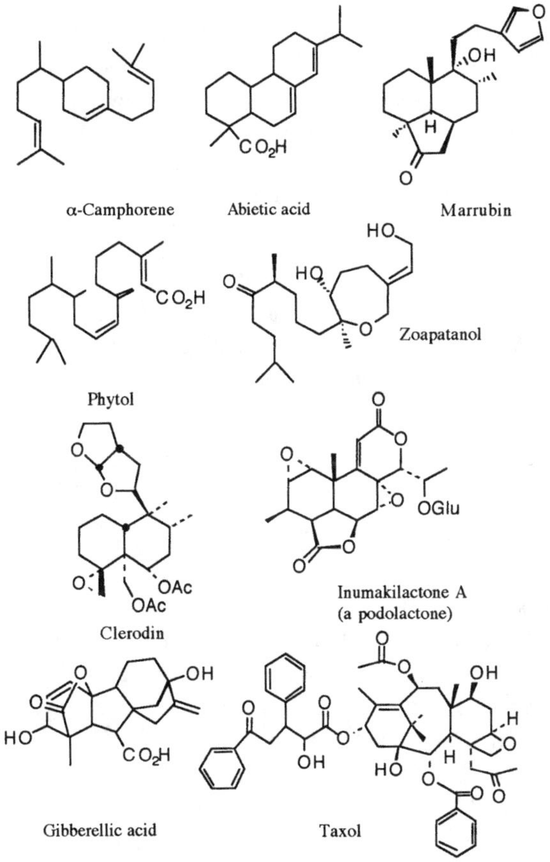

FIGURE 1.10 Some common plant diterpenes.

Many interesting examples may be mentioned. The cyclic ether **zoapatonol** is derived from the Mexican plant *Montanoa tomentosa* and has been used as an abortifacient. A number of **clerodanes** have been isolated from the *Ajuga*, *Salvia*, and *Teucrium* species, and have been found to possess insecticidal activity. A variety of cytotoxic lactones have been isolated from *Podocarpus* species. These **podolactones** have plant regulatory properties as well as antileukemic activity. The **gibberellins** comprise an important group of plant hormones. These fall into two series, including a C_{20} family represented by gibberellin A13 and a C_{19} series for which gibberellic acid is typical. **Marrubin** is a diterpene lactone from white horehound (*Marrubium vulgare*), which has been used as a bitter and choleretic in digestive and biliary complaints. **Taxol**, discovered by Wani et al. (1971) is a wholly unique antimitotic agent which binds to microtubules and stabilizes them as opposed to all other antimitotics of the tubulin-binding type, such as vincristine, the podophyllotoxins, and colchicine.

1.2.3.5   Triterpenes: C_{30}

The C_{30} terpenes are based on six isoprene units and are biosynthetically derived from squalene. They are often high-melting colorless solids and are widely distributed among plant resin, cork, and cutin. There are several important groups of triterpenes, including common triterpenes, steroids, saponins, sterolins, and cardiac glycosides. Among these is *azadirachtin*, a powerful insect antifeedant, first isolated in 1985 from Neem oil. Several triterpenes are shown in Figure 1.11.

1.2.3.5.1   Common Triterpenes

Only a few of the common triterpenes are actually widely distributed among plants. These include the amyrins and **ursolic** and **oleanic acid** which are common on the waxy coatings on leaves and as a protective coating on some fruits. Other triterpenes include the **limonins** and the **cucurbitacins**.

1.2.3.5.2   Sterols

The general steroid structure is shown in Figure 1.12. Practically all plant steroids are hydroxylated at C-3 and are in fact sterols. In the animal kingdom, the steroids have profound importance as hormones, coenzymes, and provitamins. However, the role of the phytosterols is less well understood.

1.2.3.5.3   Saponins

Saponins are high-molecular-weight triterpene glycosides containing a sugar group attached to either a sterol or other triterpene. They are widely distributed in the plant kingdom and composed of two parts: glycone (sugar) and aglycone or genin (triterpene). Typically, they have detergent properties, readily form foams in water, have a bitter taste, and are piscicidal (toxic to fish). Many of the plants that contain saponins have been used historically as soaps. These include **soaproot** (*Chlorogalum pomeridianum*), **soapbark** (*Quillaja saponaria*), **soapberry** (*Sapindus saponaria*) and **soapnut** (*Sapindus mukurossi*).

Saponins are constituents of many plant drugs and folk medicines, especially among Asian peoples. This has led to great interest in the investigation of their pharmacological properties. By 1987 over 1000 sapogenins and triterpene glycosides had been elucidated.

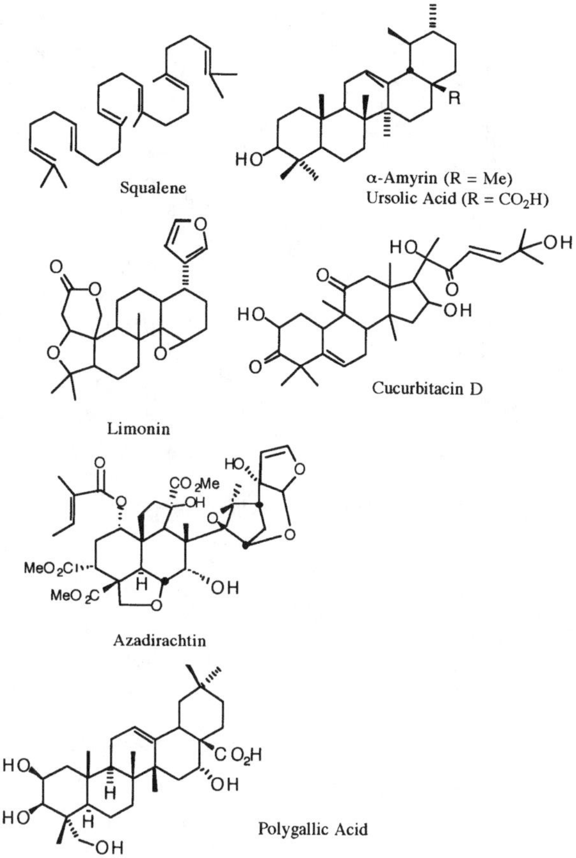

FIGURE 1.11 Some common plant triterpene natural products.

The aglycones, or genins as they are sometimes called, may be of the triterpene, steroid, or steroid alkaloid class. Saponins may be mono- or polydesmodic, depending on the number of attached sugar moieties (Figure 1.13). The most common monosaccharide groups and their corresponding abbreviations are shown in Table 1.2.

Biosynthetically, the saponins are comprised of six isoprene units and are derived from squalene. Many details, including the cyclase enzymes involved, have recently been determined. Commercially important preparations based on saponins include **sarsaparilla root** (*Sarsaparilla* spp.), **licorice** (*Glycyrrhiza* spp.), **ivy leaves** (*Hedera* spp.), **primula root** (*Primula* spp.), as well as **Ginseng** (*Panax* spp.). The structures of the ginsenosides and glycyrrhizinic acid are shown in Figure 1.14.

The ammonium and calcium salts of glycyrrhizinc acid are referred to as the **glycyrrhizins**. At 50 to 100 times sweeter than sucrose, these are the active ingredients in licorice root (*Glycyrrhiza glabra*), with expectorant, bacteriostatic, and antiviral activity. The **ginsenosides** are one of many triterpene saponins from ginseng (*Panax ginseng*) believed to be responsible for its immunostimulant activity.

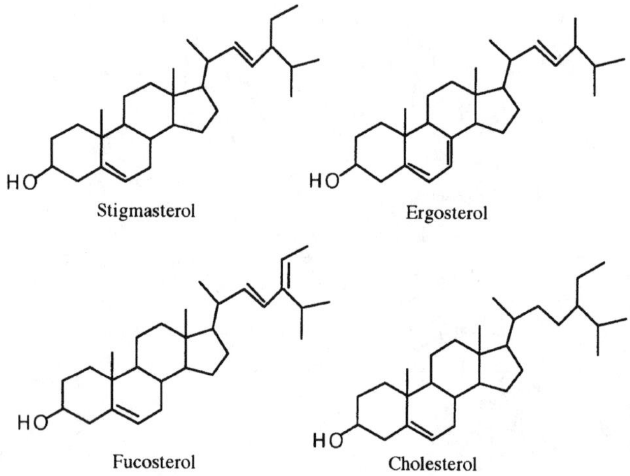

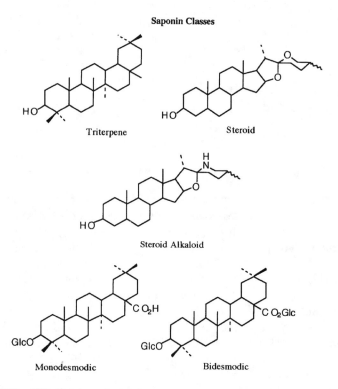

FIGURE 1.12 Some plant sterol natural products.

FIGURE 1.13 Classification of saponins.

TABLE 1.2
Common Saponin
Monosaccharide Groups

D-Glucose	Glc
D-Galactose	Gal
D-Glucuronic acid	GlcA
D-Galacturonic acid	GalA
L-Rhamnose	Rha
L-Arabinose	Ara
D-Xylose	Xyl
D-Fucose	Fuc

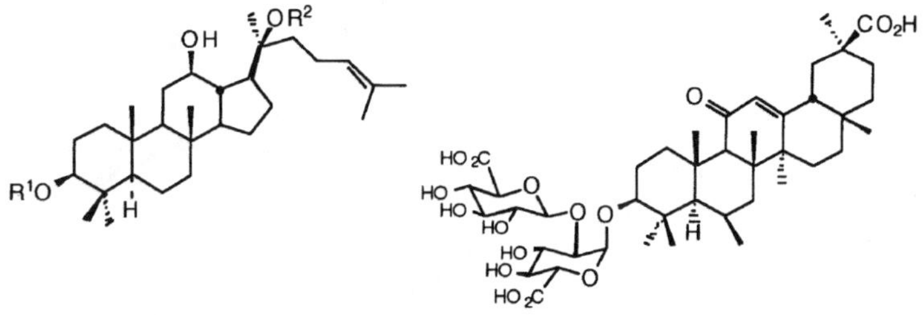

Ginsenoside Rb₁ (R¹ = R² = b-D-glucose)   Glycyrrhizinic Acid

FIGURE 1.14 Naturally occurring saponins.

1.2.3.6   Tetraterpenes: C_{40}

The most common tetraterpenoids are the *carotenoids*, a widely distributed group of C_{40} compounds. Whereas the structures of the di- and triterpenes can have a wide variety of fascinating structures, the carotenoids are generally derived from *lycopene*. Cyclization at one end gives γ-*carotene* and at both ends provides β-*carotene*. This pigment, first isolated in 1831, is by far the most common of all of these pigments and virutally universal in the leaves of higher plants. As is evident from this polyene structure, numerous double-bond isomers are possible for these basic structures, all of which can provide brightly colored pigments. Thus, in plants, carotenoids serve both as necessary pigments in photosynthesis and as coloring agents in flowers and fruits. This normally results in colors varying from yellow to red. They are also believed to protect plants from overoxidation catalyzed by other light absorbing pigments such as the chlorophylls. Some selected tetraterpenes are shown in Figure 1.15.

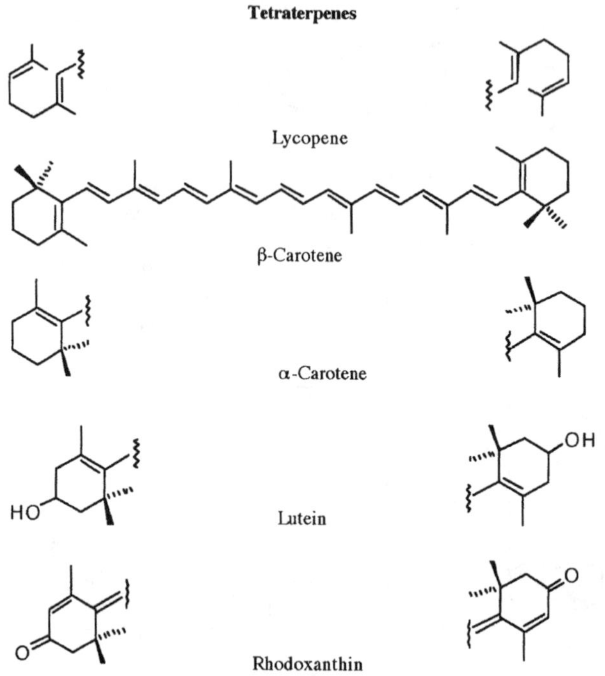

Tetraterpenes

Lycopene

β-Carotene

α-Carotene

Lutein

Rhodoxanthin

FIGURE 1.15   Representative plant tetraterpenes.

1.3   AROMATICS

Virtually all plants contain a wide variety of natural products which include an *aromatic* ring that generally contains one or more hydroxyl substituents. The vivid colors that light up the plants around us are derived generally from three sources: the tetrapyrroles, principally chlorophyll; the terpene-based carotenes, that we have already seen; and the aromatics, also referred to as the acetogenins. Several thousand aromatics are known and new structures are continuously being discovered. In some cases their functions are well known. For example, the polyphenolic *lignins* serve as structural components of the cell wall. In other cases, including the flavonols, a variety of functions have been hypothesized depending on the particular compound being investigated. Aromatic compounds are formed by several biosynthetic routes, including the polyketide and shikimate pathways, as well as from terpenoid origins. Due to the acidity of the phenol functionality (pK_a of 8 to 11 depending on substituents), phenolic substances tend to be water soluble and frequently form ether linkages with carbohydrate residues. Several individual groups exist that will be considered separately.

1.3.1   Non-Phenolic Aromatics

The vast majority of aromatic compounds of plant origin contain one or more hydroxyl functionalities. Several important natural products, however, do not contain

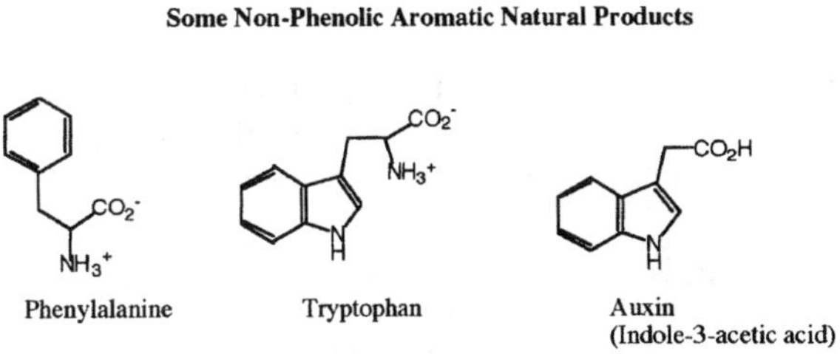

Some Non-Phenolic Aromatic Natural Products

| Phenylalanine | Tryptophan | Auxin (Indole-3-acetic acid) |

FIGURE 1.16   Some non-phenolic aromatic plant natural products.

this functionality and many will be treated in separate sections. These include the amino acids, tryptophan and phenylalanine, the indole alkaloids, and *auxin* (indole-3-acetic acid), an important plant hormone (Figure 1.16).

1.3.1.1   Tetrapyrroles

The *chlorophylls* are perhaps the most well-known plant constituents. As the primary catalysts of photosynthesis, they occur in several similar cyclic tetrapyrole forms and are located in the chloroplasts of virtually all photosynthetic plant tissues (Figure 1.17).

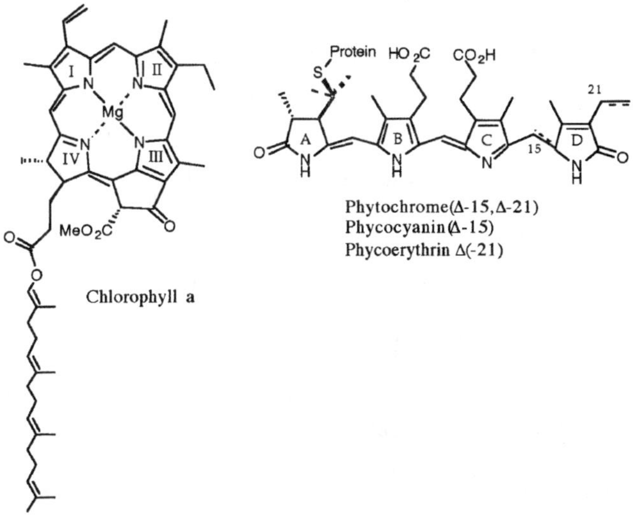

Chlorophyll a

Phytochrome(Δ-15,Δ-21)
Phycocyanin(Δ-15)
Phycoerythrin Δ(-21)

FIGURE 1.17   Cyclic and linear tetrapyrroles.

Other porphyrin pigments occur in plants in much smaller amounts. The *cytochromes*, for example, are critical components in the respiratory chain of both plants and animals. Finally, the linear tetrapyrroles, including *phytochrome*, *phycoerythrin*, and *phycocyanin*, are believed to be critical components for plant *morphogenesis*, the process by which numerous important plant developmental processes are initiated.

1.3.2 PHENOLS

The vast majority of the plant-based aromatic natural products are phenols. Numerous categories of these compounds exist; here we will consider the simple phenols, phenylpropanoids, flavonoids, tannins, and quinones.

1.3.2.1 Simple Phenols

Most of the *simple phenols* are monomeric components of the polymeric polyphenols and acids which make up plant tissues, including lignin, melanin, flavolan, and tannins. These individual components are obtained by acid hydrolysis of plant tissues. The components include *p*-hydroxybenzoic acid, protocatechuic acid, vanillic, syringic, and salicylic and gallic acids. Free phenols which do not require degradation of cell-wall polymers are relatively rare in plants. *Hydroquinone, catechol, orcinol*, and other simple phenols are found in relatively low concentrations (Figure 1.18).

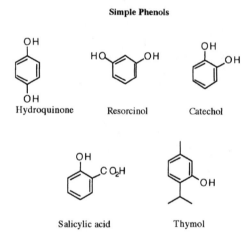

Simple Phenols

Hydroquinone   Resorcinol   Catechol

Salicylic acid   Thymol

FIGURE 1.18   Some phenolic plant natural products.

Thymol, which comprises 30 to 80% of the essential oil of thyme (*Thymus vulgaris*), has been used as an expectorant.

1.3.2.2 Phenol Ethers

Many of the phenols also exist as their methyl ethers. For illustration, a few are shown in Figure 1.19. **Khellin** and **visnagin** are the active coumarin derivatives of the ammi visnaga fruit (*Ammi visnaga*). Trans *Anethole* is chiefly responsible for the taste and smell of anise seeds (*Pimpinella anisum*). **Apiol** is a major constituent of the essential oil of parsley seed and is a powerful diuretic.

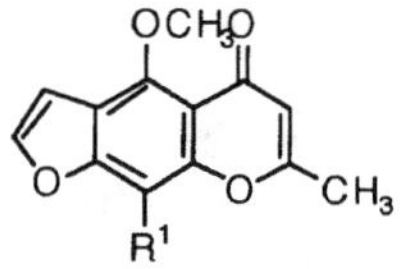

Khellin: $R^1 = OCH_3$

Visnagin: $R^1 = H$

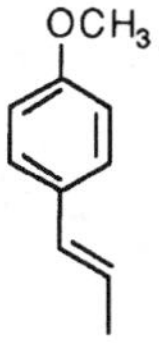

trans-Anethole

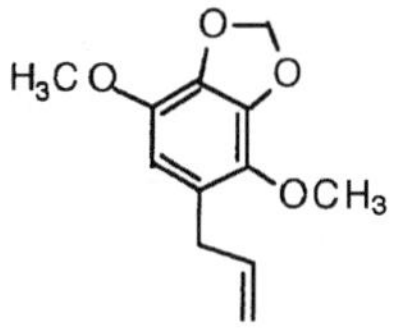

Apiol

FIGURE 1.19   Phenol ethers.

1.3.2.3   Phenylpropanoids

As the name implies, the *phenylpropanoids* contain a three-carbon side chain attached to a phenol. Common examples include the hydroxycoumarins, phenylpropenes, and the lignans. Also common are various types of hydroxycinnamic acids, including the caffeic and coumaric acids. *Coumarin* is common to numerous plants and is the sweet-smelling volatile material which is released from newly mowed hay. The phenylpropenes are not phenols at all since they lack the hydroxyl functionality. Thus, they are not water soluble but rather are essential oils and include *eugenol*, the major principle of oil of cloves. *Anethole* and *myristicin*, the principles of nutmeg, are also representative of this class of compounds (Figure 1.20).

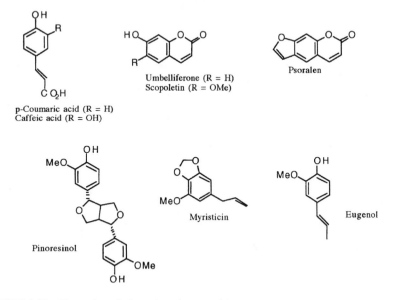

p-Coumaric acid (R = H)
Caffeic acid (R = OH)

Umbelliferone (R = H)
Scopoletin (R = OMe)

Psoralen

Pinoresinol

Myristicin

Eugenol

FIGURE 1.20   Examples of plant phenylpropanoids.

Caffeic and **p-*coumaric acid*** are hydroxycinnamic acids present in green and roasted coffee beans. ***Umbelliferone*** and ***scopoletin*** are coumarin phenylpropanoids that have been known since 1884 and are isolated from the roots of *Scopolia japonica*. The phenylpropene, ***eugenol***, has been isolated from several plant sources and has been used as a dental analgesic.

1.3.2.4 Flavonoids

The *flavonoids* have two benzene rings separated by a propane unit and are derived from *flavone*. They are generally water soluble compounds. The more conjugated compounds often are brightly colored. They are generally found in plants as their glycosides which can complicate structure determinations. Several classes are shown in Figure 1.21.

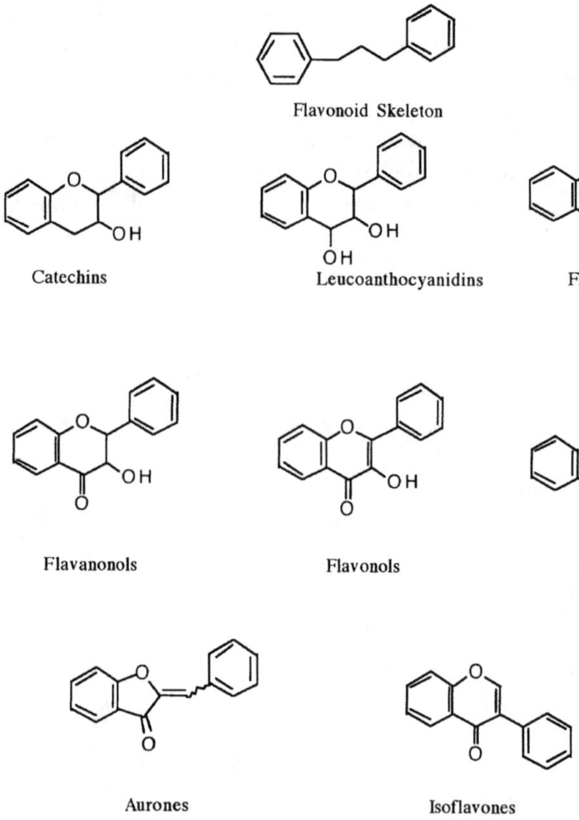

FIGURE 1.21   Common classes of flavonoids.

The different classes within the group are distinguished by additional oxygen-containing heterocyclic rings and hydroxyl groups. These include the ***catechins, leucoanthocyanidins, flavanones, flavanonols, flavones, anthocyanidins, flavonols,*** *chalcones*, *aurones*, and *isoflavones* whose general structures are shown in Figure 1.21. Other common groups include the xanthones and the condensed tannins. The *catechins* and *leucoanthocyanidins* are structurally very similar and only rarely exist as their glycosides. They polymerize to form condensed *tannins* which help give tea its color. They also are sufficiently prevalent to darken the color of streams and rivers in some woody areas.

The *flavanones* and *flavanonols* are fairly rare and normally exist as their glycosides. The *flavones* and *flavonols* are the most widely distributed of all the phenolics. The *anthocyanins* are the common red and rare blue pigments of flower petals and can make up as much as 30% of the dry weight of some flowers. They exist typically as glycosides. The *chalcones*, such as butein, lack the pyran ring found in flavonoids, although this is often subject to pH-controlled equilibria. The chalcone is more fully conjugated and normally brighly colored. Phlorizin is a strong inhibitor of apple seedling growth. The *aurones* are golden yellow pigments common in certain flowers.

1.3.2.5 Tannins

The *tannins* are common to vascular plants existing primarily within woody tissues. Tannins consist of various phenolic compounds that react with proteins to form water-insoluble copolymers. This reaction with proteins has been used industrially for the conversion of animal skins into leather. Plant tissues that are high in tannin content have a highly bitter taste and are avoided by most feeders. Tannins may be either *condensed* or *hydrolyzable*. Condensed tannins are formed biosynthetically by the condensation of *catechins* to form polymeric networks. Hydrolyzable tannins are derived from gallic acid.

1.3.2.6 Quinones

The *quinones* typically form strongly colored pigments covering the entire visible spectrum. Typically, however, they are found in the internal regions of the plant and thus do not impart a color to the exterior of the plant. Generally, quinones are derived from *benzoquinone*, *naphthoquinone*, or *anthroquinone* structures (Figure 1.22).

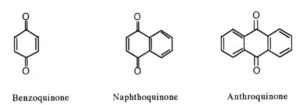

Benzoquinone   Naphthoquinone   Anthroquinone

FIGURE 1.22 Quinones.

A wide variety of phenolic natural products exist in plants many of which are of medicinal interest. A selection of these compounds is shown in Figure 1.23.

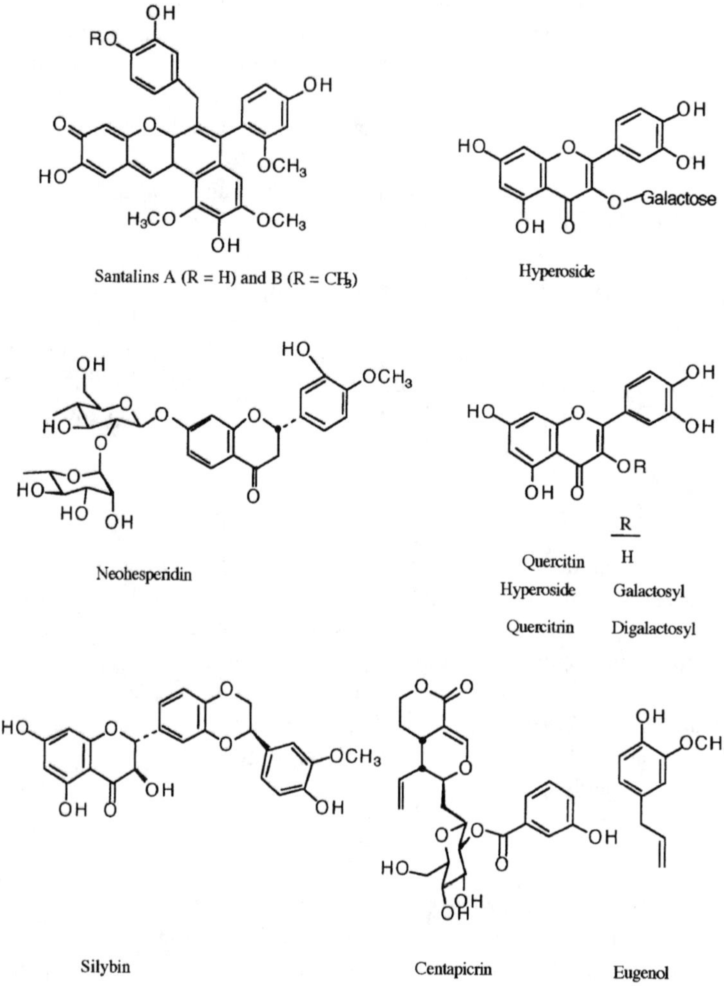

Santalins A (R = H) and B (R = CH₃)

Hyperoside

Neohesperidin

Quercitin

Hyperoside

Quercitrin

	R
Quercitin	H
Hyperoside	Galactosyl
Quercitrin	Digalactosyl

Silybin

Centapicrin

Eugenol

FIGURE 1.23 Phenols.

Many of these phenols come from familiar sources. **Santalins A** and **B** are the major pigments of red sandalwood (*Pterocarpus santalinus*). The flowers of the hawthorne tree provide **hyperoside**, one of the principle flavonoids from this source (*Crataegus laevigata*). **Neohesperidin** is responsible for the bitter taste of orange peels (*Citrus aurantium*), while **quercitin** is the active ingredient of birch (*Betula pendula*). **Silybin**, one of the silymarins, is a mixture of various flavanone derivatives (flavonolignans) and present in the fruit of the milk thistle (*Silybum marianum*). **Silymarin** is the active antihepatotoxic complex used for treatment of liver damage and increases the rate of synthesis of ribosomal ribonucleic acids. **Centapicrin** is an ultrabitter (bitterness value ca. 4,000,000) secoiridoid glycoside from the century plant (*Centaurea erythraea*). The **circuminoids** are responsible for the yellow pigment and cholagogic properties of turmeric (*Curcuma domestica*). **Hypericin**, a

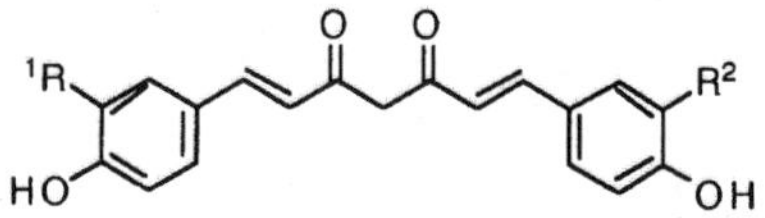

	R¹	R²
Curcumin	OCH_3	OCH_3
Desmethoxycurcumin	OCH_3	H
Bisdesmethoxycurcumin	H	H

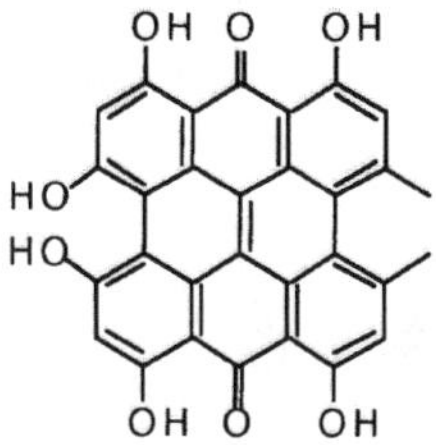

Hypericin

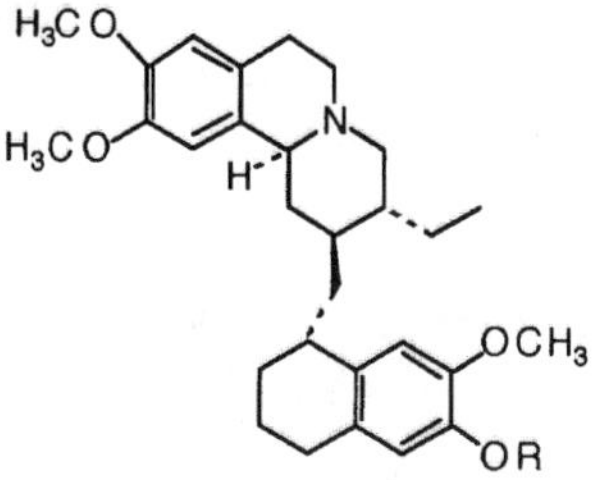

Emetine (R = CH_3), and
cephaeline
R = H)

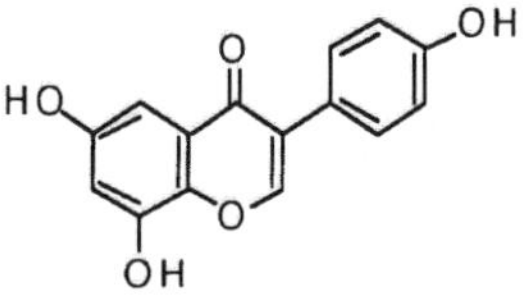

genistein

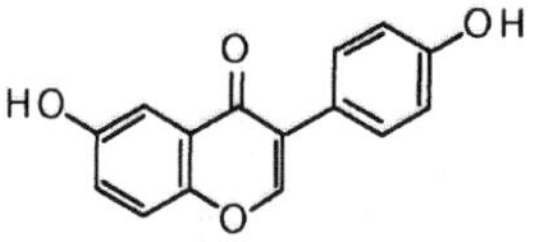

daidzein

FIGURE 1.23 (continued)

flavonoid from St. John's wort (*Hypericum perforatum*), is a monoamine oxidase inhibitor. *Emetine* and *cephaeline* are the active ingredients of syrup of ipecac, powerful emetics from ipecacuanha (*Cephaelis ipecacuanha*). The isoflavones *genistein* and *daidzein* are found in high concentrations in soybeans (*Glycine max*) as well as several other legumes. Both genistein and daidzein have been found to have anticancer activity.

1.4 CARBOHYDRATES

Sugars or *carbohydrates* are the primary products of photosynthesis and are essential as a source of energy to plants. They are stored as **starch** or **fructans**, used as **sucrose**, and polymerized to form **cellulose**, the main cell wall structural material of plants. Finally, they combine to form glycosides of many fundamental groups of natural products as we have already seen, including terpenes (to form saponins), phenols, and alkaloids.

Sugars are optically active aliphatic polyhydroxlyated compounds which are readily water soluble. This is due to the hydrophilic nature of the hydroxyl functionality, and does not involve the salt formation that we have observed for the phenolics and alkaloids. The common monosaccharides are shown in Figure 1.24. The sugars are classified into three groups depending on their size: *monosaccharides*, such as glucose; *oligosaccharides*, including sucrose; and *polysaccharides*, which include cellulose.

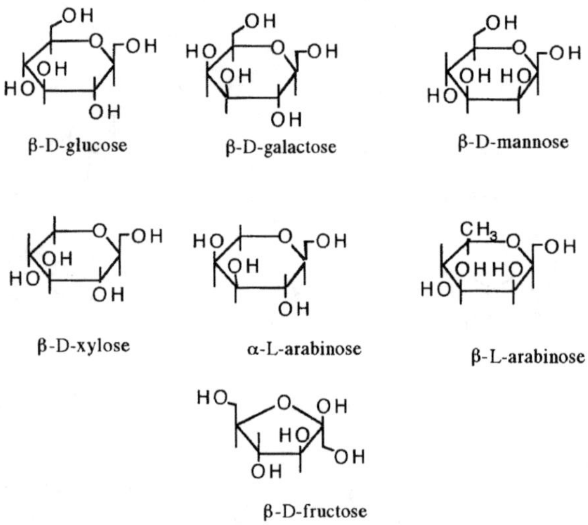

FIGURE 1.24   Monosaccharide types.

1.4.1   MONOSACCHARIDES

The most common monosaccharides are **glucose** and **fructose**. Less common are **xylose, mannose, rhamnose,** and **galactose**, as well as numerous others. In plants these sugars are normally bound as oligo- and polysaccharides or as various glycosides. The individual sugars may be physically separated either enzymatically or by treatment with acid.

1.4.2   OLIGOSACCHARIDES

The oligosaccharides normally include from two to five saccharide (or sugar) units. These are joined by any of three possible ether linkages that can complicate structure elucidation. Common oligosaccharides include **sucrose, trehalose, stachyose,** and **raffinose**.

1.4.3   POLYSACCHARIDES

Most of the carbohydrates found in plants occur as **polysaccharides** of high molecular weight. The polysaccharides (or glycans) fulfill a wide variety of functions in plants. Cellulose serves as a *structural* material, whereas in animals **keratin** and **collagen** serve similar structural roles. Cellulose is the most abundant organic compound in plants and the most abundant single polymer in the biosphere. A simple straight-chain polymer without branching is formed using β-(1,4) ether linkages and forms the main structural polysaccharides of the cell wall. *Amylose*, which is used as a storage rather than a structural glucan, uses α-(1,4) linkages. *Amylopectin* uses α-(1,4) and α–(1,6) linkages. The linkages of *cellulose* form straight ribbons that line up side by side forming polymers of high mechanical strength and limited extensibility. Other structural polysaccharides include the *polygalacturonans* (pectic polysaccharides), *xylans*, *glucomannans*, *chitins*, and the *glycosaminoglycans*.

More complex cell-wall polysaccharides act as *recognition signals* where the saccharide sequence of these heteropolysaccharides is informational, not unlike the nucleic acids. The structures of several of these polysaccharides are shown in Figure 1.25.

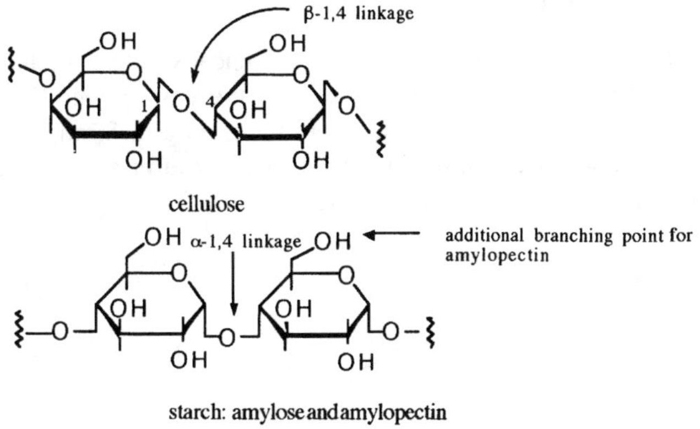

FIGURE 1.25   Common polysaccharide linkages.

1.5   AMINES, AMINO ACIDS, AND PROTEINS

Compounds that contain nitrogen as part of a chain rather than being incorporated into a ring structure generally belong to this category of natural products. For the amines, this by definition imparts a basic nature to the molecule. For the amino acids this results in molecules that may be *zwitterionic* (forming a dipolar ion) depending on the pH of the environment.

1.5.1   AMINES

The common plant amines can be subdivided into aliphatic monoamines, aliphatic polyamines, and aromatic amines. Occasionally these materials are classified as alkaloids rather than amines.

1.5.1.1   Aliphatic Monoamines

Simple aliphatic amines exist as low-boiling liquids and include most of the primary amines from methylamine, CH_3NH_2, through hexylamine, $CH_3(CH_2)_5NH_2$. These molecules typically have strong, fish-like aromas. In the case of cow parsley

(*Heracleum sphondylium*), they are believed to act as insect attractants by simulating the smell of carrion.

1.5.1.2 Aliphatic Polyamines

Common polyamines include **putrescine**, $NH_2(CH_2)_4NH_2$; **agmatine**, $NH_2(CH_2)_4NHC(=NH)NH_2$; **spermidine**, $NH_2(CH_2)_3NH(CH_2)_4NH_2$; and **spermine**, $NH_2(CH_2)_3NH(CH_2)_4NH(CH_2)_3NH_2$. Both putrescine and s-adenosylmethionine are used for the formation of spermine and spermidine. These polyamines are thought to have many functions and are invariably found complexed with nucleic acids, including both DNA and RNA.

1.5.1.3 Aromatic Amines

Many of the known aromatic amines are physiologically active. Perhaps the most well known member of this class is **mescaline**. It is the active principle of the peyote cactus, *Lophophora williamsii*, and is a potent hallucinogen. Similarly, three compounds critical to brain metabolism in animals are **noradrenaline**, **histamine**, and **serotonin**. All three occur in common plants (Figure 1.26).

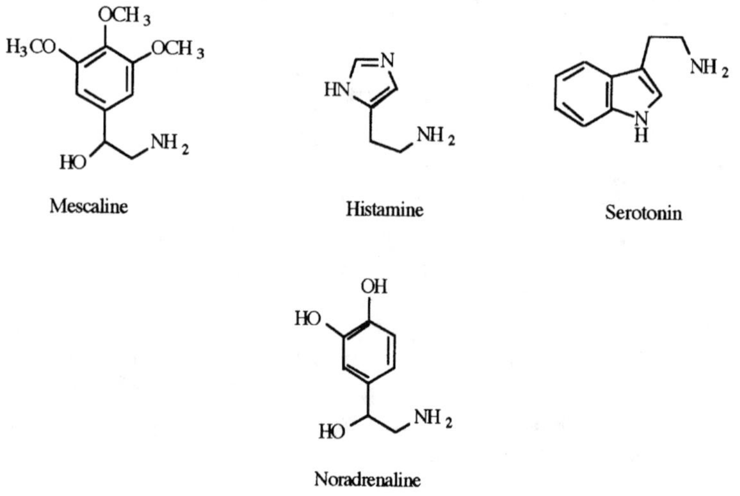

Mescaline   Histamine   Serotonin

Noradrenaline

FIGURE 1.26 Common aromatic amines.

1.5.2 Amino Acids

Much of the genetic information contained within each cell of plants and animals is expressed as **proteins**. Proteins are made up individually from large chains of **amino acids** and small oligomers comprise **peptides**. Proteins play a variety of roles. Some carry out the transport and storage of small molecules, while others make up a large part of the structural framework of cells and tissues. Perhaps the most important class of proteins are the **enzymes**, the catalysts that promote the variety of reactions that channel metabolism into essential pathways (see Chapter 2). Individual types of cells may contain several thousand kinds of proteins.

The protein amino acids are normally considered to be 20 in number for plants. The amino acids are high-melting, water soluble, zwitterionic colorless solids. Since they have both basic (amine) and acidic (acid) functionalities, the amino acids have specific pK_a's unique to each amino acid. The 20 principal amino acids are shown in Figure 1.27.

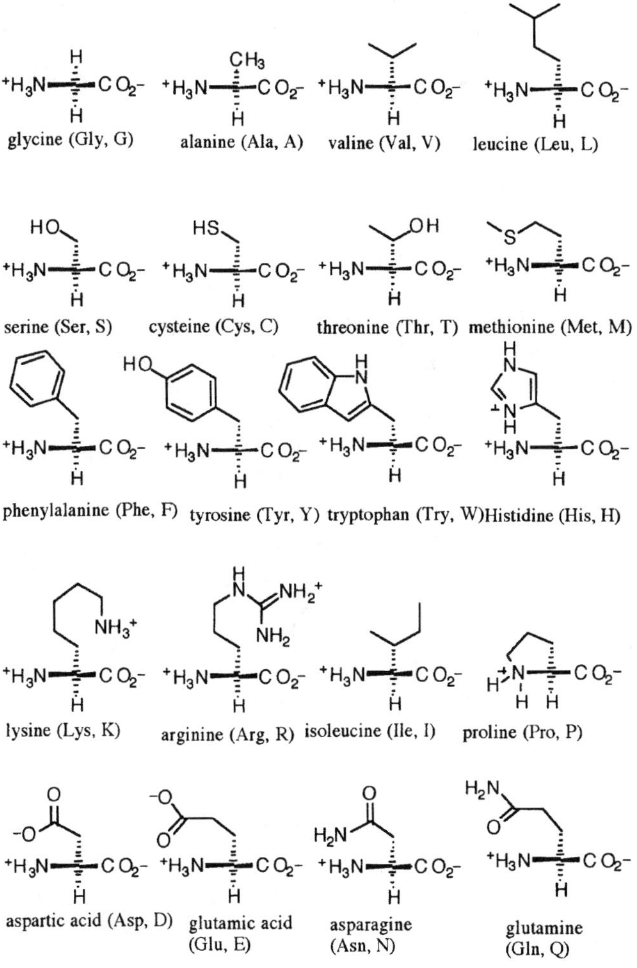

glycine (Gly, G) alanine (Ala, A) valine (Val, V) leucine (Leu, L)

serine (Ser, S) cysteine (Cys, C) threonine (Thr, T) methionine (Met, M)

phenylalanine (Phe, F) tyrosine (Tyr, Y) tryptophan (Try, W) Histidine (His, H)

lysine (Lys, K) arginine (Arg, R) isoleucine (Ile, I) proline (Pro, P)

aspartic acid (Asp, D) glutamic acid (Glu, E) asparagine (Asn, N) glutamine (Gln, Q)

FIGURE 1.27 The 20 amino acids that are incorporated into proteins.

1.5.2.1 Nonprotein Amino Acids

A nonprotein amino acid that is regularly found in plants is *d*-aminobutyric acid. Several hundred others are known, though no others have been found to be more or less ubiquitous. Additionally, atypical amino acids, peptides, and proteins exist which are constructed from nonribosomal processes that are also essential to the life of a plant. This is a more recent field that is currently popular with natural-product chemists.

1.5.3 PROTEINS

Proteins are high-molecular-weight polymers of amino acids. They are synthesized based on the triplet base code of DNA in the nucleus of a cell. Because the individual amino acids that make up proteins in plants and animals exist each as a single enantiomer, the polypeptide has a nonrandom form which gives rise to a particular three-dimensional shape, flexibility, and conformational lability. This is an active area of research.

1.6 ALKALOIDS

The *alkaloids* include those natural products that contain nitrogen, usually as part of a cyclic system. Compounds of this type are numerous among plants and are perhaps best known for their often potent pharmacological properties. Thus, many of the common drugs are alkaloid based. Relatively mild examples include *caffeine*, *quinine*, and *nicotine*. More potent examples include *cocaine*, *morphine*, and *strychnine*. Biosynthetically, they may be derived from amino acids, terpenes, or aromatics depending on the specific alkaloid structure. Because of this diversity, they are often derived from the plant source rather than being produced synthetically. In some cases, they may be grouped on the basis of the ring system present. Several common ring systems used for classification are shown in Figure 1.28.

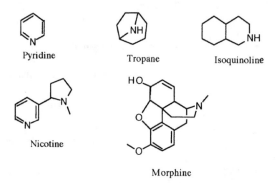

FIGURE 1.28 Classes and examples of alkaloid natural products.

Whereas the phenols are weakly acidic, the alkaloids are distinguished by the amine functionality. This makes them basic in nature. Thus, the free amines are relatively polar lipophilic substances, whereas treatment with acid forms water soluble salts. Several alkaloids are shown in Figure 1.29.

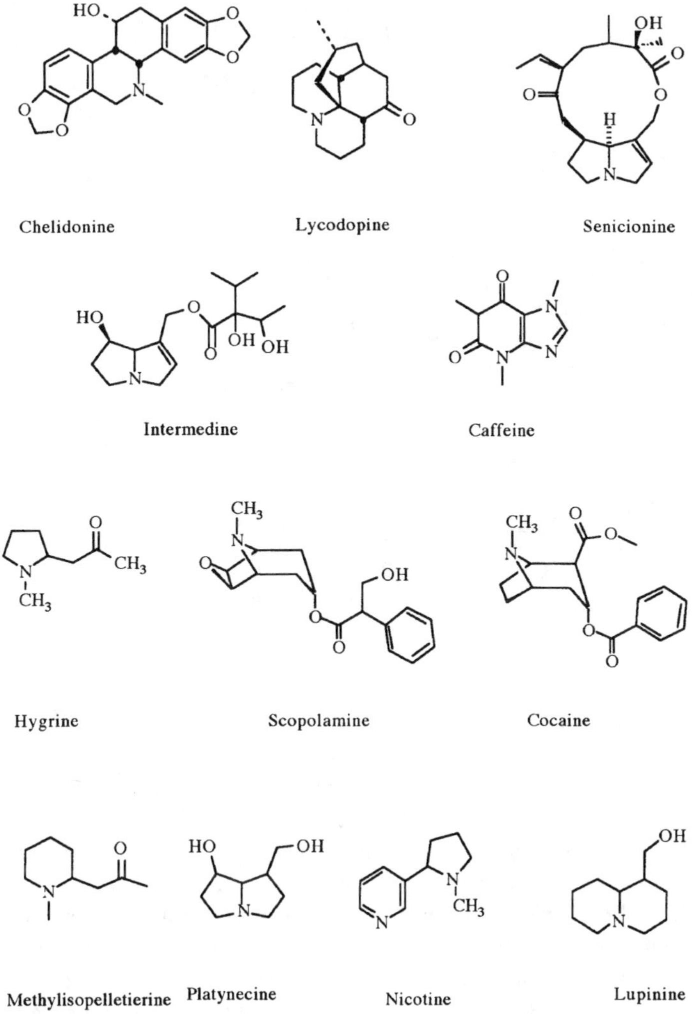

Chelidonine   Lycodopine   Senicionine

Intermedine   Caffeine

Hygrine   Scopolamine   Cocaine

Methylisopelletierine   Platynecine   Nicotine   Lupinine

FIGURE 1.29 Common alkaloids.

Many of the alkaloids are well known. **Nicotine** may comprise up to 8% of the dry weight of tobacco leaves (*Nicotiana tabacum*). **Morphine**, a powerful and addictive pain killer, may include up to 14% of the weight of high-grade opium. **Chelidonine** is one of the many alkaloids present in celandine (*Chelidonium majus*). **Quinine** is a well-known alkaloid from cinchona bark (*Cinchona pubescens*). **Lycodopine** is the principal alkaloid (with more than 100 other alkaloids) isolated from the stagshorn clubmoss (*Lycopodium clavatum*). **Senicione**, with its unique 12-membered ring, is one of several haemostyptic alkaloids from senecio (*Senecio nemorensis*). **Intermedine** is one of several pyrrolizidine alkaloids present in comfrey root (*Commiphora abyssinica*). **Caffeine**, a popular stimulant, is present in tea,

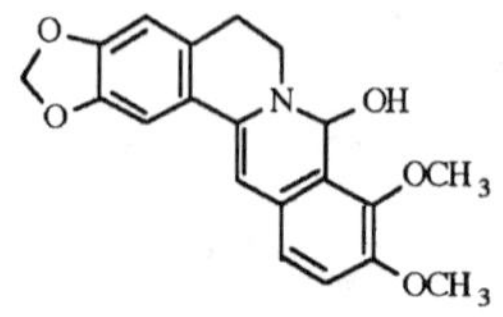

Berberine

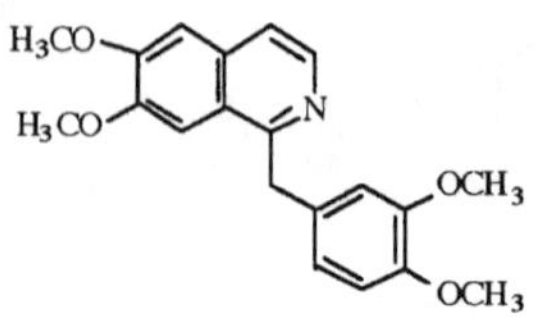

Papaverine

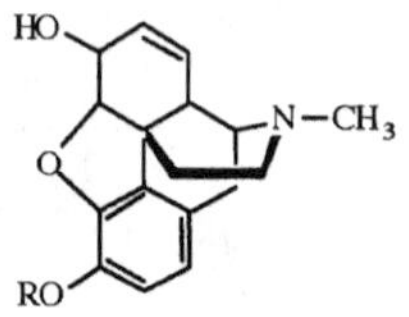

Morphine (R = H)
Codeine (R = CH₃)

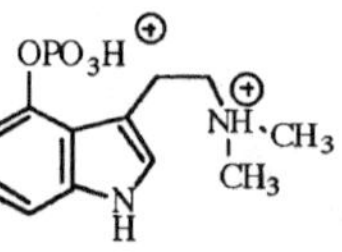

Psilocybin

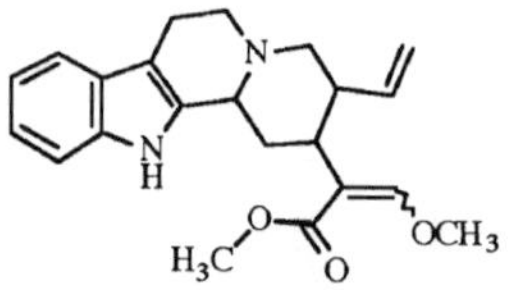

Corynantheine

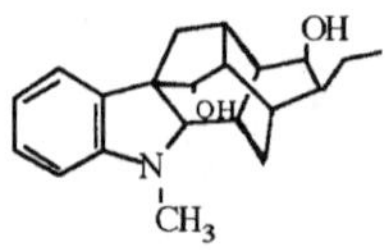

Ajmaline

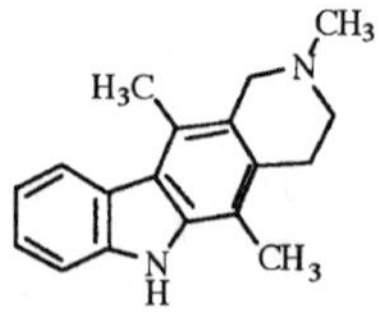

Ellipticine

FIGURE 1.29 (continued)

coffee, maté leaves, as well as guarana paste and cola nuts. *Hygrine* is a simple example of a pyrrolidine alkaloid. *Scopolamine* exists in various members of the nightshade family (Solanaceae) and can act as a powerful hallucinogen. *Cocaine* is a well-known controlled substance present in coca leaves (*Erythroxolon coca*). The yellow alkaloid *berberine*, isolated from the barberry shrub (*Berberis* spp.), shows antimicrobial and cytotoxic activity. *Papaverine*, a smooth muscle relaxant and cerebral casodilator, is also a constituent of opium. *Psilocybin* comes from the fruiting bodies of the Mexican hallucinogenic fungus, *Psilocybe mexicana*. *Colchicine* is a major alkaloid of *Colchicum autumnale*. *Strychnine*, whose structure was elucidated in 1945, is present in the seeds of *Strychnos nux-vomica* and other *Strychnos* species.

1.7. NUCLEOSIDES, NUCLEOTIDES, AND NUCLEIC ACIDS

Nucleic acids are the repositories and transmitters of genetic information for every cell, tissue, and organism. These include DNA and RNA which are polymers comprised of five different monomers: adenine, thymine (DNA only), uracil (RNA only), cytosine, and guanine. These individual monomers are composed of a sugar (ribose for RNA, deoxyribose for DNA), a base (purine or pyrimidine), and a phosphate

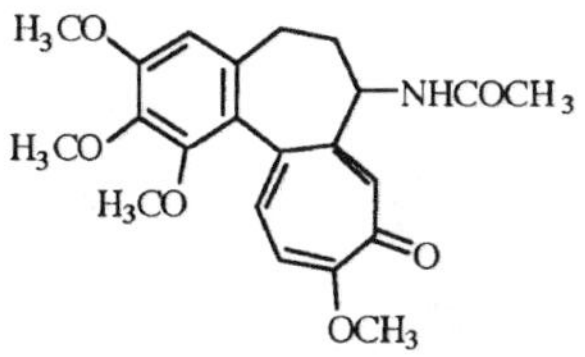

Colchicine

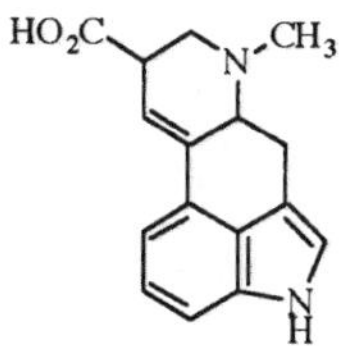

Lysergic Acid

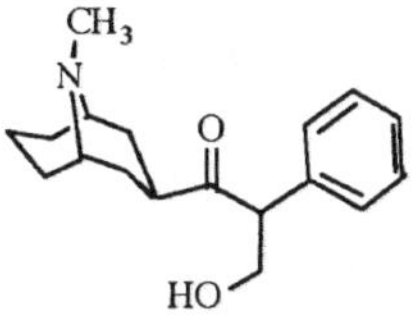

Atropine

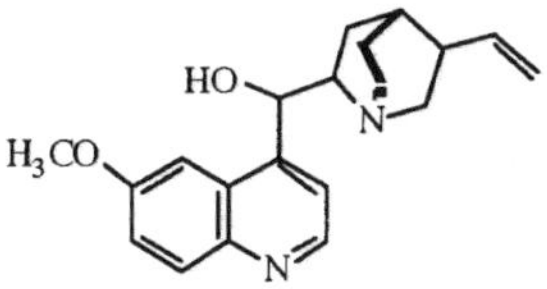

Quinine

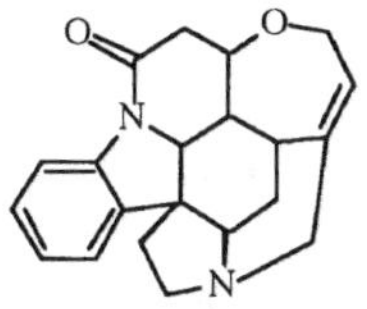

Strychnine

FIGURE 1.29 (continued)

linker. When the individual monomer contains all three components, it is referred to as a *nucleotide*, and when it lacks the phosphate, it is a *nucleoside*. These five bases can be isolated in trace amounts from plants, as well as a number of unusual bases with closely related strucutres. *5-Methylcytosine* is found in the DNA of wheat germ. The pyrimidine glycosides, vicine and convicine, have been found in certain legume seeds. The methylated purines, *theobromine* and *caffeine*, occur regularly in plants and are valued for their stimulant effects. Substituted purines constitute the *cytokinins*, which act as plant growth regulators and initiators of cell division. The purines and pyrimidines are only slightly soluble in water. The nucleosides are shown in Figure 1.30 and the purines and pyrimidines in Figure 1.31.

There are two major nucleic acids: *Deoxyribonucleic acid* (DNA) and *ribonucleic acid* (RNA). These are chain-like macromolecules that store and transfer genetic information. They are major components of all cells, comprising up to 15% of their dry weight. The monomeric units of DNA are the deoxynucleotides and for RNA, the ribonucleotides. Each of these nucleotides in turn consists of three main components: (1) a nitrogenous heterocyclic purine or pyrimidine base, (2) a pentose sugar, and (3) a molecule of phosphoric acid. DNA contains two pyrimidine bases (cytosine and thymine) and two purines (adenine and guanine). RNA has the same

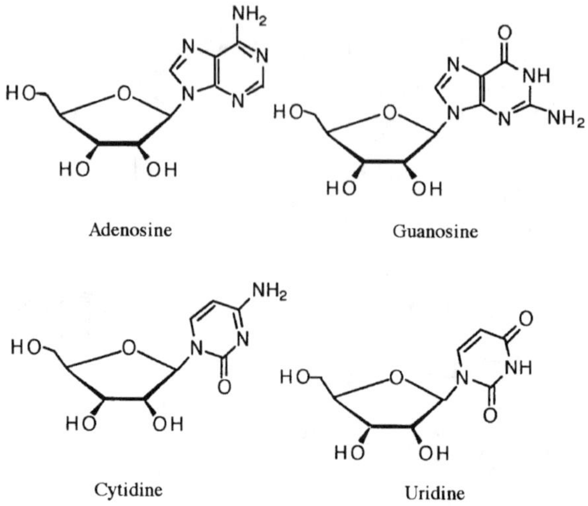

FIGURE 1.30   The nucleosides that can be obtained from RNA.

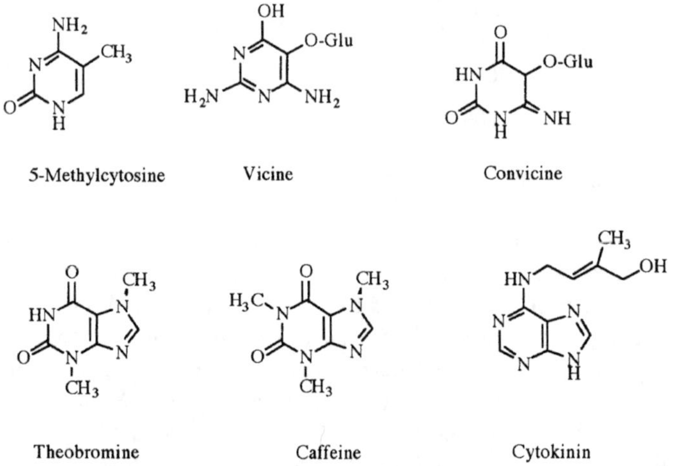

FIGURE 1.31   Naturally occurring purines and pyrimidines.

nitrogenous bases except that uracil replaces thymine. When the phosphate group of a nucleotide is absent, the remaining structure is called a nucleoside. Like the free purines and pyrimidines, free nucleosides occur only in trace amounts in most plant cells. The nucleotides can be present in significant amounts.

   In addition to the common bases listed above, numerous other purine and pyrimidine derivatives also have been isolated. The function of many of these rare bases is not well understood though it has been found that transfer RNA may contain up to 10% of these minor components.

BIBLIOGRAPHY

Journals devoted to the study of the chemistry of natural products include the *Journal of Natural Products*, *Natural Products Reports*, and *Phytochemistry*. New compounds may also be published in other journals, including *Journal of the American Chemical Society*, *Journal of Organic Chemistry*, and *Tetrahedron Letters*, among others.

NATURAL PRODUCTS

Atta-ur-Rahman, *Advances in Natural Products Chemistry*, Harwood Academic Publishers, Geneva, 1992.
Barker, R., *Organic Chemistry of Biological Compounds*, Prentice Hall, Englewood Cliffs, NJ, 1971.
Bisset, N. G., Ed., *Herbal Drugs*, CRC Press, Boca Raton, 1994.
Colegate, S. M. and Molyneux, R., *Bioactive Natural Products*, CRC Press, Boca Raton, FL, 1993.
Harborne, J. B., *Phytochemical Methods*, 2nd ed., Chapman and Hall, New York, 1984.
Hendrickson, J. B., *The Molecules of Nature*, W. A. Benjamin, New York, 1965.
Hostettmann, K. and Lea, P. J., Eds., *Biologically Active Natural Products*, Clarendon Press, Oxford, 1987.
Robinson, T., *The Organic Constituents of Higher Plants*, Burgess Publishing, Minneapolis, MN, 1964.

MARINE NATURAL PRODUCTS

Scheuer, P. J., *The Chemistry of Marine Natural Products*, Academic Press, New York, 1973.
Scheuer, P. J., Ed., *Marine Natural Products: Chemical and Biological Perspectives*, Vols. 1-5, Academic Press, New York, 1978-1985.

LIPIDS

Deuel, H. J., *The Lipids: Biochemistry*, Vol. 3, Interscience, New York, 1957.
Hawthorne, J. N. and Ansell, G. B., Eds., *New Comprehensive Biochemistry, Phospholipids*, Vol. 4, Elsevier, Amsterdam, 1982.
Stumpf, P. K., *The Biochemistry of Plants, Lipids: Structure and Function,* Vol. 4, Academic Press, New York, 1980.

TERPENES

Adams, R. P., *Identification of Essential Oil Components by GC/MS*, Allured Publishing, Carol Stream, IL, 1995.
Britton, G. and Goodwin, T. W., *Carotenoid Chemistry and Biochemistry*, Pergamon, New York, 1981.
Hanson, J. R., *The Tetracyclic Diterpenes*, Pergamon, New York, 1968.
Harborne, J. B. and Tomas-Barberan, F. A., Eds., *Ecological Chemistry and Biochemistry of Plant Terpenoids*, Clarendon Press, Oxford, 1991.
Newman, A. A., *Chemistry of Terpenes and Terpenoids*, Academic Press, New York, 1972.
Pfander, H., et al., Eds., *Key to the Carotenoids,* Verlag, Boston, 1987.
Taylor, W. I., *Cyclopentanoid Terpene Derivatives*, Marcel Dekker, New York, 1969.

Templeton, W., *An Introduction to the Chemistry of the Terpenoids and Steroids*, Butterworths, London, 1969.

AROMATICS

Dhar, D. N., *The Chemistry of the Chalcones and Related Compounds*, Wiley, New York, 1981.
Markham, K. R., *Techniques of Flavonoid Identification*, Academic Press, New York, 1982.

CARBOHYDRATES

Aspinall, G. D., *Polysaccharides*, Pergamon Press, New York, 1970.
Banks, W. and Greenwood, C. T., *Starch and Its Components*, Edinburgh University Press, Edinburgh, 1975.
Binkley, R. W., *Modern Carbohydrate Chemistry*, Marcel Dekker, New York, 1988.
Bols, M., *Carbohydrate Building Blocks*, Wiley, New York, 1996.
Collins, P. and Ferrier, R., *Monosaccharides: Their Chemistry and Their Role in Natural Products*, Wiley, New York, 1995.
El Khadem, H., *Carbohydrate Chemistry*, Academic Press, New York, 1988.
Kennedy, J. F. and White, C. A., *Bioactive Carbohydrates in Chemistry, Biochemistry, and Biology*, Wiley, New York, 1983.

AMINES, AMINO ACIDS, AND PROTEINS

Bailey, R. D., *An Introduction to Peptide Chemistry*, Wiley, New York, 1990.
Dence, J. B., *Steroids and Peptides*, Wiley, New York, 1980.
Devenyi, T. and Gergely, J., *Amino Acids, Peptides, and Proteins*, Elsevier, New York, 1974.
Gottschalk, A., *Glycoproteins*, Elsevier, New York, 1966.

ALKALOIDS

Dalton, D. A., *The Alkaloids: The Fundamental Chemistry*, Marcel Dekker, New York, 1979.
Kametani, T., *The Chemistry of the Isoquinoline Alkaloids*, Elsevier, New York, 1969.
Pelletier, S. W., *Chemistry of the Alkaloids*, Van Nostrand Reinhold, New York, 1970.
Swan, G. A., *An Introduction to the Alkaloids*, Wiley, New York, 1967.

NUCLEOSIDES, NUCLEOTIDES, AND NUCLEIC ACIDS

Blackburn, G. M. and Gait, M. J., *Nucleic Acids in Chemistry and Biology*, Oxford University Press, New York, 1990.
Townsend, L. B., *Chemistry of Nucleosides and Nucleotides*, Plenum, New York, 1988.
Townsend, L. B. and Tipson, R. S., *Nucleic Acid Chemistry*, Wiley, New York, 1991.
Ulbricht, T. L. V., *Purines, Pyrimidines, and Nucleotides*, Pergamon, New York, 1964.

2 How and Why These Compounds are Synthesized by Plants

Leland J. Cseke and Peter B. Kaufman

CONTENTS

0-8493-3134-X/99/$0.00+$.50
© 1999 by CRC Press LLC

2.1   INTRODUCTION

In Chapter 1, we presented a compilation of the many types of chemical compounds that plants produce. Now the question arises: How do plants synthesize these compounds and why do plants synthesize such a vast array of compounds? These are the primary topics of this chapter and in the process of exploring the answers, we hope to shed some light on the factors that drive the evolution of the biosynthetic pathways that produce these compounds. For example, the simple fact that plants have roots results in very different selective pressures than those driving the evolution of animal metabolism. After all, very few plants have the ability to run away when another organism sees them as food. Consequently, plants have evolved ways to repel or in some cases attract other organisms. Their lack of movement also allows them to produce very rigid compounds (such as cellulose or lignin) which, among other things, allows them to grow upward into new environmental niches.

To make such compounds as sugars, waxes, lignin, starch, pigments, or alkaloids, plants utilize very specific enzymes, each of which catalyzes a specific metabolic reaction. The enzymes are proteins called *organic catalysts*. They are coded by specific genes in the plant's DNA and are made via processes we call transcription (conversion of DNA to RNA via the enzyme RNA polymerase) and translation (conversion of RNA to protein via enzymatic reactions associated with complex structures called ribosomes). When there is a series of enzymatically catalyzed reactions in a well-defined sequence of steps, we have what is termed a *metabolic pathway*. Some enzymes may be involved in metabolic pathways requiring just a few, as in synthesis of starch from the sugar nucleotide, adenosine diphosphate (ADP)-glucose or many enzymatic steps, as in the synthesis of gibberellin hormones from mevalonic acid. Some enzymes may be involved in pathways that break down compounds (as in the hydrolysis of starch to sugars by α- and β-amylases). Still other enzymes may be involved in making storage forms of given compounds, such as glucosides, amides, or esters of the plant hormone indole-3-acetic acid (IAA). It is these different enzymatic pathways involved in the synthesis, breakdown, and creation of storage forms of a compound that regulate the level of the given compound. The regulation of each pathway and each of its enzymes is, however, extremely complicated. More will be said about this and other modes of regulation of enzyme activities in particular metabolic pathways in Chapter 3. Please note that not all proteins are enzymes. Many proteins within a given cell may be purely structural in function.

2.2   PRIMARY METABOLIC PATHWAYS IN PLANTS

To make some sense out of the various "highways and by-ways" of plant metabolism, we have put together the scheme shown in Figure 2.1. It depicts the interrelationships between the major metabolic pathways that occur in plants. Similar schemes have been produced for the major pathways for mammalian and microbial metabolism. Some pathways are unique to plants, such as the carbon reduction cycle in photosynthesis and the shikimic acid pathway which produces, among other things, essential amino acids (like tryptophan) that animals cannot live without. These aromatic amino acids are also required for the production of many plant-specific, nitrogen-containing, and phenolic compounds. Microbes and mammals also have their own unique pathways such as those involved in steroid hormone production, but common to plants, microbes, and mammals are the pentose phosphate pathway, glycolysis, and the tricarboxylic acid (TCA) cycle that are concerned with aerobic respiration and adenosine triphosphate (ATP) biosynthesis — the key energy molecule of the cell.

The scheme shown in Figure 2.1 for plant metabolic pathways will be an essential reference when we discuss individual metabolic pathways and sites where they are known to occur in plant cells. This scheme does not indicate where these pathways occur in plant cells; that will be covered in the next section. It also does not show the individual enzymatic steps that occur in each of the pathways shown. What it does show, however, is (1) the major kinds of metabolites produced by plants (most of these are indicated around the right and bottom fringes of this scheme); (2) the interrelationships between each of the major metabolic pathways; and (3) the molecule, carbon dioxide, which when fixed in photosynthesis leads to the formation of all the other kinds of molecules shown in the diagram. We have tried to show that the larger categories of all plant products are few in number. The majority of all essential products are made from sugars, acetyl CoA, (coenzyme A) or amino acids (which make up all the proteins in the plant including the all-important enzymes involved in each biochemical pathway). The same holds for products having somewhat less of an impact on the growth and development of the plant. These are generally considered to fall into three categories — terpenoids, nitrogen-containing compounds, and phenolic compounds. Some of these compounds require the addition of some soil nutrients such as nitrogen or sulfur, and many are the building blocks for higher organisms and thus are absolutely necessary for life on this planet. Therefore, the fact that plants can utilize the energy of the sun to convert carbon dioxide into more complex compounds is the primary factor that makes plants so essential and so interesting.

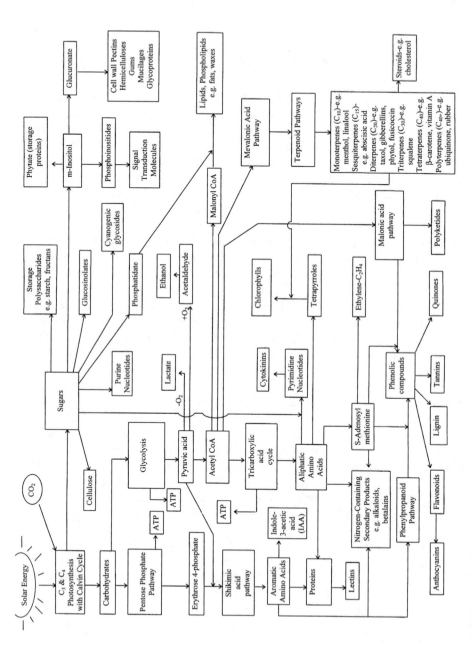

FIGURE 2.1 Illustration of the primary metabolic pathways in plants.

2.3 GENERALIZED VIEW OF A PLANT CELL AND ITS SUBCELLULAR COMPARTMENTS

Before considering individual compartments within plant cells where plant metabolites are synthesized and stored, we must first examine how a typical plant cell is organized and how its various components are related to one another. For this purpose, we shall refer to the cell illustrated in Figure 2.2. The "jacket" that encloses this cell is the cell wall. This is the primary site for polymerization of cell wall polysaccharides (cellulose, hemicellulose, and pectic polysaccharides), lignin, and amorphous silica gel in plants that accumulate this polymer. Within the cell wall is the plasma membrane that encloses the nucleus, cytoplasm, and various cellular organelles [chloroplasts or other plastids, mitochondria, endoplasmic reticulum, golgi apparatus (also called dictyosome), peroxysomes or glyoxysomes, and vacuoles]. The nucleus is the information center of the cell. It is surrounded by a double lipid membrane and contains genetic information (DNA) needed to create proteins within the cell. Thus, within the nucleus of each cell is all the information needed to create the entire organism. The cytoplasm is the liquid phase of the cell that contains (1) the majority of the ribosomes involved in protein synthesis (the other main location of ribosomes is on the endoplasmic reticulum); (2) microtubules and microfilaments that provide a physical skeleton for the cell and also act in cellular trafficking of proteins and organelles; and (3) all the soluble enzymes of the cell not found within organelles or cellular membranes.

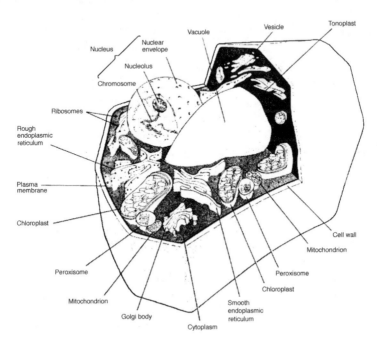

FIGURE 2.2 Diagrammatic representation of a plant cell and its constituent organelles.

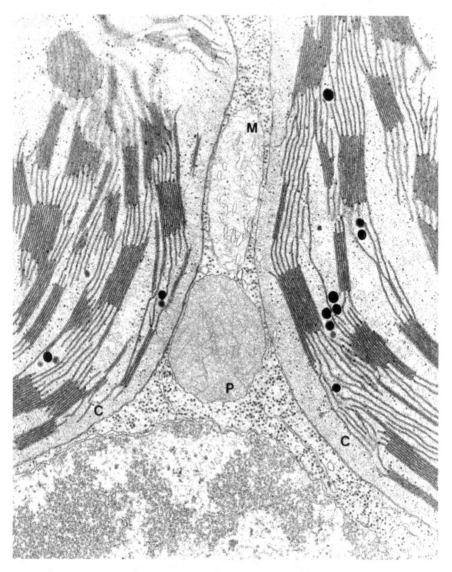

FIGURE 2.3 Transmission electron micrograph (TEM) illustrating leaf peroxysome (P) and mitochondrion (M) between two chloroplasts (C) in timothy grass (*Phleum pratense*). (Photo courtesy of Eldon H. Newcomb, University of Wisconsin.)

Each kind of organelle has many biochemical functions but the generally accepted function of each major class of organelle is as follows. The chloroplasts (Figure 2.3) of a plant cell are organelles bounded by a double lipid membrane which contains the enzymes and pigments (such as chlorophyll) that perform photosynthesis. There are, however, other types of plastids, such as those found in the petals of flowers (chromoplasts or leucoplasts) that do not develop the "machinery" to perform photosynthesis, yet still act as locations for the production of many plant

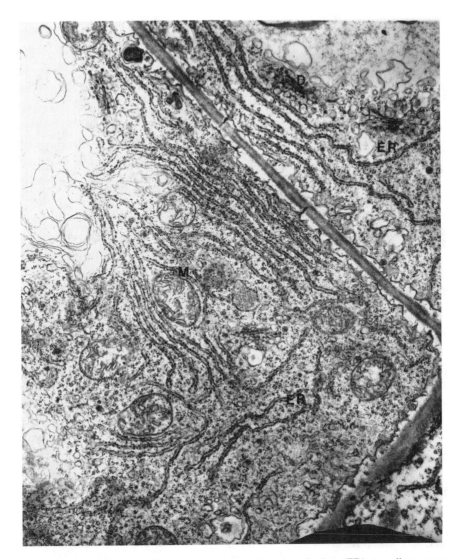

FIGURE 2.4 TEM photo illustrating rough endoplasmic reticulum (ER), two dictyosomes (D) in upper cell and mitochondrion (M) in lower cell of radish (*Raphanus sativus*) root hair. (Photo courtesy of Eldon H. Newcomb, University of Wisconsin.)

products. Mitochondria (Figure 2.4) are also surrounded by two lipid membranes. These organelles are the location of the TCA cycle, the respiratory chain, and oxidative phosphorylation all of which produce ATP. The endoplasmic reticulum (ER) (Figure 2.5) is a system of membrane-bound tubes and flattened sacs that spread throughout the cell and work in conjunction with dictyosomes (on Golgi bodies) to produce and secrete various compounds as well as to deliver specific proteins and membrane lipids to their proper locations within the cell. Peroxysomes (Figures 2.3 and 2.6) are microbody organelles that have the very important function

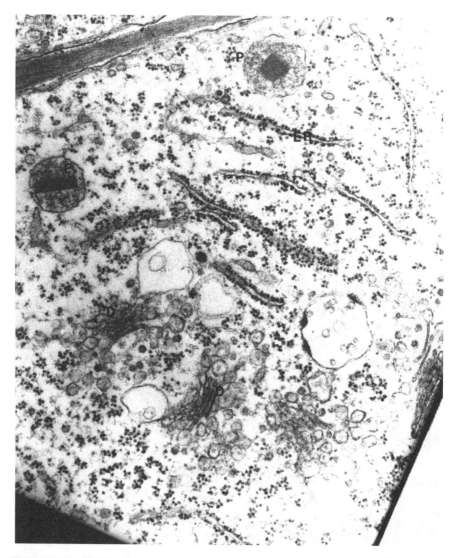

FIGURE 2.5   TEM photo illustrating rough endoplasmic reticulum (ER), two peroxysomes (P) containing calcium oxylate crystals, and dictyosomes (D) in young root cell of oats (*Avena sativa*). (Photo courtesy of Eldon H. Newcomb, University of Wisconsin.)

of housing the formation of toxic peroxides that are necessary for other metabolic mechanisms but would otherwise kill the cell. Peroxysomes also break down fats and participate in photorespiration (an important metabolic pathway coupled to photosynthesis). Glyoxysomes are specialized peroxysomes found only in the early stages of plant development. They contain the enzymes necessary for the conversion of stored lipids to carbohydrates in such processes as seed germination where photosynthesis is not yet possible. Finally, the vacuole is a liquid-filled cavity in the plant cell enclosed by a single membrane. Vacuoles play a wide variety of roles in

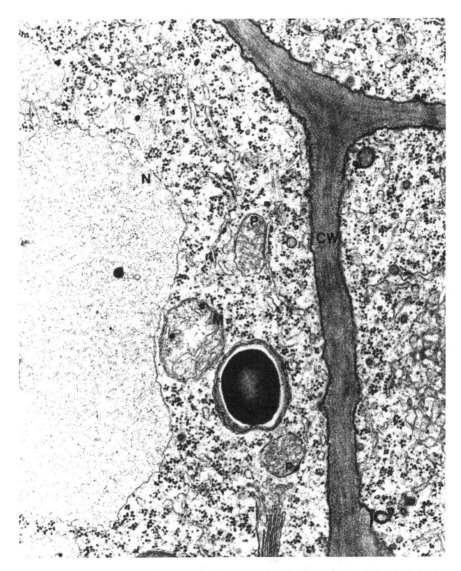

FIGURE 2.6 TEM photo with mitochondrion (M), two peroxysomes (P), an amyloplast (A) with large starch grain in it, nucleus (N), and cell wall (CW) in root tip of garden bean (*Phaseolus vulgaris*). (Photo courtesy of Eldon H. Newcomb, University of Wisconsin.)

cellular metabolism, some being digestive chambers, some storage chambers, some "waste bins", but they also play a very important role as a support structure. Water is pumped into the vacuole of each plant cell causing a build up of pressure, called turgor, which allows non-woody plants to remain standing. Without an adequate supply of water a plant will wilt.

Remember that what you see in Figure 2.2 represents only a static view of the cell at only one point in time. Most of the cell's contents are in a continuous state of motion, called *cyclosis*, and each molecule within the cell is, at a molecular level, experiencing **Brownian movement**. So, each cell of even the most solid looking of plants is actually a dynamic system of complex biochemical pathways which, when linked together, not only result in the organisms that we see but also define and regulate the interaction that the plant has with its environment. We shall now take a look at the individual components of a plant cell and some of the kinds of biosynthetic pathways known to occur in each structure.

2.4 HOW AND WHERE SOME OF THE WELL-KNOWN PLANT METABOLITES ARE SYNTHESIZED IN PLANT CELLS

2.4.1 LIPIDS, PROTEINS, AND NUCLEOTIDES

All living organisms produce three major categories of compounds: (1) lipids, which make up both the plasma membrane and the membranes of all internal compartments and organelles; (2) proteins, which make up both structural units of the cell such as microtubles and all the enzymes of every biochemical process; and (3) nucleic acids and nucleotides, which code for all proteins, act as metabolic energy molecules such as ATP and biochemical regulators such as GTP or cAMP, and in some cases work in conjunction with proteins to produce certain specific activities. Ribosomes, for example, consist of both protein and RNA, the combination of which allows the production of all other proteins. Since all organisms produce these compounds, their synthesis in plants will not be considered in detail. We refer the interested reader to any modern biochemistry or cell biology text.

Lipids are highly hydrophobic compounds produced by a partnership between plastids and the ER. Most lipids have a fatty acid portion made from acetyl-CoA and malonyl-CoA in a reaction whose repetition produces longer molecules. Malonyl-CoA is simply the carboxylated form of acetyl-CoA. In animals, fatty acid biosynthesis takes place in the cytosol, but in plants it occurs in plastids (chloroplasts in green tissue, proplastids in non-green tissue). In higher plants and animals, the predominant fatty acid residues are those of the C_{16} and C_{18} species, palmitic, oleic, linoleic, and stearic acids (Figure 2.7). However, there are many different forms of lipids. Membrane lipids such as phospholipids and glycolipids are made from a combination of glycerol, fatty acids, and hydrophilic compounds such as serine, choline, inositol, or various sugars. The many varieties of phospholipids and glycolipids are made from phosphatidate, a phosphorylated sugar derivative which acts as the precursor for the polar heads of these lipids. Vesicles that bud off of the ER or Golgi apparatus carry specific phospholipids to their proper location in the plasma membrane or organelles. Other than the typical lipid cell components, plants also have different metabolic pathways that produce waxes (Tables 2.1 and 2.2) which make up the protective cuticle of epidermal cells (see Cellulose Biosynthesis) and terpenes which are lipids synthesized from acetyl CoA via the mevalonic acid pathway. Terpenes produced in the terpenoid pathway serve a variety of functions in photosynthesis (see Carotenoid Biosynthesis), hormone controlled development (gibberellins and abscisic acid), and flower coloration and scent (see Section 2.6.6),

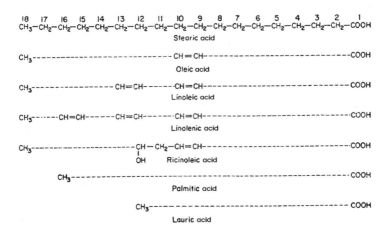

FIGURE 2.7 Illustration of the most common fatty acids in oils derived from plant seeds that are used for non-food purposes.

TABLE 2.1
Some Long-Chain Saturated Acids and Alcohols Found Free
or Esterified in Plant Waxes

Number of carbons	Acid	Alcohol
24	Lignoceric acid	Lignoceryl (*n*-tetrocosanol)
26	Cerotic acid	Ceryl (*n*-hexacosanol)
28	Montanic acid	Octacosyl (*n*-octacosanol)
30	Melissic acid	*n*-Myricyl (*n*-triacontanol)
32	Lacceroic acid	*n*-Lacceryl (*n*-dotriacontanol)
34	*n*-Tetratriacontanoic acid	Tetratriacontyl (*n*-tetratriacontanol)

to name a few. For humans, they are a source for rubber, essential oils (perfumes), and medicinal drugs such as taxol (an anticancer drug). Plants do produce very important storage forms of lipids (fats and oils) as energy reserves in fruits and seeds

TABLE 2.2
Some Common Components of Plant Cuticular Waxes

Compound type	Structural formula	Usual range of chain lengths
n-Alkanes	$CH_3(CH_2)_nCH_3$	C_{25}–C_{35}
Iso-alkanes	$\overset{\displaystyle CH_3}{\underset{\displaystyle \vert}{CH_3CH(CH_2)_nCH_3}}$	C_{25}–C_{35}
Alkenes	$CH_3(CH_2)_nCH{=}CH(CH_2)_mCH_3$	C_{17}–C_{33}
Monoketones	$CH_3(CH_2)_n\overset{\displaystyle O}{\overset{\displaystyle \Vert}{C}}(CH_2)_mCH_3$	C_{24}–C_{33}
β-Diketones	$CH_3(CH_2)_n\overset{\displaystyle O}{\overset{\displaystyle \Vert}{C}}CH_2\overset{\displaystyle O}{\overset{\displaystyle \Vert}{C}}(CH_2)_mCH_3$	C_{31}–C_{31}
Secondary alcohols	$CH_3(CH_2)_n\overset{\displaystyle OH}{\underset{\displaystyle \vert}{CH}}(CH_2)_mCH_3$	C_{20}–C_{33}
Wax esters	$CH_3(CH_2)_n\overset{\displaystyle O}{\overset{\displaystyle \Vert}{C}}{-}O(CH_2)_mCH_3$	C_{30}–C_{60}
Primary alcohols	$CH_3(CH_2)_nCH_2OH$	C_{12}–C_{36}
Normal fatty acids	$CH_3(CH_2)_n\overset{\displaystyle O}{\overset{\displaystyle \Vert}{C}}OH$	C_{12}–C_{36}
ω-hydroxy acids	$\overset{\displaystyle OH}{\underset{\displaystyle \vert}{CH_2}}(CH_2)_n\overset{\displaystyle O}{\overset{\displaystyle \Vert}{C}}OH$	C_{10}–C_{34}

such as the fats and oils found in avocados, olives, soybeans, sunflower seeds, and peanuts. In some cases, these reserves also may serve as rewards for animals that disperse the plant's seeds. These stored lipids are often found in the cytoplasm of either cotyledon or endosperm cells in organelles known as spherosomes (also called lipid bodies) which, like vesicles, bud off of the ER.

The production of proteins is completely dependent on the presence of nucleotides because every protein is coded by nucleic acids which are made from nucleotides. In eukaryotic cells, most proteins are initially produced in the cytosol and then transported to their final destination in the cell where they will perform their specific function. Organelles, such as chloroplasts and mitochondria, can also make proteins specific to these organelles. We have already mentioned that proteins may be enzymatic or structural in function, but plants do produce storage forms of proteins, like phytate, to provide a reserve of amino acids and energy especially in the process of seed germination. Some of these storage proteins can be **lectins** which are highly toxic and serve as herbivore deterrents (see Section 2.6.5), but their ability to bind sugars gives them function in recognition of symbionts, pathogens, and species-specific pollen grains as well. The purine and pyrimidine nucleotides that allow the synthesis of nucleic acids (DNA or RNA) are made in the cytoplasm from sugars and aliphatic amino acids. Purine nucleotides are made from ribose-5-phosphate, a modified ribose sugar, while pyrimidine nucleotides also require glutamine. So, a nucleotide is simply one of several different nitrogen-containing ring compounds linked to a five-carbon sugar (either ribose or deoxyribose) that carries a phosphate group. Nucleotides are also salvaged within the cell from the degradation or breakdown of nucleic acids (usually RNA). Please remember that all biochemical processes are ultimately controlled by the timing of the expression of the genes encoded by DNA.

2.4.2 CELLULOSE AND CELLULOSE BIOSYNTHESIS

Cellulose is the world's most common naturally synthesized polymer. It makes up the majority of all the biomass on the planet and is the primary component of all plant cell walls (Figure 2.8). This homopolymer is made from the glucose molecules produced by photosynthesis and is organized as glucan chains of β-1,4-linked glucose units in which every other glucose unit is rotated 180° with respect to its neightbor.[1] The glucan chains in primary walls of growing plant cells aggregate into fibers called cellulose microfibrils. In secondary walls, laid down after cell growth has ceased, the cellulose microfibrils are organized into macrofibrils or bundles.[1] Cells which expand more or less equally in all directions have cellulose microfibrils oriented in a random pattern; in contrast, cells which expand by elongation growth (e.g., fibers, pollen tubes, root hairs, and conducting cells of the vascular system) have cellulose microfibrils oriented parallel to each other, lying at right angles to the direction in which the cell elongates. These patterns of orientation of cellulose microfibrils help govern the specific function of a given cell and can be determined microscopically by the use of crossed polarizers and a red filter placed diagonally to the crossed polarizers.

The synthesis of cellulose occurs at the plasma membrane which is located at the interface between the cell wall and the cytoplasm. The monomeric unit that donates glucose units to a growing cellulose chain is UDPG (uridine diphosphate glucose). The glucose in UDPG comes from the hydrolysis of the disaccharide sugar, sucrose, catalyzed by the enzyme, SuSy (sucrose phosphate synthase). An elegant hypothetical model of a cellulose synthase complex in the plasma membrane is provided in the excellent review article on cellulose biosynthesis.[1]

The cell wall in plants provides structural support for the plant. This structural support is provided not only by cellulose but by other polymers such as hemicellulose and pectic polysaccharides. Like cellulose, these are chains of sugars, but their many varieties differ from cellulose in the kinds of sugars present, how they are linked together, and how many branches they have in their chains. Hemicellulose and pectins are not made at the plasma membrane. Instead, they appear to be made in the secretory system of the ER and Golgi. Then they are transported to the cell wall

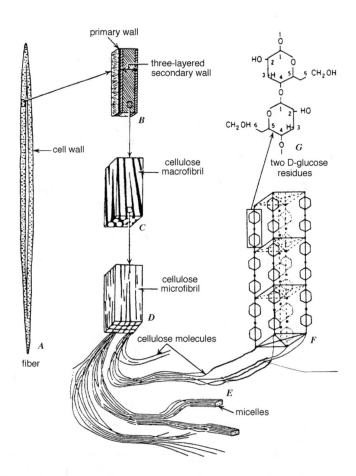

FIGURE 2.8 Interpretation of plant cell-wall structure in a fiber cell (A). It shows the structure of the primary wall in B, a cellulose macrofibril in C, a cellulose microfibril in D, a crystalline micelle of cellulose in E, the molecular architecture of repeating, ordered D-glucose units that make up cellulose in F, and two D-glucose residues connected by β-1,4-glucosidic bonds in G. (Modified from Esau, K., *Plant Anatomy*, 2nd ed., John Wiley & Son, New York, 1965.)

via vesicles. In woody plants such as vines, trees, and shrubs, the cell walls become lignified through deposition of the polymer, lignin (see Section 2.4.3). Cellulose combined with lignin is the primary plant product involved in support and provides the physical structure that allows such plants as trees to grow very tall. In vascular plants such as grasses, sedges, and scouring rushes (*Equisetum* spp.) as well as diatoms (one type of algae), the cell walls become infiltrated with amorphous silica gel which, like lignin, provides structural support (see Section 2.4.4). In epidermal cells of plant shoots, the cell walls can also become infiltrated and covered on their outer surfaces with a waxy lipid coating called the **cuticle**, made up of a polymer called cuticular wax or cutin made and secreted by the ER of epidermal cells combined with a wide variety of saturated and unsaturated acids as well as many forms of alcohols (Tables 2.1 and 2.2). This "water-proofing" of the surface of the shoot (leaves and stems, flowers, and fruits) prevents excess water loss from the plant and consequent desiccation. In some species, such as those living at high altitudes, the cuticle is very white which helps reflect damaging ultraviolet (UV) light. Cellulose infiltrated with lignin and/or cuticular wax also provides a physical barrier which greatly deters most potential herbivores because of the toughness of the polymers. In contrast, roots do not produce a cuticular wax layer on their outer surfaces, but they do synthesize a wax known as *suberin* in an interior layer of cells called the *endodermis* that prevents leakage of ions and metabolites out of the vascular cylinder in the center of the root.

In commerce, cellulose is important in fabric made from cotton or other plant fibers, in softwood fibers (derived from conifers) that make up paper and cardboard, and in purified or modified forms as a matrix used in column and thin layer chromatography to purify compounds such as plant pigments and enzymes (e.g., DEAE cellulose = diethylaminoethyl cellulose). Obviously, it is also a major structural component of wood derived from trees used to make lumber. Figure 2.9 illustrates a cross-section of the trunk of a California redwood tree (*Sequoiadendron sempervirens*) whose wood (secondary xylem tissues) is mostly composed of cellulose, but which also is lignified (see the following section).

FIGURE 2.9   Photo of Peter Kaufman and Mike Messler examining a cut stump of a California redwood (*Sequoiadendron sempervirens*) tree whose wood (secondary xylem) is made primarily of lignified cellulose. Diameter of this tree at cut surface is about 5 m. (Photo courtesy of Casey Lu, Humboldt State University, Arcata, CA.)

2.4.3   LIGNIN AND LIGNIN BIOSYNTHESIS

Lignin is a complex polymer (Figure 2.10) that exists as a 3-dimensional matrix around the polysaccharides of secondary cell walls found in plant fibers and in the tracheids and vessel elements of secondary xylem (wood). It is composed of varying amounts

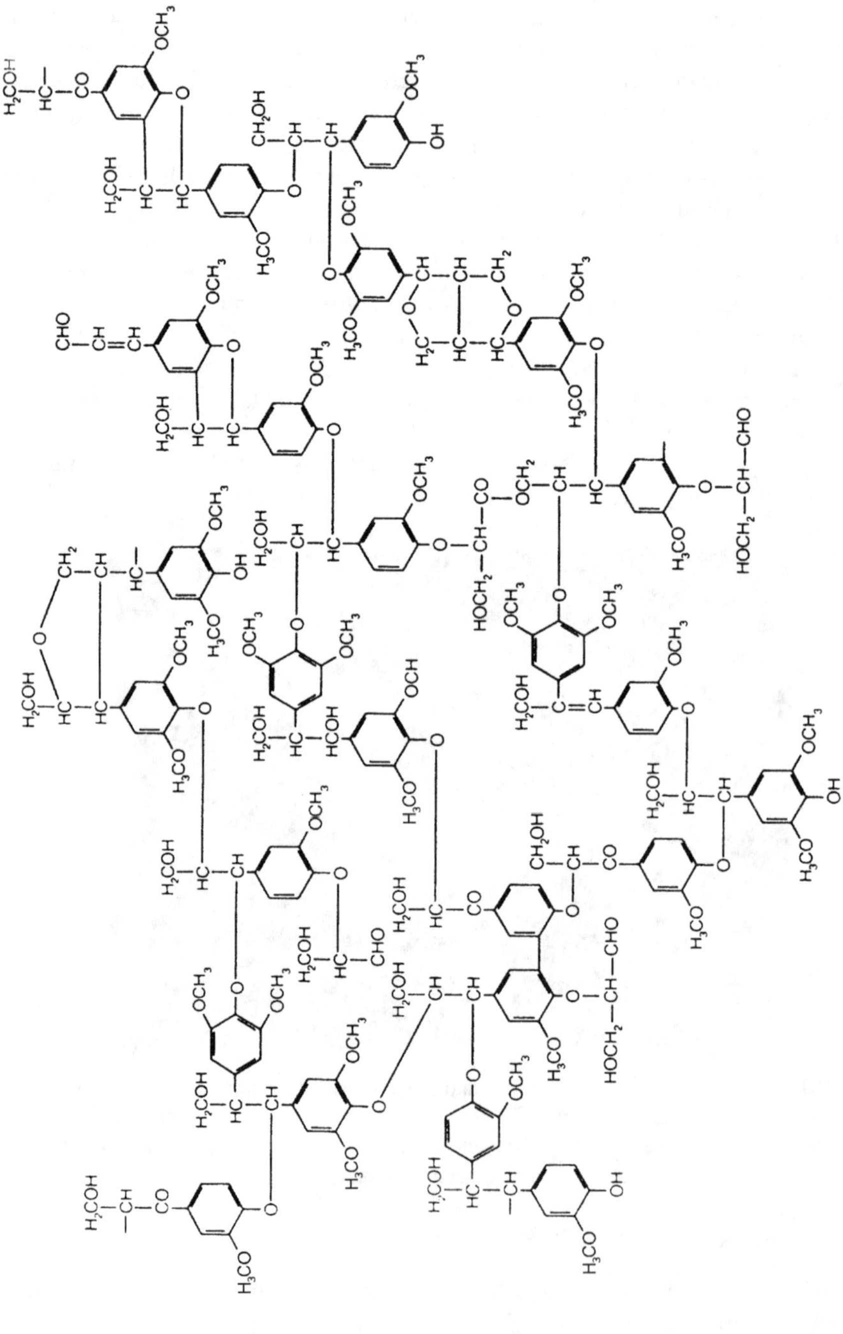

FIGURE 2.10   Partial polymeric structure of a lignin molecule made up of phenylpropane (C_6-C_3) monolignol alcohol units (see Figure 2.11).

of the aromatic phenylpropanoid subunits (monolignols), para-coumaryl alcohol, coniferyl alcohol, and sinapyl alcohol made via the shikimic acid pathway (Figure 2.1). These monolignols are usually synthesized from the amino acid l-phenylalanine, although tyrosine can also be used. Subsequent steps in the monolignol biosynthetic pathway are shown in Figure 2.11. These monolignols appear to be made in the ER and Golgi bodies, but the polymerization of lignin itself occurs outside the plasma membrane.

Lignin makes up between 15 and 35% of the dry weight of woody tissue,[2] and it acts to provide additional rigidity and compressive strength to cell walls. Because lignin is hydrophobic, it also makes cell walls that become lignified impermeable to water.[3] In plants, there are simple histochemical tests available which allow one to test for the presence of lignin in cell walls. They basically involve the use of phloroglucinol/HCl or para-rosaniline HCl. In either case, the cell walls stain deep reddish brown in color. We have used these reactions to demonstrate that, in cereal grasses, bundles of fibers associated with vascular bundles show excellent lignin staining in mature stems and leaf sheaths that make up stiff straw. In contrast, it is totally absent in strands of fiber-like collenchyma cells associated with vascular bundles in the swollen leaf sheath bases of cereal grass shoots which are sites for upward bending (negative gravitropic curvature) of lodged shoots prostrated by the action of wind, torrential rain, or hail.[4] So, while lignin may provide support, help prevent water loss, and even resist herbivores, it is not a benefit to plant tissues that need to grow or to bend.

2.4.4 BIOGENIC SILICA AND SILICIFICATION

Some plants have developed the ability to absorb inorganic constituents from their environment and use them towards their benefit. Biogenic silica is a polymer of biological origin that is characteristically found in the cell walls of diatoms, scouring rushes (*Equisetum* spp.) or horsetails, grasses (all members of the Poaceae or grass family), members of the rush family (Juncaceae), and members of the sedge family (Cyperaceae). Silica found in these silica-accumulating plants has its origin from silicates found in soil minerals. It is taken up as monosilicic acid, $Si(OH)_4$ via the roots (or cell membrane in the case of the single-cell diatoms) from which it moves up the plant in the xylem-conducting elements. This upward movement of monosilicic acid with water and other mineral compounds occurs as a result of "transpirational pull" mediated by transpirational loss of water from stomates (pores) located in epidermal tissues of leaves and stems. Once monosilicic acid arrives in stems and leaves where transpiration is occurring, it irreversibly polymerizes as amorphous silica gel, $SiO_2 \cdot nH_2O$, mostly in cell walls which are hydrogen bonded to cellulose molecules. However, in grasses, within specialized silica cells located in the epidermis of leaves and floral bracts, it can also polymerize directly in the cytoplasm after breakdown of all cell organelles has occurred.[5] Silica secretions can also result in specialized structures such as the needles on nettles.

The annual scouring rush, *Equisetum arvense*, can produce up to 20% of its dry weight as silica. A classical experiment done at California Institute of Technology[5] showed that these plants, grown in silicon-free hydroponic nutrient solutions became

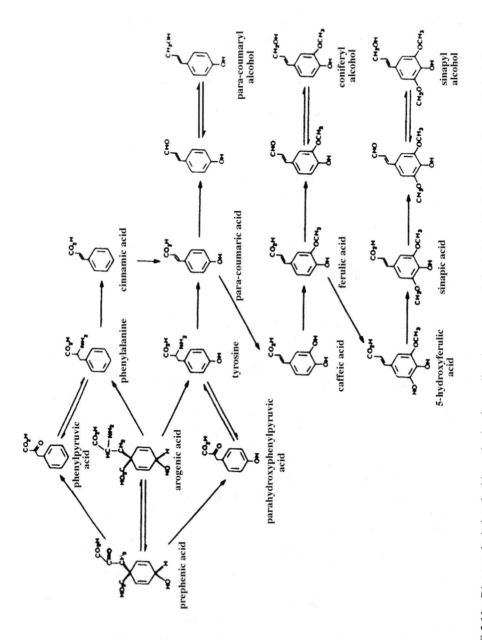

FIGURE 2.11 Diagram depicting the biosynthesis of monolignols, *p*-coumaryl alcohol, coniferyl alcohol, and synapyl alcohol.

very weak and appeared collapsed. Additions of silicon, as sodium metasilicate, to the hydroponic nutrient solution at only 80 ppm yielded plants whose shoots were upright and appeared strong and robust. This indicated that silica provided direct support for the shoot and, hence, is considered an essential element for normal growth and development in these types of plants. So, the primary role of silica in the cell wall is to provide support to the shoot in addition to that provided by cellulose and lignin. Aside from providing support to shoots of grasses, sedges, rushes, and *Equisetum* spp., amorphous silica gel that gets deposited in outer cell walls of epidermal tissue of leaves and stems forms very hard and often very sharp structures which can deter attack by predacious animals, insects, and disease-causing fungi. In fact, the mandibles of many insects that attack rice plants (e.g., green and brown leaf hoppers that transmit tungro virus pathogens) get worn down and rendered ineffective in piercing the leaves of the rice plants. Likewise, the teeth of sheep get worn down significantly by eating high silica-containing pasture grasses. Fortunately, these animals can replace their worn-down teeth with new teeth.

It also should be noted that silica is not the only inorganic constituent that plant roots can absorb. Marine algae can absorb calcium in the form of calcium carbonate that they deposit on their surfaces as crusty support compounds, much like biogenic silica, which seem to prevent the plants from getting damaged by crashing waves. Some plants can absorb toxic elements such as selenium (Se) or bromine which help ward off herbivores. For example, *Astragulus* (loco weed) accumulates Se and incorporates it into certain amino acids and proteins. The plant itself can distinguish if a protein has Se, so there is no toxic effect to the plant. However, the animal metabolism cannot distinguish proteins that contain Se from those that do not and the effect is quite toxic. The fact that Se is toxic to most other plants also allows *Astragulus* to avoid competition in soils that contain Se. These soils often occur around uranium deposits, so *Astragulus* has been used as an indicator species in botanical prospecting. Other plants such as alpine penny-cress (*Thalspi caerulescens*) will take up elements such as zinc and cadmium making them very useful when planted in polluted areas needing bioremediation.

2.4.5 STARCH AND STARCH BIOSYNTHESIS

Starch is the most common storage polysaccharide found in plants and serves as a primary food source for humans, domestic animals, birds, insects, and microbes. Starch is essentially made up of monomers of sugars linked end to end in long chains through α-1,4 linkages along the chain with α-1,6 linkages as branch points. Martin and Smith[6] present an excellent review of how starch is synthesized in plants. A scheme from their review article depicting amylopectin (branched chains of starch molecules with α-1,4- and α-1,6-linked glucan), starch granule formation, and the biosynthetic pathway for starch biosynthesis is shown in Figure 2.12. The other configuration of the starch molecule is that of amylose which is unbranched. It is composed exclusively of α-1,4-linked glucan. In rice gains, for example, one encounters varying amounts of amylose (straight-chain starch) and amylopectin (branched chain starch), depending on the cultivar. In sticky, short-grain rice (Japonica cultivars), amylopectin predominates. Such rice is better for soups and for eating with

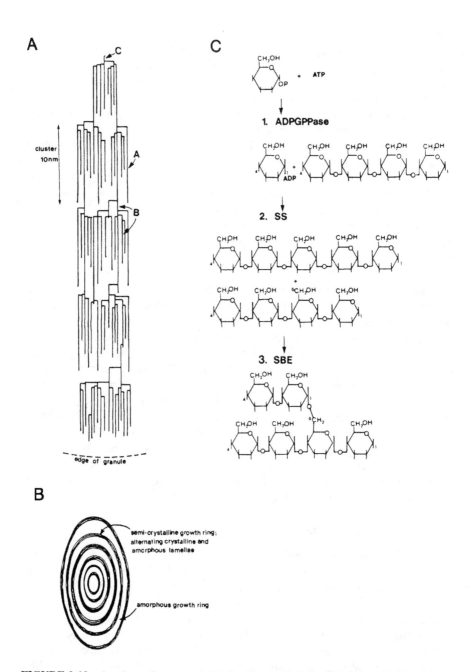

FIGURE 2.12 Amylopectin structure (**A**), starch granule form (**B**), and starch biosynthesis (**C**). 1. ADPGPPase – adenosine diphosphate glucose pyrophosphorylase 2. SS – starch synthase 3. SBE – starch branching enzyme (From Martin, C. and Smith, A. M., *The Plant Cell,* 7, 971–985, 1995. With permission.)

chopsticks. In non-sticky, long-grain rice (Indica cultivars), amylose starch predominates. Both forms of starch are made in the chloroplasts of mesophyll and bundle sheath cells from triose phosphate, a product of the Calvin cycle.

A large amount of free sugar in a cell will cause the cytosol to become thick and syrupy. This causes a hypertonic osmotic condition in the cell which will result in excessive water uptake and potential damage. So, one of the primary benefits of producing starch is to make sugars osmotically inactive by making them insoluble within the cell. The starch produced by chloroplasts is, in most species, the primary storage form that is mobilized for translocation to other plant parts during night periods. It often aggregates into starch grains which typically occur as several granules lying between grana membrane stacks inside the chloroplasts. This starch can be hydrolyzed to D-glucose that can be used for ATP synthesis via aerobic respiration to maintain turgor pressure in growing cells via its osmotic effects and for synthesis of cellulose and other polysaccharides in the cell wall. Translocated sugars and starch are also important in the development of storage organs, such as the above rice grains.

Large quantities of starch can be found in storage organs such as tubers and tuberous roots, taproots, stems located above ground, as well as seeds. These tissues are termed "sinks" by physiologists and agronomists. They allow the plant to survive on stored energy for long periods of winter or drought. In potatoes, under warm weather conditions starch typically gets hydrolyzed to sugar used for growth of new shoots. Starch also occurs in the root caps located at the tips of growing roots in the soil. It is stored in specialized colorless plastids in rootcap cells called *amyloplasts*. The starch-filled amyloplasts are dense, heavy bodies which fall downwards in the rootcap cells when a root is placed horizontally (gravistimulated). These serve as gravisensors that trigger signal transduction events resulting in asymmetric growth of the root downward. Why is this so? It has been shown that if the rootcaps are removed from corn roots, the roots will not curve down when placed horizontally and thus will not grow into the soil where nutrients are located. When the rootcaps are replaced, gravisensitivity is restored. The gravitropic curvature response is much lower[4] in *Arabidopsis* mutants that have a lesion in starch biosynthesis that results in poorly formed, small starch grains in the amyloplasts. The starch grains in chloroplasts (Figure 2.13), as in amyloplasts of root caps, can serve as gravisensors in prostrated stems of plants. When this starch is depleted artificially by placing the shoots in the dark for 4 to 5 d, the stems no longer respond to gravity; but, when fed sucrose, starch is resynthesized and gravisensitivity is restored.[4]

Humans eat starch in products such as potatoes, cereal grains, taro, and tapioca. It is also important in beer brewing as a "modified barley substrate" used in secondary fermentations. What happens here is that starch in barley is hydrolyzed to maltose (a disaccharide) and eventually to D-glucose. This hexose is used as a substrate (food) for beer fermenting yeast which, under anaerobic conditions, convert the sugar to ethyl alcohol (ca. 3.5 to 4.5%) and carbon dioxide. Starch is easily visualized in storage organs, such as potatoes, or in swollen joints (pulvini) of cereal grass stems by the use of a simple histochemical test. Fresh sections of plant tissue are placed in a 1% solution of iodine-potassium iodide (1:1) and the resulting stained starch grains appear blue-black in the light microscope (Figure 2.13).

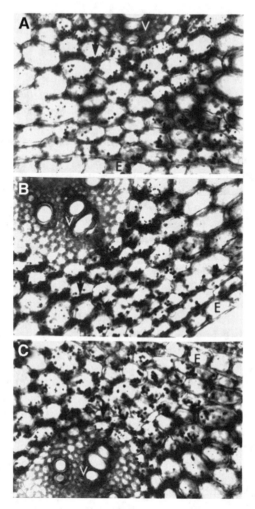

FIGURE 2.13   Starch grains stained with I₂KI in chloroplasts of oat cells located in the graviresponsive swollen leaf sheath pulvini of the shoot. Arrows indicate the direction of the gravity vector; E = epidermis; V = vascular bundle. A, B, and C × 100; D × 200. (Photo courtesy of Casey Lu and Peter Kaufman.)

2.4.6   Fructans and Fructan Biosynthesis

Fructans are soluble storage polysaccharides found in the vacuoles of cells of plants which are known to be fructan accumulators. They are made predominantly from the sugar D-fructose (hence the name), but D-glucose molecules may also be present in the chain. Classic examples include temperate zone monocots such as grasses, lilies, onions, irises, and dicots such as dahlias and Jerusalem artichokes. A complete compilation of families of monocots and dicots in which fructans are known to occur is cited in Reference 7. In dahlia and Jerusalem artichoke, the fructan is referred to as *inulin*, where the polysaccharide is composed mostly or exclusively of 2–1

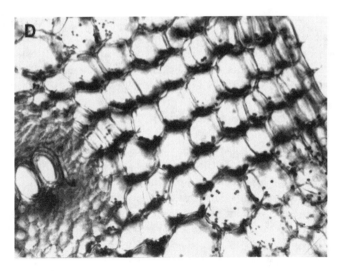

FIGURE 2.13 (continued)

fructosyl-fructose linkages (a D-glucose molecule is allowed but not necessary). In the temperate zone grasses, the fructan is termed either ***graminan***, which has both 2–1 and 2–6 fructosyl-fructose linkages, or ***phlein*** which contain mostly or exclusively 2–6 fructosyl-fructose linkages.[7] As with inulin, D-glucose is allowed in the chain but is not necessary in the structure of graminan and phlein-type fructans.

The basic pathway of synthesis of inulin-type fructans in plant cells has been summarized by Edelman and Jefford.[8] It involves the following three steps, as summarized by Suzuki[7]

- Conversion of sucrose to trisaccharide in the cytosol by the enzyme sucrose-sucrose 1-fructosyltransferase (SST)
- Transfer of the terminal fructosyl moiety of this trisaccharide in the cytosol to sucrose in the vacuole by the enzyme B(2–1′) fructan:B(2–1c) fructan 1-fructosyltransferase (FFT), which is possibly located on the tonoplast membrane surrounding the vacuole
- Continued transfers by FFT of terminal fructosyl groups from the resulting molecules of trisaccharide (e.g., kestose or isokestose) in the vacuole to the extending fructan chain resulting in the formation of inulin molecules

Fructan's primary role in plants is that of a reserve carbohydrate similar to starch. Temperate, cold-tolerant grasses like oats, barley, wheat, and rye typically contain fructans and sucrose as the primary carbohydrate reserves. Tropical, warm-loving and cold-intolerant grasses such as maize contain starch and sucrose as the primary reserve carbohydrates. It is interesting that in the shoots of temperate zone grasses and in the tubers of the Jerusalem artichoke, fructan synthesis accelerates under low temperature conditions of autumn; then the stored fructans become hydrolyzed through the action of fructan hydrolase in the spring when temperatures warm and shoot and root growth begin. This appears to provide the plant with a source for energy for a head-start on growth in the early spring.

Jerusalem artichoke (*Helianthus tuberosus*) tubers are frequently eaten by humans as a potato substitute (but **not** starch substitute). Humans cannot digest the inulin fructan present in these tubers because of the absence of the gene that makes the fructan-specific hydrolase in humans. Furthermore, the ubiquitous intestinal colon bacterium, *Escherichia coli*, cannot hydrolyze fructan. This would make one think that these tubers would be perfect food for dieters. However, there is recent evidence from Japanese studies that *Bifidobacteria*, found in intestinal microflora, can digest fructan; in fact, when fructans are eaten, populations of this microbe in the large intestine increase significantly. This being the case, enrichment of the human diet with fructans from plants such as rye, onions, Jerusalem artichoke tubers, and garlic may be beneficial, not because they are hydrolyzed in the small intestine, but because they are hydrolyzed in the large intestine. There is also evidence that fructans from plant sources may also be beneficial in the diets of swine and poultry (see "Fructans in Human and Animal Diets" by Farnworth in Susuki and Chatterton[7]).

2.4.7 Gum, Mucilage, and Dietary Fiber

There are other forms of polysaccharides that plants produce. Some of these are gums and mucilage which are highly branched heteropolysaccharides (related to hemicellulose and pectic polysaccharides) that contain acidic residues thus making them very hydrophobic, insoluble within the plant cell, and often difficult for animals to digest. One benefit to the plant that produces indigestible polymers is that it reduces the reward for herbivores. In other words, the animal spends its time eating yet gets nothing out of the process. For the plant, these polysaccharides can function as a storage reserve for carbohydrates, but they are also found as part of the matrix that surrounds the walls of some cells. This matrix is called the **glycocalyx** and is mostly seen on the surfaces of roots where it may serve to protect the plant against microbial invasion. Glycocalyx secretions are not unique to plants. They are also found in bacteria and animals where, as in plants, they act in cell-cell recognition of symbionts or pathogens. Another function of mucilage is seen in carnivorous plants like Sundew (*Drosera* spp.) where a substance called mucin is produced to catch unwary insects in nutrient-poor environments. Gums are also useful in sealing wounds in leaves and stems. For example, when a cherry tree is injured, it will produce a thick substance called gum arabic that fills in the wound thus preventing infection. This also acts as a human cosmetic.

Cellulose, pectin, lignin, waxes, gums, and mucilages are some of the many types of dietary fiber. Fiber is simply the insoluble polymers of plants and most come from cell walls. Fiber stimulates the gastrointestinal tract and acts as a laxative. Fiber containing pectins reduces blood cholesterol by adsorbing cholesterol molecules. Fiber, in general, appears to inhibit many cancers, especially colon cancer, by binding the carcinogens and preventing them from entering the body while they pass through the system. One problem, however, is that fiber may also adsorb vitamins thus carrying them out of the body before they can be absorbed. So, a balance of fiber in the diet is essential.

2.4.8 CHLOROPHYLL AND CHLOROPHYLL BIOSYNTHESIS

There are three main locations of pigments within the cell: (1) plastids, (2) vacuole, and (3) the cell wall. The chemistry of the pigments varies with the location. Chlorophyll is a porphyrin that constitutes the primary photoreceptor pigment for the process of photosynthesis in plants. It is produced in the chloroplasts and is responsible for the green appearance of leaves and stems, aerial and prop roots, many kinds of floral bracts, and green fruits before they ripen. The chlorophyll molecule is made up of four pyrrole rings, made from alliphatic amino acids, (designated I to IV) that are ligated to form a *tetrapyrrole* ring with a magnesium atom in its center; ring IV is esterified with a hydrophobic long chain *phytol* molecule ($C_{20}H_{39}$) made in the terpenoid pathway.[9] For light harvesting, plants use two forms of chlorophyll, a and b. Chlorophyll a is in all plants and is the only chlorophyll at the reaction centers. It has a methyl group at C3, while chlorophyll b, found in most plants, has a formyl group at this position and, like other accessory pigments, functions to absorb the energy from wavelengths of light that differ from chlorophyll a (Figure 2.14). The chemical structure of chlorophyll a is shown in Figure 2.15.

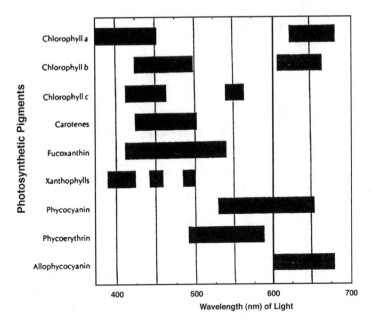

FIGURE 2.14   Absorption of different wavelengths of light by various photosynthetic pigments in plants.

The biosynthesis of the chlorophylls is quite complex. It starts with the synthesis of 5-aminolevulinic acid from glutamic acid. The porphyrin ring containing conjugated double bonds is assembled in the chloroplast from eight molecules of 5-aminolevulinic acid. Subsequent steps lead to the formation of protochlorophyllide, addition of the phytyl tail, and insertion of a Mg^{2+} atom in the center of the

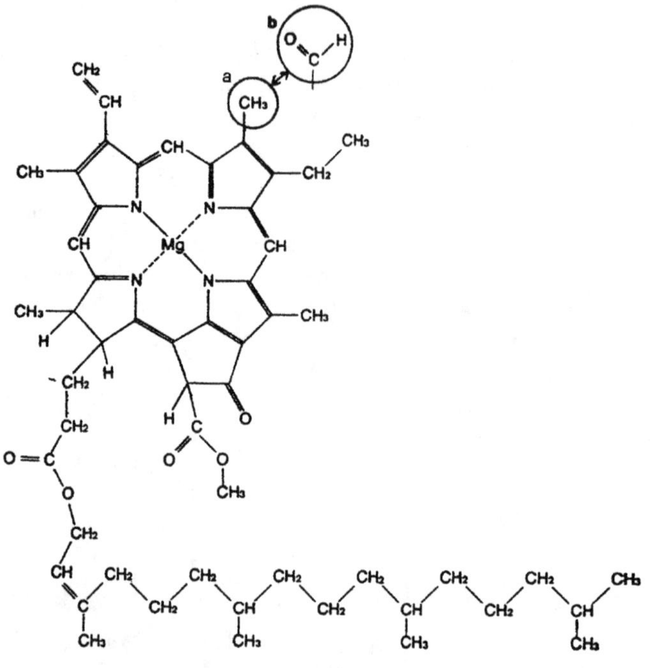

FIGURE 2.15 Chemical structure of a chlorophyll a molecule. Note the tetrapyrole ring with a Mg atom in its center at the top and the long chain phytol "tail" at the base. Note also the formyl group substitution at the methyl group in upper right corner of the chlorophyll a molecule; such a substitution gives one the structure of chlorophyll b.

tetrapyrrole ring. These are illustrated in the reviedw by vonWettstein et al.[9] In the chloroplasts, the chlorophyll pigments are bound to proteins of the photosynthetic membranes (stacks of thylakoid membranes inside chloroplasts called *grana* stacks). These proteins, called chlorophyll a/b binding proteins, are arranged into large complexes with many other proteins, cytochromes, and quinones to form the photosynthetic electron-transport chain whose primary function is to produce the ATP required to fix carbon dioxide. It is the pigment chlorophyll that absorbs the energy of the sun and shuttles resulting free electrons to this all-important series of chemical events.

Chlorophyll absorbs photons of light energy from the sun or from artificial lamps (e.g., incandescent lamps, high pressure lamps, light-emitting diodes) in the red and blue portions of the electromagnetic spectrum with peaks of maximal absorption occurring at 660 and 450 nm, respectively. This is called its **absorption spectrum**. Absorption spectra are commonly used to characterize pigment types. Maximal rates of photosynthesis (measured by the rate of CO_2 uptake or O_2 evolution) also occur in the red and blue portions of the electromagnetic spectrum. This is called its **action spectrum**. When the action spectrum peaks, like those for photosynthesis, match the absorption peaks for a given pigment(s), like those for chlorophylls, one can deduce that this pigment(s) is the one that is essential for absorption of light for the particular process under consideration. Another type of proof is to find plants that lack the pigment of interest and determine which processes are functional. For example, albino mutants and parasitic plants such as Indian pipe (*Monotropa* spp.), which are devoid of chlorophyll pigments, cannot carry out photosynthesis. Also, please note that not all plants photosynthesize. Parasitic plants feed off the nutrients and sugars provided by their hosts.

2.4.9 CAROTENOID BIOSYNTHESIS

Plant carotenoids (Figure 2.16) are responsible for the red, orange, and yellow pigments found in fruits and roots such as tomatos, red peppers, pumpkins, and carrots. They can be seen in the petals of many flowers and are the primary pigments responsible for the fall coloration of deciduous trees. Carotenoids are synthesized in the terpenoid pathway as C_{40} tetraterpenes derived from the condensation of eight isoprene units starting with isopentenyl diphosphate (Figure 2.17).[10,11] There are two basic types of carotenoids: (1) carotene which contains no oxygen atoms and (2) xanthophyll which does contain oxygen (Figure 2.16). At the center of each carotenoid molecule, the linkage order is reversed, resulting in a molecule which is symmetrical. A set of double bonds in the molecule is responsible for the absorption of light in the visible portion of the spectrum.[10] As mentioned above, this has an important impact on the absorption of a wider range of light wavelengths for use in photosynthesis (see Figure 2.14). Consequently, in photosynthetic organisms, carotenoids are an integral structural component of photosynthetic antenna and reaction center complexes, but they also protect against the harmful effects of photooxidation processes.[10] So, like chlorophyll, carotenoids are found in the thylakoids of green leaves and stems. In fruits and flowers they are also found in plastids, but these plastids have structural differences and are referred to as chromoplasts to indicate that they contain pigments other than chlorophyll.

β-carotene is the orange pigment in carrot roots, pumpkin fruits, leaves of deciduous trees, and some flower petals. Zeaxanthin and violaxantin are also found in autumn-colored leaves and flower petals and are responsible for the bright yellows that are sometimes seen. Coloration of flowers is very important to the survival success of the plants producing them. The color of the flowers is one of the primary factors involved in attracting pollinators. For more information on the attraction of pollinators, see Section 2.6.6. For humans, β-carotene is important in our diet because of its purported anticancer activity, its use as a food coloring, and as an important source of vitamin A that is synthesized from β-carotene and other carotenoids. Vitamin A produced by animals is in turn converted to the pigment, *retinal*. This pigment is one of the essential components in the light receptors of the eye that allow us to see.

Carotenoid pigments can also act in fruit and seed dispersal by attracting animals which in turn spread the seeds. Most fruits also produce odor compounds such as monoterpenes to help attract these organisms, and the sugars produced and stored in the fruits act as a positive reward. In ripening fruits, as in leaves turning color in the autumn, chlorophyll pigments gradually break down in chloroplast thylakoid membranes, revealing the carotenoid pigments that were masked by the chlorophyll pigments. During ripening there is also significant synthesis of new carotenoid pigments.

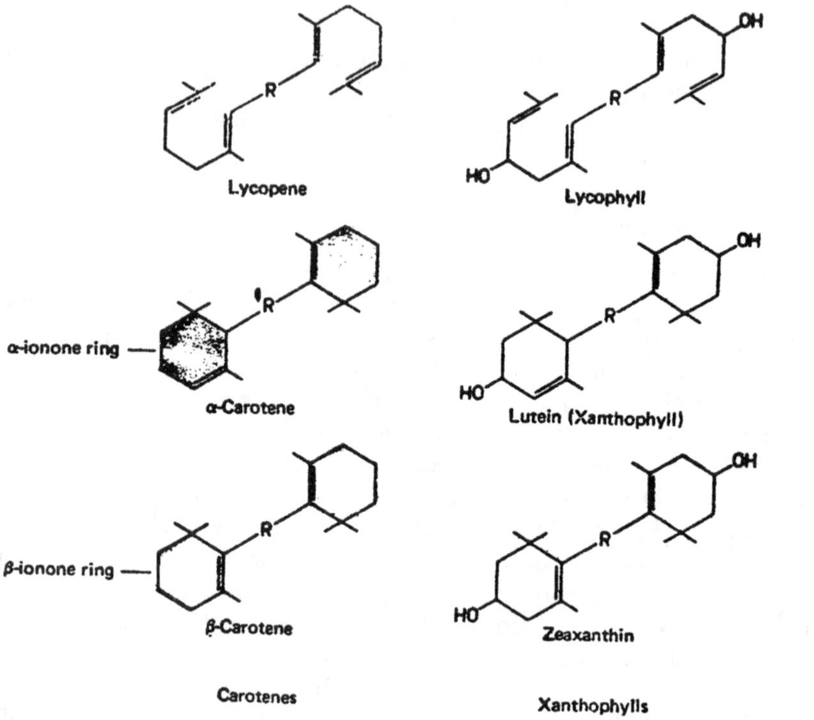

FIGURE 2.16 Examples of carotenoid and xanthophyll pigments from plants.

In the case of ripening tomatoes and peppers, for example, the unripe fruits are characteristically bright green. As ripening progresses (triggered by the plant hormone, ethylene) various carotenoid pigments appear and, newly synthesized, account for the color of the ripe fruits. Lycopene is the red pigment seen in mature tomato and red pepper fruits. Tomatoes can have both red and yellow fruits depending on the genotype of the parent. In some peppers, we encounter ripe fruits which are green at maturity (bell peppers). Here, chlorophyll pigments do not breakdown during the ripening process. Other ripe peppers may be yellow or red depending again on the genotype of the parents. Similar types of color changes occur in fruits of ripening cucurbits (squash, gourds, pumpkins) and in the fruits of egg plant.

Some plants such as red maple (*Acer rubrum*) trees produce red-colored flowers and/or leaves, yet they are wind pollinated. It's obvious that these plants do not have to attract pollinators. So, why the color? One interesting theory behind why they spend their energy to do this is that the pigments help to warm the flowers or leaves during early spring or late fall. This extra heat would greatly aid seed development and photosynthetic processes in early spring allowing the plant to get a head start on growth over other plants as well as provide a longer period to produce energy reserves in the fall. Red coloration in many plants is not due to carotenoids, but rather, anthocyanin pigments.

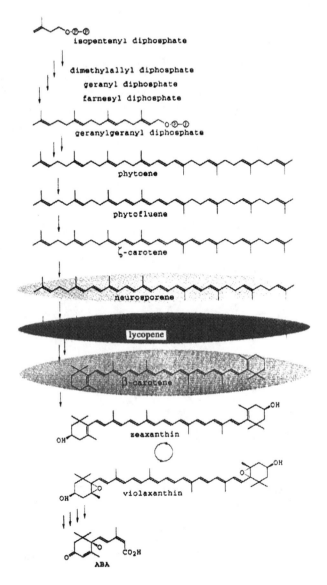

FIGURE 2.17 Carotenoid biosynthesis pathway in plants. (From Bartley, G. E. and Scolnik, P. A., *The Cell Plant,* 7, 1027–1028, 1995. With permission.)

2.4.10 ANTHOCYANIN BIOSYNTHESIS

Anthocyanins are flavonoid-type compounds responsible for most of the red, pink, purple, and blue pigments found in roots, stems, leaves, flowers, seeds, and fruits. Examples include the red anthocyanins in red radish, the red leaves of some Norway maple cultivars (e.g., *Acer saccharum* cv. '*Schwedleri*'), the red fruits of some peppers, apples, and Acerola cherry (*Malpighia glabra*, said also to contain the highest content of vitamin C and ascorbic acid of any fruit), and the red, pink, purple, and blue flowers of *Rhododendron*, *Hibiscus*, and *Fuchsia* to name a few. Anthocyanin pigments occur in the vacuoles of plant cells. They are synthesized from the aromatic amino acid, phenylalanine, in the phenylpropanoid pathway (Figure 2.18). This is the same pathway that is responsible for the synthesis of tannins, flavonones (Figure 2.19) like naringenin, flavonols, flavonoids, isoflanonoids like genistein and daidzein, lignin, lignans, and coumarin.[13]

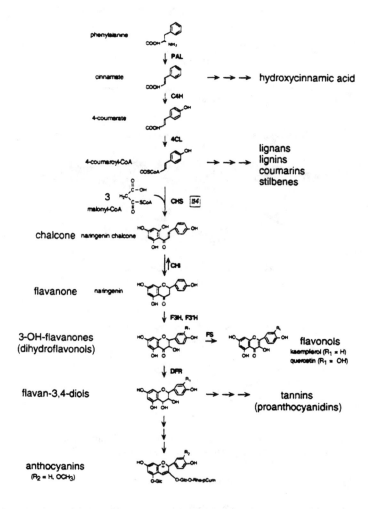

FIGURE 2.18   Diagram of the general phenylpropanoid and flavonoid branch pathways. (From Burbulis, I. E., Iacobucci, M., and Shirley, B. W., *The Plant Cell*, 8, 1013–1025, 1996. With permission.)

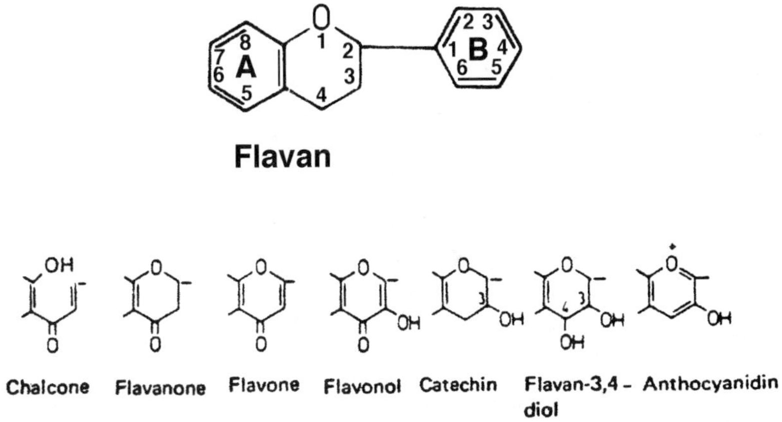

Flavan

Chalcone   Flavanone   Flavone   Flavonol   Catechin   Flavan-3,4-   Anthocyanidin
                                                                          diol

FIGURE 2.19   Survey of several flavan derivatives (below) based on the basic flavan skeleton (above).

The primary enzyme that commits the pathway to biosynthesis of the anthocyanin pigments is chalcone synthase (CHS). There is a whole gene family of CHS genes within most plants. Some of the genes are expressed in very specific tissues. CHS(A), for example, is only expressed in the petals and stamens of flowers that produce anthocyanins. This and subsequent enzymes in the pathway have been well characterized. In petunia, genetic loci controlling the synthesis of most of these enzymes have been located with the exception of 5GT (5-glucosyl transferase).[12] The different colored anthocyanins arise from precursors that include dihydrokaempferol (a precursor of the orange-to-red anthocyanin, pelargonidin), dihydroquercetin (a precursor of the purplish red anthocyanin, cyanindin), and dihydromyricetin (a precursor of the bluish purple anthocyanin, delphinidin) (Figure 2.20). All of these anthocyanidins are converted to their glucosides such as pelargonidin-3-glucoside, cyanidin-3-glucoside, and delphinidin-3-glucoside which allows them better solubility in the aqueous solution of the vacuole.

The glucosyl moieties are typically glucose and rhamnose sugars. The color of anthocyanins is affected by the number of hydroxyl and methoxyl groups in the B ring of the anthocyanidin, but apart from structure, color is also affected by the presence of chelating metals such as iron and aluminum, the presence of flavone or flavonol co-pigments, and the vacuolar pH where these pigments are stored (Table 2.3).[2] As one example, in Hydrangea flowers, where the vacuolar pH is acidic, the flower petals appear blue; where it is alkaline, they appear pink. So the vast variety of coloration of many leaves, flowers, and fruits is often the result of several different pigments — chlorophylls, carotenoids, and anthocyanins.

Anthocyanins serve many diverse functions in plants, including attraction of insect and bird pollinators to flowers and dispersal of seeds and fruits by birds and mammals. In some cases, they are feeding deterrents and, like other flavonoids, can also protect the plant against damage from UV irradiation.[12]

Anthocyanins have great economic importance in expression of the wide array of flower colors in plants grown as ornamentals. In fact, attempts to obtain blue

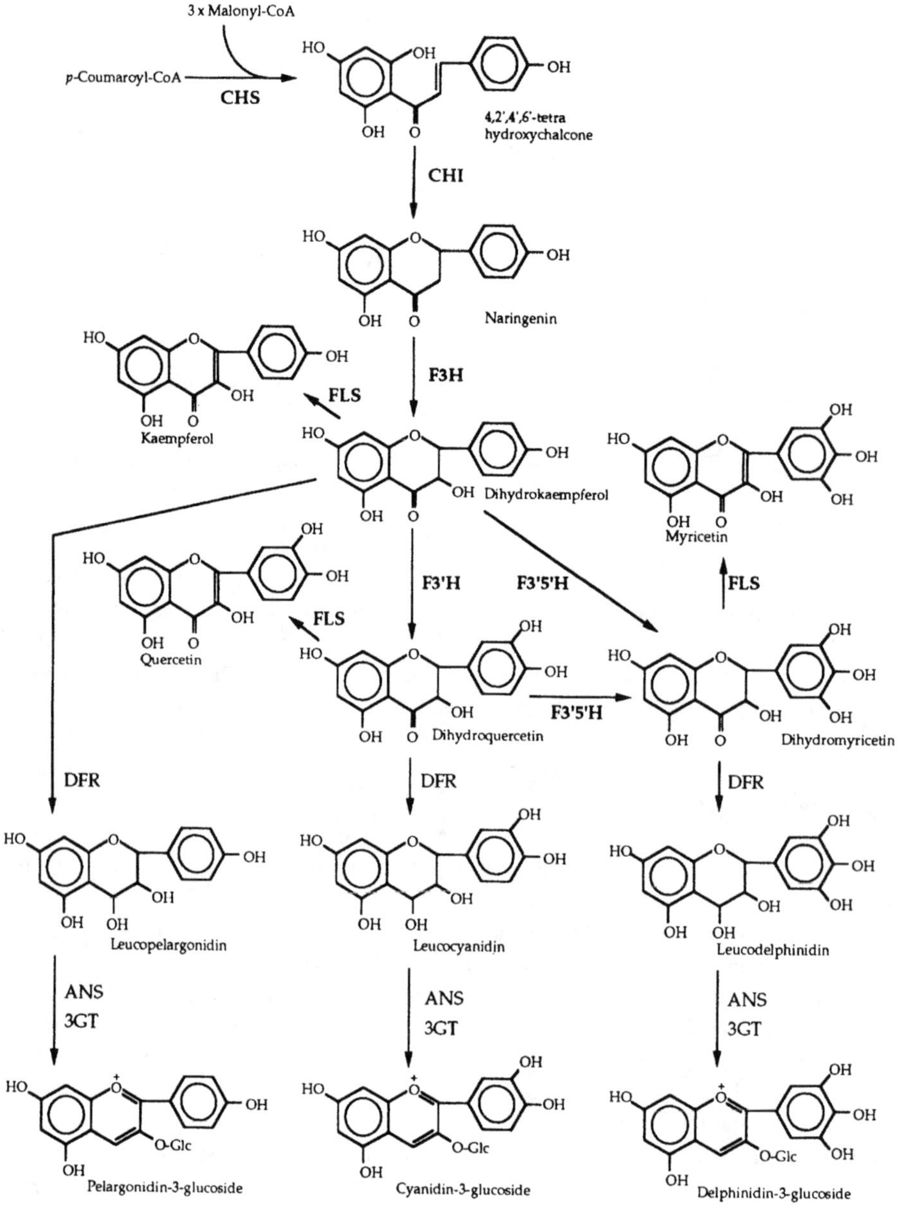

FIGURE 2.20 Anthocyanin and flavanol biosynthetic pathway. (From Holton, T. A. and Cornish, E. C., *The Plant Cell,* 7, 1071–1083, 1995. With permission.)

roses, chrysanthemums and carnations are now possible with transgenic plants. In these plants, synthesis of the blue pigment, delphinidin-3-glucoside, does not normally occur because the 3′,5′-hydroxylase is not normally expressed. In the transgenic plants, this gene, obtained from other plants like petunia, is expressed, resulting

TABLE 2.3
Factors Controlling Cyanic Color in Flowers

[a]Hydroxylation pattern of the anthocyanidins (i.e., based on pelargonidin, cyanidin, or delphinidin)

Pigment concentration

Presence of flavone or flavonol co-pigment (may have blueing effect)

Presence of chelating metal (blueing effect)

Presence of aromatic acyl substituent (blueing effect)

Presence of sugar on B-ring hydroxyl (reddening effect)

Methylation of anthocyanidins (small reddening effect)

Presence of other types of pigment (carotenoids have browning effect)

[a] In approximate order of importance. There are other minor factors, including pH, physical phenomena, etc.

in the synthesis of the blue delphinidin-3-glucoside anthocyanoid pigment. One interesting application in the use of naturally occurring anthocyanin pigments is the one present in the red roots of radish, *Raphanus sativus*. This water-soluble pigment is extracted from these roots and is currently used to dye Maraschino cherries bright red instead of using a synthetic red dye as was done previously. This process was developed by horticulturists at Oregon State University in Corvallis, OR.

2.4.11 ALKALOID BIOSYNTHESIS

So far we have tried to touch upon each of the major categories of products produced by plants in general. We have discussed the biosynthesis of the major cellular components found in the majority of plants including primary storage compounds and key compounds which start the carbon fixation process (chlorophylls). We have used carotenoids to demonstrate the production of terpenoids and anthocyanins to give examples of phenolic compounds. Now we will say a few words about nitrogen-containing compounds which will be represented by the *alkaloids*.

Most of these products are not considered to be essential to the growth and development of the plant, but some, such as pyrimidine nucleotides and tetrapyrroles are absolutely essential. This is why we have separated these compounds from the rest of the nitrogen-containing compounds in Figure 2.1.

There are literally thousands of different plant products that have nitrogen in their structures. Perhaps the most diverse of these types of compounds (found in 20 to 30% of vascular plants) are the alkaloids which, like most other nitrogen-containing compounds, are synthesized from amino acids. Alkaloids are especially interesting because they are quite toxic to both herbivores and humans; yet they have some very important medicinal properties for humans. The nitrogen atom in these substances is almost always part of a heterocyclic ring whose origin is found innately in the structure of the amino acids from which they came or are the result of the circularization of the given amino acid. This is the case with aspartic acid which combines with glyceraldehyde-3-phosphate in the production of nicotinic acid (a precursor of the alkaloid, nicotine) in plants such as tobacco. Nicotine is well known as a toxic component of tobacco smoke. There are many categories of alkaloids, including pyrrolidine, tropane, piperidine, pyrrolizidine, quinolizidine, isoquinoline, and indole alkaloids. Much of the carbon skeleton of some of these alkaloids is derived from the mevalonic acid pathway, but it is beyond the scope of this chapter to go into the details of the biosynthesis of all types of alkaloids. Figure 2.21 shows the major alkaloid classes and the biosynthetic precursors.

Alkaloid Class	Structure	Biosynthetic Precursor	Examples
Pyrrolidine		Aspartic acid	Nicotine
Tropane		Ornithine	Atropine, cocaine
Piperidine	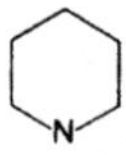	Lysine (oracetate)	Coniine
Pyrrolizidine	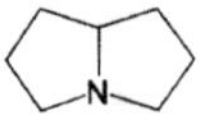	Ornithine	Retrorsine
Quinolizidine	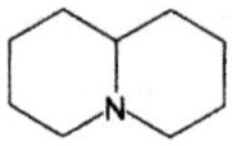	Lysine	Lupinine
Isoquinoline	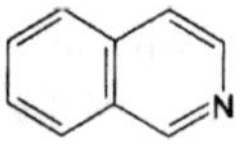	Tyrosine	Codeine, morphine
Indole	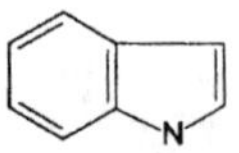	Tryptophan	Psilocybin, reserpine, strychnine

FIGURE 2.21 Major classes of alkaloids, their chemical structures, their biosynthetic precursors, and well-known examples of each class.

An old idea about the function of alkaloids in plants depicted them as waste products of plant metabolism. However, plants are, energetically, very efficient organisms. They simply don't waste their energy in the production of compounds that they don't need — there always seems to be a reason for their production. The predominant activity of alkaloids in plants seems to be the deterrence of herbivores. Many livestock deaths are caused by ingestion of alkaloid-containing plants such as lupines (*Lupinus* spp.), larkspur (*Delphinium* spp.), and groundsel (*Senecio* spp.).

They also have been shown to be toxic to insects, bacteria, and fungi. Alkaloids are not solely defensive substances. Some red and yellow-colored alkaloids called beta-lains, like carotenoids and anthocyanins, act as attractants in flowers and fruits of plant species such as beets and cacti. It is interesting to note that plant families that contain betalain pigments never contain anthocyanins. Some pyrrolizidine alkaloids also act as attractants by mimicking such compounds as the sexual pheromones normally produced by some insects like butterflies. These compounds trick the insect into visiting the flower and spreading the plant's pollen, but alkaloids in general are toxic. When taken in sufficient quantity, alkaloids are dangerously toxic to humans, but at lower doses, many are helpful — morphine, codeine, atropine, and ephedrine to name a few. Other alkaloids, including nicotine, caffeine, and cocaine find pop-ularity as nonmedicinal stimulants or sedatives, but they too have their toxic effects.

2.5 SYNTHESIS OF PLANT METABOLITES IN SPECIALIZED STRUCTURES OR TISSUES

Plants do not always produce their products in every cell of the organism. Often plants have developed tissue-specific locations for synthesis of certain compounds which not only accentuates the compound's specific function but perhaps avoids the toxic effects that the compound may have on the plant itself. Indeed, this is the case for all plant products within each cell of every plant, but the plant as a whole must have a system for dealing with potentially hazardous substances. The following are a few examples.

2.5.1 SYNTHESIS OF MONOTERPENES IN LEAVES OF PEPPERMINT (MENTHA PIPERITA)

It has been shown by Croteau and Winters[14] at Washington State University that leaves can synthesize a variety of monoterpenes from geranyl pyrophosphate (GPP), as shown in Figure 2.22. GPP production in the terpenoid pathway is the universal precursor of all monoterpenes. Monoterpenes, as well as some sesquiterpenes, in general serve as antiherbivore agents that have significant insect toxicity while having negligible toxicity to mammals. Mixtures of these low molecular weight volatiles, called essential oils, are what give plants such as peppermint, lemon, basil, and sage their characteristic odors, and many are commercially important in flavoring foods and in making perfumes.

Of particular interest in peppermint, is the pathway of l-menthone metabolism illustrated in Figure 2.23. The branch of this pathway at the top of the figure shows the biosynthesis of l-menthol and l-menthyl acetate from l-menthone. These substrates and the enzymes that lead to their biosynthesis occur in the glandular hairs that arise from leaf epidermal tissue. The products are stored in a modified extracellular space between the cuticle and the cell wall. Well known to repel insects, menthol at the very surface of the leaves (in hairs) seems to deter herbivores before they even get a chance to take a trial bite. In contrast the branch in the pathway at the bottom of the figure that leads to the synthesis of d-neomenthol and d-neomenthyl glucoside occurs not in the epidermal hairs, but rather, in the photosynthetic mesophyll tissue of the leaves

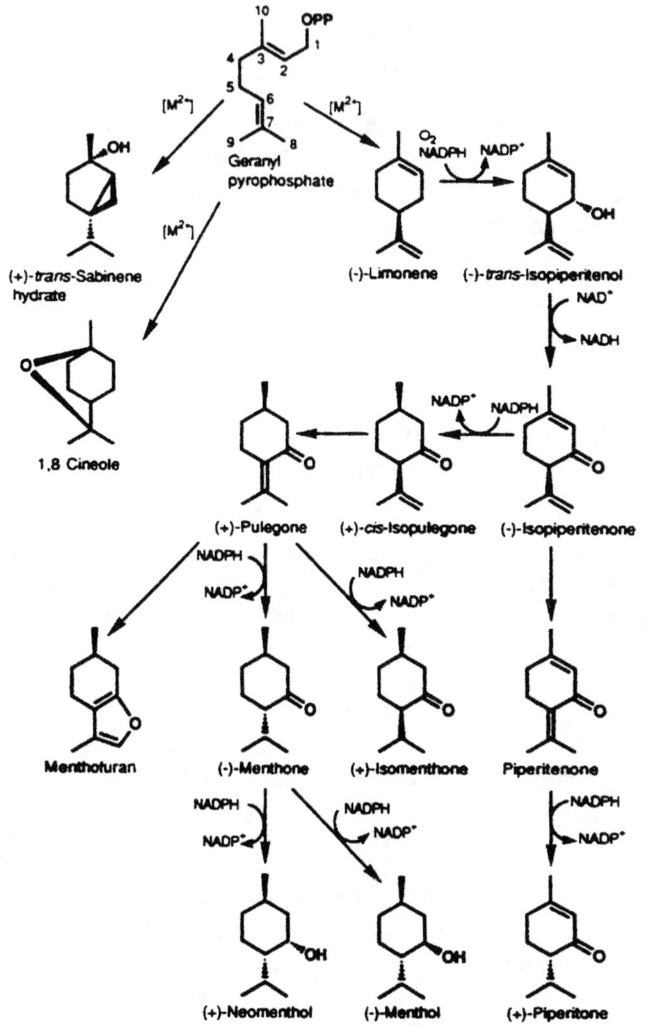

FIGURE 2.22 Major pathways and cofactor requirements for monoterpene biosynthesis in peppermint (*Mentha spicata*). [M^{2+}] is the divalent metal ion cofactor (either Mg^{2+} or Mn^{2+} required by monoterpene cyclases). (From McCaskill, D., Gershenzon, J., and Croteau, R., *Planta*, 187, 445–454, 1992. With permission.)

that lies inside the epidermis. The ultimate product, *d*-neomenthyl glucoside, is then translocated from the leaf mesophyll tissue to the phloem in the leaf vascular bundles, and from there to the roots of the plant where it is stored. This difference in cell/tissue compartmentation for monoterpene biosynthesis in peppermint leaves is of particular interest to biochemists, physiologists, and cell biologists as a model for the control of gene expression in different tissues and for the study of translocation of compounds within the plant. For the plant, it most likely evolved this bifurcation in the *l*-menthone

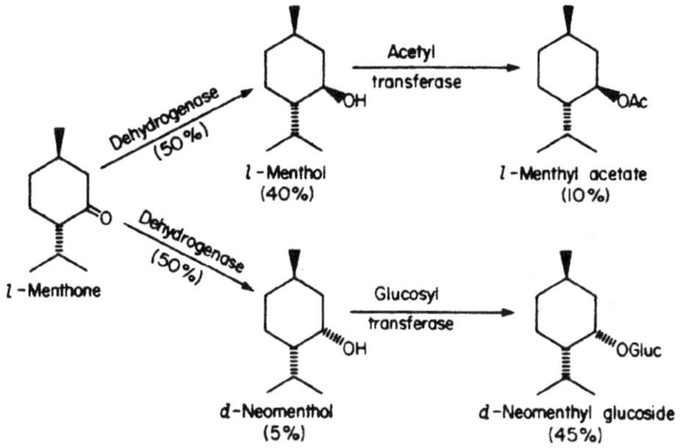

FIGURE 2.23 Pathways of *l*-menthone metabolism in peppermint (*Mentha spicata*). The percentages indicate the approximate distribution of the products derived from *l*-menthone in mature leaf-tissue. (From Croteau, R. and Winters, J. N., *Plant Physiology*, 69, 975–977, 1982. With permission.)

biosynthetic pathway in response to predation pressures by insects and herbivores which prey on both the leaves/stems and roots of these plants. However, not all monoterpenes and sesquiterpenes are repellents. Sometimes their primary function is to attract.

2.5.2 SYNTHESIS OF MONOTERPENES IN THE FLOWERS OF *CLARKIA BREWERI* AND OTHER SPECIES

Apart from the coloration factors discussed above and in Section 2.6.6, the flowers of many plant species attract pollinators by producing different complex mixtures of volatile compounds within the various floral organs (i.e., stigma, style, ovary, filaments, petals, or sepals). It is the combinations of the constituents of this scent mixture that give each flowering plant species a unique fragrance.[15,16] The fact that insects can distinguish between these different floral scent mixtures is the key to the reason that many specific plant species often have specific pollinator species. For example, plants that make flowers which produce linalool (a monoterpene) very often attract moth pollinators during the night, while species that may look very similar and live in the same area but do not produce linalool do not attract moths.[17] They are pollinated by other insects, usually bees or butterflies during the daytime. Thus, the components of a floral scent have important implications for the pollination success of the plants that produce them.[18,19-21]

Although floral scent production is crucial, it has been virtually ignored by the biochemical and molecular biology disciplines. Few of the biochemical pathways that produce the vast array of scent compounds have been elucidated, and although many of these compounds are monoterpenes, only one of the enzymes that directly produces a monoterpenoid floral scent compound has been identified at this time. This enzyme, linalool synthase (LIS), catalyzes the conversion of GPP directly to linalool (Figure 2.24). Linalool is a common acyclic monoterpenoid floral scent

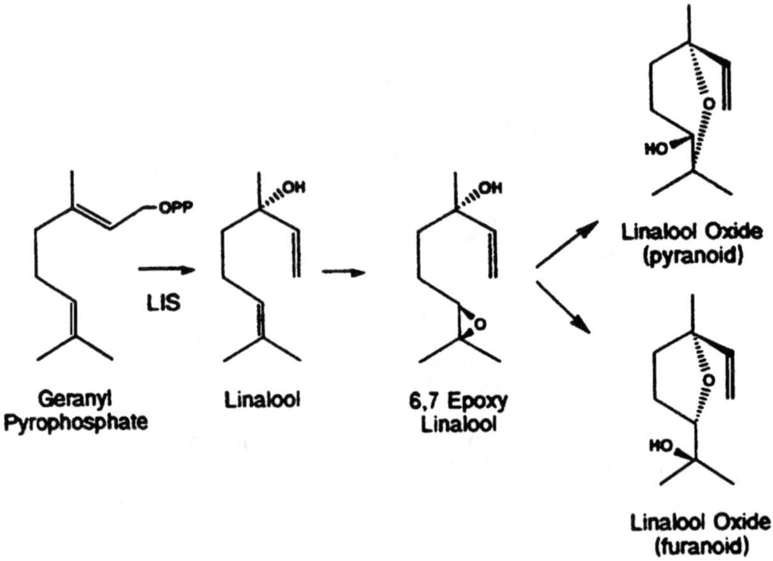

FIGURE 2.24   The linalool and linalool oxides pathway. (Courtesy of Eran Pichersky and Leland Cseke).

compound produced by the flowers of many plant species.[18,20-24] In *Clarkia breweri* plants (a small annual plant native to California and the only species where LIS activity is well characterized), it is produced predominantly by the epidermal cells of the petals which are responsible for the majority of linalool emission from the flower. Linalool also has its oxide forms that are produced through a suspected epoxide intermediate by an as-yet unidentified epoxidase (see Figure 2.24). These oxides are produced predominantly in the transmitting tissue of the stigma and style of each flower where pollen tubes grow during pollination. The oxides, however, are a minor component of the floral scent mixture. Both linalool and it's oxides are only produced when the flower is open, beginning as soon as the flower opens and ending just after the flower is pollinated. This timing has a distinct advantage for the plant since it avoids wasted energy by the production of compounds when they are not needed. Linalool is known to be toxic to some insects such as fleas. There is also some evidence through transgenic studies that linalool production is toxic to young plant tissue. Thus, producing linalool only when a more mature tissue, such as a flower, has developed may avoid other toxic effects within the plant. In any case, the primary activity of linalool itself seems to be to attract a specific moth pollinator (a hawkmoth) that lives in the same regions as *C. breweri*. The oxides may also play a part in this role, but it seems likely from their expression pattern that linalool oxides have potential roles (1) in directing the visiting insect specifically to the stigma where it is most advantageous for the plant to have pollen placed or (2) in the inhibition of pollen tube growth of other species or the stimulation of pollen tube growth from the same species. The true function of the oxides, however, is not known. Like other monoterpenes, linalool is also important in industry as a starting material in the production of perfumes and as a flavoring compound in food and drink.[25] So, its study not only helps us understand how plants communicate with insects but may also benefit industry and agriculture — especially with the potential for the modification of scent production through transgenic plants of crop plants that are grown outside of their natural pollinator's living range and thus suffer from lower crop yields.

Another interesting part of the *Clarkia* project deals with the general question of how the ability to produce linalool changes over evolutionary time. As mentioned above, species that produce linalool are generally pollinated by moths, while species that do not produce linalool are pollinated predominantly by bees and butterflies. This part of the study focuses on the differences in the molecular genetics and biochemistry of scent production between *Clarkia* and *Oenothera* (evening primrose) species that determines the differences in primary pollinators. *Oenothera* and *Clarkia* are in the same family (Onagraceae) and are thus very closely related. Most *Oenothera* produce scent including linalool, yet only two species within the *Clarkia* genus, *C. concinna* and *C. breweri*, produce any linalool at all.[22-24] Flowers of *C. concinna*, like those of all other *Clarkia* species, are odorless to the human nose. However, linalool and its pyranoid and furanoid oxides have been detected in *C. concinna* stigmas using gas chromatography/mass spectrometry (GC-MS), but at levels a thousand-fold less than in *C. breweri*. Additionally, chromosomal, morphological, and genetic data suggest that *C. breweri* has evolved relatively recently from *C. concinna*.[20,24] These observations raise at least two questions: (a) What is the function of the linalool pathway in nonscented plants such as *C. concinna*; and (b) What is the mechanism of evolution that allows the scent trait to be switched off and on over evolutionary time? This evolution could occur through several mechanisms — enzymatic, morphological, or genetic — but research so far has narrowed the possibilities for differential scent production between *C. breweri* and *C. concinna* to control at the level of transcription.[26] The project aims to determine if the same type of mechanism is involved in the biosynthesis and emission of linalool in *Oenothera* species. It is generally accepted that *Oenothera* and *Clarkia* species share a common ancestor, yet they show a surprising diversity in the ability to produce linalool. By characterizing the expression and regulation of the gene encoding linalool synthase, the project promises to uncover how scented species, represented by *Oenothera*, evolve into nonscented species, represented by most *Clarkia*, and yet retain the ability to evolve into scented species again — represented by *C. breweri*.

2.5.3 SYNTHESIS OF OLEORESIN TERPENES IN CONIFERS

Oleoresin is a mixture of terpenoid compounds in the tissues of many species but is best characterized in conifers. Oleoresin from pine trees, also known as "pitch", is composed mainly of monoterpene olefins (turpentine) and diterpene resin acids (rosin).[27] So-called constitutive oleoresin is synthesized in epithelial cells surrounding resin ducts in the needles and stem[28] as well as in resin blisters on the bark of the tree trunk. In contrast, induced resin arises from nonspecialized cells located adjacent to site(s) of injury that are not normally associated with oleoresin biosynthesis.[29] This resin is secreted in response to physical wounding and/or attack by fungal pathogens and insects such as bark beetles. Resins, however, are not all related to gum which may have the same function in other species. This defense reaction by conifers is adaptively important to the survival of conifers in natural habitats because the oleoresins are antifungal and toxic to bark beetles. Wounded areas in the bark of a tree trunk or branch physically become sealed by the solidification of the resin acids after the terpentine has evaporated.[29] This may also serve to prevent loss of water.

The biosynthetic pathway for the synthesis of monoterpene olefins and abietic acid (the primary diterpenoid resin of grand fir, *Abies grandis*) is shown in Figure 2.25.[27] Note that the starting substrate in the pathway is acetyl-CoA. From it, oleoresin biosynthesis proceeds stepwise via mevalonate, isopentenyl pyrophosphate, dimethylallyl pyrophosphate, farnesyl pyrophosphate, and geranylgeranyl pyrophosphate (GGPP) in the same biochemical processes that produce the precursors of menthol and linalool. The GPP leads directly to synthesis of monoterpene olefins such as α- and β-pinene, 3-carene, β-phellandrene, and limonene catalyzed by monoterpene cyclases. The substrate, GGPP, leads to the synthesis of the diterpenoid resin, abietic acid, via four enzymatic steps involving a single cyclase, two hyroxylases, and a dehydrogenase. Each of these enzymes has been isolated and assayed for the production of respective products by liquid scintillation spectrometry, using [1(2)-^{14}C] acetic acid as the starting substrate.[27]

2.5.4   SECRETION OF SODIUM AND POTASSIUM CHLORIDE FROM SALT GLANDS OF PLANTS THAT GROW IN SALINE ENVIRONMENTS (HALOPHYTES)

Over evolutionary time, plants have developed mechanisms that allow them to survive in a given environment which has specific conditions. Some times these environmental conditions are quite harsh. A number of plants which are tolerant of and grow in saline environments actually secrete salts from their leaves using specialized salt glands. In fact, when one tastes the leaves, they taste salty because of these saline secretions. One such plant is salt grass, *Distichlis spicata*, which grows in such areas as the "playas" or salt flats near the Great Salt Lake and the Bonneville Salt Flats in Utah or in the saline soils of the Sacramento Valley of California. Scanning electron micrographs of the surfaces of the leaves of saltgrass reveal glands and toothpaste-like secretions that emanate from these glands. If one makes X-ray analysis maps for sodium, potassium, and chlorine of the same area photographed with the scanning electron microscope, the images seen on the CRT (cathode ray tube) will reveal bright-dot images over each toothpaste secretion. This tells researchers which elements are present in the secretions. From this information, it was shown that these secretions are potassium chloride and sodium chloride, corresponding to the predominant salts in the soil in which these plants grow. So, to avoid possible damage due to the osmotic effects of salt, these plants simply secrete the salt that is taken up by the roots thus keeping it out of the plant's cells.

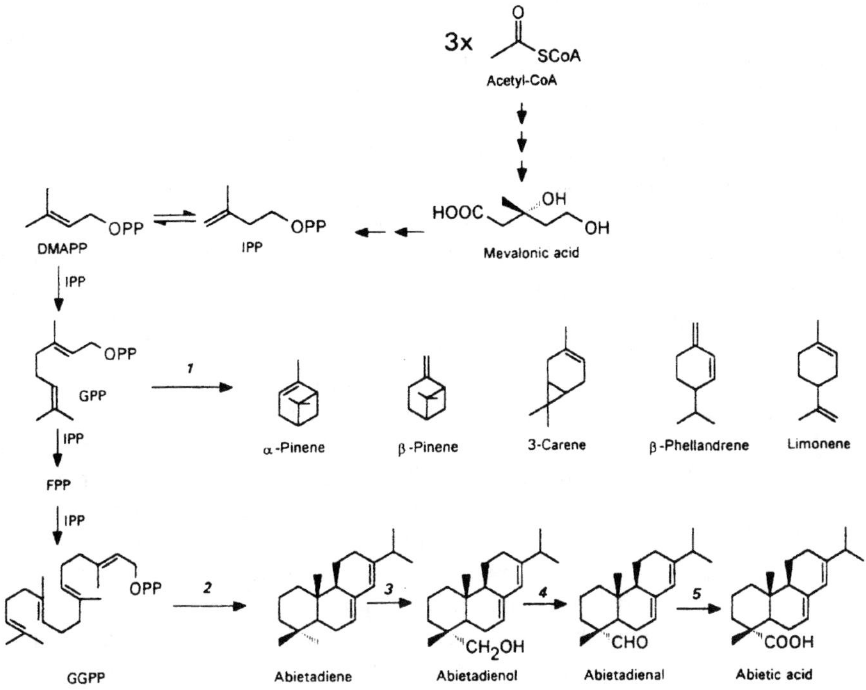

FIGURE 2.25 Outline of the biosynthesis of monoterpene olefins and abietic acid, the principal diterpenoid resin of grand fir (*Abies grandis*) oleoresin. IPP, isopentenyl pyrophosphate; DMAPP, dimethylallyl pyrophosphate; GPP, geranyl pyrophosphate; FPP, farnesyl pyrophosphate; GGPP, geranyl-geranyl pyrophosphate; 1, monoterpene cyclases; 2, abietadiene cyclase; 3, abietadiene hydroxylase; 4, abietadienol hydroxylase; 5, abietadienal dehydrogenase. (From Funk, C. et al., *Plant Physiology*, 106, 999–1005, 1994. With permission.)

2.6 ADAPTIVE FUNCTIONS OF METABOLITES IN PLANTS

2.6.1 SOURCES OF METABOLIC ENERGY AND ENERGY TRANSFER

Without a source of metabolic energy or the ability to transfer the energy obtained from the environment through metabolic pathways, a living organism will die. This is why plants devote so much of their time and energy towards the production of pools of compounds that ultimately store the energy of the sun. Plants have several such sources of metabolic energy derived from stored metabolites or from ATP. The stored metabolites include starch (universal in green vascular plants), fructan [in grass family (Poaceae), lily family (Liliaceae), amaryllis family (Amaryllidaceae), aster family (Asteraceae), and in other families], other polysaccharides (gums and mucilage), and stored lipids and proteins (as in the fats, oils, and protein bodies of seeds and fruits). Each of these polymers may be broken down by specific enzymes when the need for energy arises such as during the night when sunlight is not available or during seed germination. The units of these polymers (sugars, amino acids, or acetyl CoA) then enter the mainstream of the plant's metabolism where they can once again help produce ATP. ATP is produced in the electron transport cascade during photosynthetic photophosphorylation in chloroplasts and oxidative phosphorylation in mitochondria.

Plants have evolved two major pathways of photosynthetic carbon fixation. In C-3 plants (which are represented by most plant species), the primary product is phosphoglyceric acid (PGA) which is used for synthesis of 4-, 5-, 6-, and 7-carbon sugars in the Calvin cycle. In C-3 plants, typically 30% of the fixed carbon is lost as carbon dioxide through photorespiration (a process which liberates CO_2 and constitutes a significant energy drain). In contrast, in C-4 plants (such as sugarcane, corn, and many fast growing weeds) the primary products are both the 3-carbon acid, PGA, as well as the 4-carbon acids, malate and aspartate; the former is produced in chloroplasts in leaf mesophyll tissue, whereas the latter are produced in chloroplasts of vascular bundle sheath cells. What is especially interesting is that there is little or no photorespiration in C-4 plants, so that total carbon fixed is, on average, 30% higher than in C-3 plants. The reason for this difference is that in the process of shuttling carbon from mesophyll to bundle sheath cells, a much higher concentration of carbon dioxide is generated in the bundle sheath cells. It is this elevated CO_2 partial pressure in bundle sheath cells that suppresses RUDP (ribulose *bis*phosphate) oxygenation. This is the first step in photorespiration. The enzyme involved here (RUBISCO) has both carboxylase and oxygenase catalytic activity. So, the higher level of CO_2 inhibits this enzyme's oxygenase activity.

2.6.2 CELLULAR BUILDING BLOCKS AND STRUCTURAL SUPPORT

By cellular building blocks, we are referring primarily to the polysaccharides that make up the cell walls of plants — cellulose, hemicellulose (xyloglucans in dicot flowering plants and arabinoxylans in monocot flowering plants), and pectins (polygalacturonans, based on polymers of galacturonic acid coupled to different sugar moieties such as rhamnose and fucose). As mentioned above, it is these polysaccharides that constitute the majority of all plant biomass on this planet. Animal cells do not have cell walls; each cell is circumscribed by a plasma membrane alone. It is this structural support provided by cell walls of plants along with the additional structural support provided by such processes as lignification and/or silicification of these cell walls that turns the plant into a type of scaffold upon which to hang its photosynthetic tissues (leaves or stems) in the best possible orientation to absorb carbon dioxide and the energy of the sun. Without this support terrestrial plants would not be able to support the weight of their leaves, and consequently the leaves would not get good exposure to the sun for photosynthesis. Interestingly, many aquatic plants such as algae do not have this problem. They are supported by the water itself, and hence, usually do not produce additional support compounds such as lignin. In some cases, these cellular building blocks allow the development of massive plant bodies. This is most dramatically exemplified by giant redwood and sequoia trees. Please remember, however, that the other major contributors to the structure of plants and their cells alike are lipids (especially membrane lipids such as phospholipids) and proteins (such as those in membranes, microtubules, and microfilaments).

2.6.3   SOURCES OF GENETIC INFORMATION

One finds the genetic information (DNA, RNA) of plants residing in their nuclei, chloroplasts, mitochondria, and ribosomes. All of the proteins of the cell including both structural proteins and enzymes are encoded by these nucleic acids. Most of the proteins synthesized in plant cells are encoded by nuclear DNA; on the other hand, many of the proteins that occur in mitochondria or in chloroplasts are synthesized on ribosomes within these respective organelles. Some of these proteins are structural components of membranes and membrane channels, others are enzymatic. Many proteins that occur in organelles are also coded for by DNA in the nucleus. How do nuclear encoded proteins find their way to the proper cell organelle? "Signal peptides" (specific amino acid sequences also encoded by DNA) occur on these proteins to target the proteins to the membrane(s) of specific organelles, such as peroxysomes, glyoxysomes, Golgi (dictyosomes), mitochondria, or plastids. Once the protein gets targeted to the proper organelle, it is then transported into the organelle across the membrane(s) enclosing that organelle (this involves different mechanisms for targeting and for transport). In most cases the signal peptide gets cleaved off by a specific peptidase which produces a functional protein that may act as a monomeric enzyme, may become associated with other proteins to form multimeric complexes, or may become a structural component of a membrane. Some proteins exist as glycoproteins or as lipoproteins which have carbohydrate or lipid components attached. These may also be involved in targeting and/or intercalation of the protein onto the inner or outer surface of a given membrane such as the tonoplast membrane that surrounds the vacuole or the plasmalemma that surrounds the cytoplasm and lies just inside the cell wall.

The point here is that all the information for production, localization, and functionality of every protein is ultimately contained on a strand of DNA. The ability to pass this information onto offspring is one of the key factors that determines if a species of organism will survive in a given environment. The fact that genetic information can change (mutate) over evolutionary time is what allows organisms in general to adapt to ever changing environments. This change (evolution) is always occurring and produces new combinations that may or may not work in that environment, but only the individuals that have the combinations that do work will survive to the next generation. This variation between individuals is very important to the survival of each species of plant (or animal). Thus, plants have evolved various methods of sexual reproduction, such as pollination, that allow the sharing of genetic information between individuals or a given species. This holds the benefit of spreading combinations of enzymatic reactions that work throughout a population.

2.6.4 CATALYSTS OF METABOLIC REACTIONS

By now, it has become quite apparent how important enzymes are in catalyzing metabolic reactions in different compartments of plant cells. It is these proteins, coded for by the plant's genetic information and placed in the proper locations within cells of tissues held in the correct positions by the plant's cellular building blocks, that allow not only the production of but the utilization of the metabolic energy compounds that run the biochemical reactions that control the processes of life. In such reactions, binding of the substrate to the active site of the enzyme to form the enzyme-substrate complex is a prerequisite to catalytic action of the enzyme. Enzymes act to lower the amount of free energy required to make a reaction proceed to the formation of the product of the reaction which is released following separation of the enzyme from the enzyme-substrate complex. Without this interaction of enzyme with substrate, the reaction would proceed very slowly or not at all under normal conditions of temperature and pressure. So, enzymes act as organic catalysts by speeding up the rate of a given metabolic reaction.

Some of these enzymes act to cause hydrolysis of substrates and are called *hydrolases*, like amylase which hydrolyzes starch, invertase which hydrolyzes sucrose, and fructan hydrolase which hydrolyzes fructan. Other enzymes, called *synthases* or *transferases* are involved in synthesis, as for example, cellulose synthase that makes cellulose or callose (depending on concentrations of Mg^{2+} co-factor and substrate concentration) or starch synthase, one of the enzymes involved in starch synthesis. Still others are involved in cyclicization reactions and are called *cyclases*. They make linear molecules circular as in the conversion of GPP to cyclic monoterpenes. In photosynthesis, you remember the substrate, RUDP; it is acted upon by a single enzyme (RUBISCO) that has *carboxylase* as well as *oxygenase* activity connected with photosynthetic carbon fixation from CO_2 and with photo-respiration, respectively. There are important enzymes involved in signal transduction processes related to hormone action in plant and animal cells. These include *phosphorylases*, *phosphatases*, and many kinds of *protein kinases*. Then, there are enzymes called *dehydrogenases*, such as mannitol dehydrogenase, which catalyzes the formation of D-mannose from mannitol. *Chaperones* are a group of enzymes which promote the folding of proteins into their correct (i.e., active) forms, hold proteins which are to be transported to organelles in an unfolded form, and help maintain protein integrity during heat stress and thus prevent denaturation. These are the main classes of enzymes, but the list does go on.

Finally, we need to mention the concept of *isozymes*. These refer to the same type of enzyme which (1) may exist in different cellular compartments, (2) have different pH optima for the same substrate, or (3) at the molecular level, have different nucleotide sequences for the signal peptides of the enzyme. A good example is invertase (a β-fructofuranosidase) which hydrolyzes sucrose to D-glucose and D-fructose. There are several known isozymes of invertase: (1) intracellular soluble invertase located in the cell vacuole, and possibly the cytosol, with pH optima from slightly alkaline (pH 7.5) to acidic (pH 4.5) and (2) insoluble forms ionically bound to the cell wall with pH optima of 4.0 and 5.3.[30,31]

2.6.5 DETERRENCE OF PREDATORS AND PATHOGENS VIA POISONS AND VENOMS

Plants have evolved a vast array of chemical defenses which effectively deter herbivores and pathogens from attacking them. These have obvious selective and survival value for the plants because plants are almost always confined to one spot and thus can fall easy prey to wandering animals out to steal plant nutrients. This brings up several questions. What is the nature of these chemical defense strategies? How do they work? Which came first, the chemical deterrent evolution in different groups of plants or the predator/pathogen-dictated selective pressure for plants to evolve new chemical defense strategies? How effective are human-designed chemical defense strategies, as in transgenic plants, as compared to the multifaceted strategies plants have evolved and continue to evolve to deter predators or pathogens? These questions are addressed in four excellent references.[2,32-34]

Here, we will consider some of the more important ways plants defend themselves against attack by insect predators, herbivores, pathogenic fungi, bacteria, and viruses. These methods are based on the classification scheme of Becerra[34] as follows:

- **Structural defense strategies.** These include lignification, silicification, callose formation, and wax deposition. We have alluded to these processes in more detail in the preceding sections of this chapter. The chemical polymers act as a sort of armor and present fungi, bacteria, or virus with a physical barrier through which to penetrate or present insects or herbivores a hard surface through which to chew.
- **Chemical defense strategies.** These include almost all compounds that, based on their chemical nature, deter attack. There are many fascinating stories behind the mechanisms of each of these compounds, but let it suffice to say that each of these compounds can interfere (usually in a species-specific manner) with at least one critical biochemical pathway within the attaching organism thus killing or making sick this organism. There are literally thousands of examples of chemical defense including
  - Alkaloids (e.g., nitrogen-containing, heterocyclic ring compounds)
  - Active oxygen species such as H_2O_2, O_2^- (superoxide anion), and OH (hydroxyl radical)
  - Proteins, including cell wall glycoproteins (hydroxyproline-rich, proline-rich, and glycine-rich glycoproteins); inhibitory proteins (many are induced and endogenous antiviral proteins, antifungal lipid transfer proteins, antibacterial a-thionins); lectins (which are carbohydrate-binding proteins); antioomycete pathogenesis-related protein, and antifungal defensin proteins; extracellular hydrolases (e.g., cellulases, pectinases, chitinases, ribonucleases, proteases, and lipid acyl hydrolases such as patatins); and ribosome-inactivating proteins such as trichosanthin in the Chinese cucumber plant, *Trichosanthes kiriliowii*)

- Saccharides and polysaccharides such as callose and pectins, effusive gums, mucilage, cardiac glycosides, cyanogenic glyosides and gluco-sides of organic nitrogen-containing compounds consisting of a sugar moiety linked to a cyanide or nitrite, respectively
- Phenolics and coumarins
- Polyphenolics such as suberins, lignins, and tannins [both hydrolyzable and condensed (see Figure 2.26)]

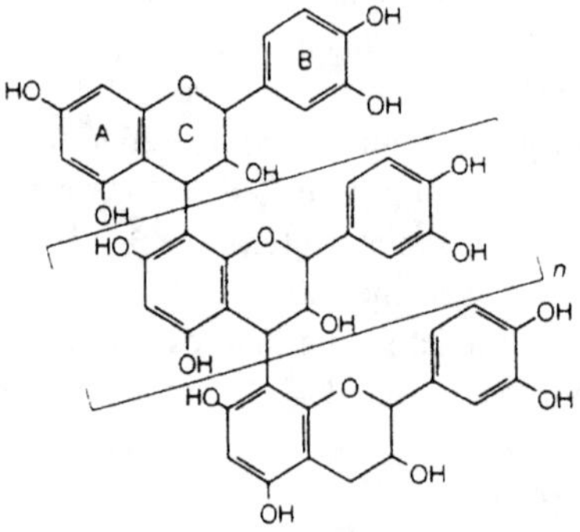

(A) Condensed tannin

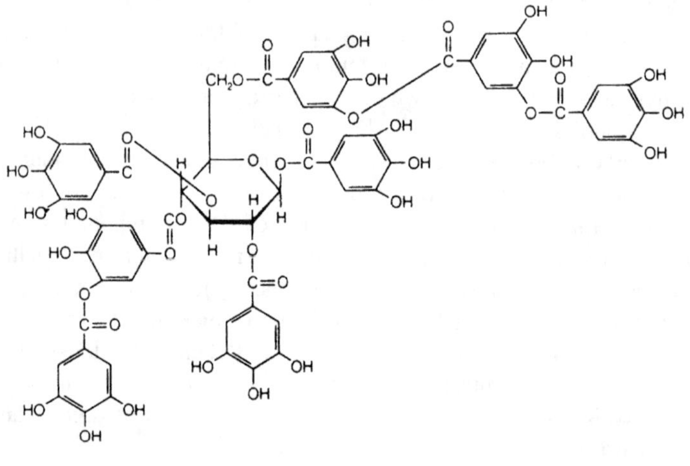

(B) Hydrolyzable tannin

FIGURE 2.26   Chemical structures of condensed (A) and hydrolyzable (B) tannins.

- Flavonoids and isoflavonoids, quinones and isoquinones
- Terpenoid/steroid compounds such as cardiac glycosides, leguminous saponins (often glycosylated), gossypiol-related terpenoids, aphid alarm pheromones, brassinosteroids (insect hormone-mimicking com-pounds), and phytoecdysones (insect molting hormone mimics)
- Cyanide-releasing compounds, which release hydrogen cyanide on ingestion and block electron transport during respiration, and include

cyanogenic glycosides and the glucosinolates (mustard oil glycosides which release isothiocyanates)
- Organic acids, including the salts of oxalic acid (as found in aroids such as *Dieffenbachia*, *Symplocarpus*, and *Monstera*, as well as the leaf blades of rhubarb, *Rheum* spp.), monofluoroacetic acid, and L-DOPA (3,4-dihydroxyphenylalanine)
- Long-chain carbon compounds such as antimicrobial polyacetylene, antifungal alkenales, anti-mammalian polyacetylene toxins, and fatty acid/lipid-containing waxes, oils, and cutin

The point here is that defense is very diverse and often very complex. It is also important to note that not all toxins act in an acute or immediate manner. Some act as chronic toxins, having a noticeable affect only after a long period of time.

There is an interesting connection between tannins and their possible role in deterring attack by the chestnut blight pathogen (*Endothia parasitica*) that has caused the near demise of the American chestnut tree in Eastern North America. It has been shown by Hebard and Kaufman[35] that in callus cultures of five clones of chestnut, the callus clones of "resistant" American chestnut trees (*Castanea dentata*) and of resistant chestnuts (*C. crenata* and *C. mollissima*), as compared with susceptible *C. dentata*, had much higher levels of hydrolyzable tannins (galloyl esters and ellagitannins) in the clones from resistant trees than in clones from susceptible trees. Challenging the respective callus cultures with virulent strains of the fungal pathogen showed that calli from susceptible chestnuts were overgrown by the pathogen, while calli from resistant strains were not affected and remained healthy. The conclusion from these findings is that the levels of hydrolyzable tannins in chestnut trees is correlated with resistance to the American chestnut blight fungal pathogen, but the correlation does not prove that condensed tannins are responsible for resistance to the pathogen either in culture or in chestnut trees. A question to ponder is this: What experiment(s) are necessary to show this kind of proof? (Hint: one must find a way to prevent the synthesis of tannins in cells that normally produce them.)

2.6.6   ATTRACTION AND DETERRENCE OF POLLINATORS

As with the example of linalool production in *Clarkia breweri* plants seen in Section 2.5.2, many species of flowering plants have evolved the ability to produce various compounds that appeal to the visual, olfactory, and taste senses of insects or animals. Since many flowering plants are strictly dependent on a mobile organism to visit its flowers and pass its pollen to another plant of the same species, there is a distinct adaptive advantage to the plant that can attract a pollinator that will visit the same plant species over and over rather than spreading pollen around at random. One must remember that the pollinators too are evolving the ability to distinguish the plant species that provide the best rewards (food) over those that do not. This is in their best interest. So, a system of reward plays a critical role in the plant's pollination success.

Attraction of pollinators to flowers is achieved by several mechanisms. As discussed in Sections 2.4.9 and 2.4.10, coloration is a critical factor for attracting insects and animals that come out during the day. The color of flowers may be due to carotenoids in biomembranes (as in chromoplast membranes), brown phlobaphenes and black melanins in the cell walls, or red, yellow, pink, blue, and deep violet flavonoids, betacyanins, and betaxanthines in the cell vacuole (Table 2.4). Many of these colors are dependent on possible complexes with Fe^{3+} and Al^{3+} as well as on pH. Different pollinators are attracted to different colors (Table 2.5). Birds are generally attracted to red. Moths are attracted to white or light yellow flowers because these flowers are more visible at night when the moths are active. Flies prefer greens and browns. Butterflies tend to visit brightly colored flowers — yellow, blue, reddish — while bees prefer yellow and blue. Bees do not usually visit red flowers. This is most likely due to the fact that a bee's spectrum of vision includes very little red. It is shifted towards the ultraviolet range. Consequently, bees preferentially pollinate flowers that produce ultraviolet nectar guides (usually present on petals) that are invisible to the human eye and are the result of the biosynthesis of specific phenolic compounds (certain flavonoids) in specific patterns which are apparently discernible to different insects.

TABLE 2.4
Chemical Basis of Flower Color in Angiosperms (Flowering Plants)

Color	Pigments responsible	Examples
White, ivory, cream	Flavones (e.g., luteolin) and/or flavonols (e.g., quercetin)	95% of white flowered spp.
Yellow	Carotenoid alone	Majority of yellows
	Yellow flavonol alone	*Primula, Gossypium*
	Anthochlor alone	*Linaria, Oxalis, Dahlia*
	Carotenoid + yellow flavenoid	*Coreopsis, Rudbeckia*
Orange	Carotenoid alone	*Calendula, Lilium*
	Pelargonidin + aurone	*Antirrhinum*
Scarlet	Pure pelargonidin	Many, incl. *Salvia*
	Cyanidin + carotenoid	*Tulipa*
Brown	Cyanidin on carotenoid background	*Cheiranhus,* many *Orchidaceae*
Magenta, crimson	Pure cyanidin	Most reds, incl. *Rosa*
Pink	Pure peonidin	Peony, *Rosa rugosa*
Mauve, violet	Pure delphinidin	Many incl. *Verbena*
Blue	Cyanidin + copigment/metal	*Centaurea*
	Delphinidin + copigment/metal	Most blues, *Gentiana*
Black (purple black)	Delphinidin at high concentration	Black tulip, pansy
Green	Chlorophylls	*Helleborus*

Odiforous substances that attract insects, birds, and mammals to flowers are usually produced as soon as the flower opens and help potential pollinators find the flower during both the day and night. These compounds include monoterpenes (e.g., linalool, limonene, geraniol), sesquiterpenes (e.g., β-ionone and α-(−)-*bis*abolol),

TABLE 2.5
Color Preferences of Different Pollinators

Animal	Flower color preferences	Comments
Bats	White or drab colors, e.g., greens and pale purples	Mostly color blind
Bees	Yellow and blue intense colors, also white	Can see in UV, but not sensitive to red
Beetles	Dull, cream or greenish color	Poor color sense
Birds	Vivid scarlets, also bicolors (red-yellow)	Sensitive to red
Butterflies (Lepidoptera)	Vivid colors, including reds and purples	—
Moths (Heterocera)	Reds and purples, white or pink	Mostly pollinate at night
Flies	Dull, brown, purple or green	Checkered pattern may be present
Wasps	Browns	—

aromatics (e.g., vanillin, eugenol, methyl eugenol), aliphatics (e.g., pentadecane, *i*-octanol), monoamines (e.g., methylamine, ethylamine, propylamine, butylamine, amylamine, hexylamine), diamines (e.g., putrescine and cadaverine), and indoles (e.g., indole and skatole). It is the various amines and indoles just listed that have unpleasant odors and attract pollinators such as flies and fungal gnats. Some plants such as Skunk Cabbage and Voodoo Lily actually benefit from photorespiration because, while the plant loses stored energy, it creates heat which better volitilizes the amines allowing them to be released more quickly and with a stronger odor. The other compounds, in general, produce pleasant odors which attract pollinators such as bees, butterflies, moths, and bats. The chemical attractants may be produced in special scent glands (called osmophores) produced by various organs of flowers, by epidermal cells along the upper sides of the petals, or in some cases by glandular hairs on leaves. Excellent discussions of these different pollination attractant syndromes are found in Larcher[32] and Harborne.[36] As mentioned in Section 2.5.2, it is the mixture of these chemicals produced by flowers that allows insects to distinguish between different species of plants, but there can be one specific scent component that determines which pollinator will pollinate a specific species of plant.

The rewards for pollination in plant flowers usually come in the form of sugar-rich solutions (sucrose, fructose, and glucose are the most common) that are secreted into the nectaries of flowers. They act in much the same way as the sugars, fats, and proteins found in mature fruits act to reward animals to disperse plant seeds (good examples here include squirrels that "forget" where they buried the acorn and birds that spread seeds all over your freshly washed car). Nectaries are located in different locations in different species, but they are almost always located at the junction between two different flower organs such as petals and ovary. The nectar held in the nectaries is not usually just sugar and water. It may also contain pigments such as anthocyanins, scents such as monoterpenes, and in some cases toxins.

Some compounds in flowers that are known to attract certain pollinators can also repel other potential pollinators. Indole, for example, can deter bees from pollinating alfalfa (*Medicago sativa*) flowers.[37] Skatole, monoamines, and the offensive smelling diamines (putrescine and cadaverine) seem to serve similar functions in other flowers. Why would a plant "want" to repel a potential pollinator? The answer may lie in the fact that some insects and animals can cue in on plant attractants (odor for example) and take the reward produced by that plant without dispersing the plants pollen. To repel such a visitor would save the plant's energy in producing rewards. So, flowering plants undergo very distinct selective pressures to produce the specific compounds that will attract the best pollinators living in a specific environment.

2.6.7 ALLELOPATHIC ACTION

Allelopathy refers to plants which give off chemical substances that are injurious to other plants or prevent other plants from becoming established in the vicinity of the plant which gives off the allelopathic chemicals (also called **allomones**).[32] Such chemicals have an obvious advantage to the plant that produces them by preventing the growth of other plant species that may compete for soil nutrients, carbon dioxide, or sunlight. Allelopathic chemicals include short-chain fatty acids, essential oils, phenolic compounds, alkaloids, steroids, and derivatives of coumarin. A classic example is the compound naphthalene glucoside produced by leaves and roots of walnut (*Juglans* spp.). This compound itself is not allelopathic; it must undergo hydrolysis and oxidation by soil microorganisms to produce hydrojuglone, and finally, the active compound, juglone. Juglone prevents the germination of seeds of many, but not all, plant species. This is why it's a bad idea to plant a wild flower garden in the same area as walnut trees. Another good example is the release of carboxyphenolic acids and hydroxycinnamic acids by heath family (Ericaceae) members which grow in such places as Scotland. Scotland was once covered with pine trees, but it was stripped to provide fuel for the growing industrial revolution. Now there is a problem with attempts at reforestation because the heath plants inhibit the association of mycorrhizal fungi with young pine roots. These fungi are essential symbionts for pines (see Section 2.6.8), so the seedlings eventually die. There are many more examples including cases with *Calluna* and *Arctostaphylos* (bearberry) that inhibit the growth of grasses (Poaceae family) and herbs; the release of terpenes and water-soluble phenolics by plants which inhabit steppes and arid shrub communities (*Parthenium* or guayule, *Encelia*, and *Artemisia* or sagebrush in the Asteraceae family and members of the Lamiaceae, Myrtaceae, Rutaceae, and Rosaceae families); and the release of orcinol depsides and usnic acid by lichens (plants that have algal and fungal partners living in association mutualistically) which exert an alleopathic effect on conifer seedlings and have an antibiotic effect on fungi which again may be symbionts.

2.6.8 ATTRACTION OF SYMBIONTS

Not all plants are capable of getting enough nutrients out of the soils in which they live. Bacteria and fungi are sometimes much better at absorbing and/or producing some of the nutrients that plants require. Consequently, it is often the case that plants will elicit a partnership between themselves and specific bacterial or fungal symbionts. A classic case illustrating this concept is that of the establishment of a mutualistic association between a host plant and nitrogen-fixing bacteria such as *Brachyrhizobium* species. The bacteria become associated with the roots of the host plants and trigger the formation of nodules which provide the bacteria with a safe place to live as well as some plant nutrients and water while they supply the host plant with reduced nitrogen in the form of NH_4^+ that is derived from atmospheric nitrogen, N_2. The reduced nitrogen is then used by the plant for synthesis of amino acids via amination reactions. At the start of this scenario, flavonoids are synthesized in significant amounts within the root systems of leguminous plants (e.g., the isoflavonoid, daidzein, in soybean, *Glycine max*; and the flavonoid, luteolin, in clovers, *Trifolium* spp.). These flavonoids play a key role in the establishment of the infection of host roots by nitrogen-fixing bacteria signaling the bacteria to bind to the plant roots after recognition of specific factors contained in the root glycocalyx (see Section 2.4.7). During the infection process, the flavonoids produced by the host plant upregulate the expression of so-called *nod* genes in the bacterial cells. These *nod* genes are required for three key steps in the infection process: (1) synthesis of a lipooligosaccharide molecule that induces root hair curling (root hairs are the sites for entry of the bacteria into the host root system), (2) the formation of an infection thread of bacterial cells in the host root hairs (from this thread the bacteria enter the cortical tissue of the root), and (3) the cell divisions in root cortical cells which give rise to root nodules.

Plants other than legumes can also develop symbiotic relationships with nitrogen-fixing organisms. The deciduous tree alder (*Alnus* spp.) can produce similar nodules upon infection. Grasses can form associations with soil bacteria, but they do not produce root nodules. Here, the bacteria seem to be anchored to the root surfaces. Fungi are also "recruited" for nutritional help. As mentioned, this is very common in the root system of pine and other trees which require an interaction with mycorrhizal fungi. Each of these examples has its own series of communicational signaling events between the plant and its specific symbiont.

2.6.9 FOOD FOR POLLINATORS, SYMBIONTS, HERBIVORES, PATHOGENS, AND DECOMPOSERS

It would not be right if we did not say something about the adaptive value of plant metabolites to other organisms. As we have emphasized throughout this chapter, it is the plant's ability to fix carbon from CO_2 into more complex storage forms of metabolic energy that makes plants crucial to the survival of all other organisms including humans. Quite simply, plants provide organisms with most of the food necessary for their growth and reproduction. Witness the following examples:

- Pollinators foraging in flowers to find food rewards in the form of sugars produced in nectaries located near the sites of insertion of the floral organs.
- Other pollinators, e.g., the blastophaga wasp in fig fruits (synconia) which lay their eggs inside the developing fruit and whose larvae hatch out to

use the inside portion of the fleshy fruit as a food source before they metamorphose into adult wasps.

- Symbiotic associations that benefit bacteria or fungi as well as the plant itself (e.g., algal/fungal partners in lichens, nitrogen-fixing bacteria in nodules of leguminous and other plants, nitrogen-fixing blue-green algae such as Nostoc and Anabaena in fronds of ferns such as *Azolla* spp.).
- Birds that devour whole fruits and regurgitate the flesh to their young while dispersing seeds along the way.
- Cows grazing on grasslands and later providing milk which builds strong bones in human offspring.
- Fungal and bacterial pathogens that invade plant cells and cause all sorts of plant diseases including blights.
- Shelf fungi, edible and toxic mushrooms, slime molds, and soil bacteria that feed off the plants even after they have long ceased to fix carbon.

These are only the very beginning of the diversity that we see due to the food supplied by plants.

2.7 CONCLUSIONS

Plants synthesize thousands of metabolites that are used for their growth and development, reproduction, defense against attack by many different kinds of organisms, and survival in often harsh and ever-changing environments. It all starts with photosynthetic carbon fixation using carbon dioxide and energy supplied by the sun. The synthesis of the various metabolites proceeds along metabolic pathways located in one or more cell compartment(s) (e.g., cell walls, membrane systems, the cytosol, and various cellular organelles) within tissues that are often specialized for particular tasks. Most metabolites produced by these pathways never leave the plant, but occasionally plant compounds, some of which attract and some of which repel, are the basis for a complex type of communication between plants and animals. The specific enzymes that catalyze the respective steps in each metabolic pathway are encoded in nuclear, chloroplast, and mitochondrial genomes by specific genes. We will explore in the next chapter what factors influence the expression and regulation of these genes.

REFERENCES

1. Delmer, D. P. and Amor, Y., Cellulose biosynthesis, *The Plant Cell*, 7, 987, 1995.
2. Taiz, L. and Zeiger, E., *Plant Physiology*, Benjamin/Cummings, Redwood City, CA, 1991.
3. Whetten, R. and Sederoff, R., Lignin biosynthesis, *The Plant Cell*, 7, 1001–1013, 1995.
4. Kaufman, P. B., Wu, L.-L., Brock, T. G., and Kim, D., Hormones and the orientation of growth, in *Plant Hormones, Physiology, Biochemistry and Molecular Biology*, Davies, P. J., Ed., Kluwer Academic, Dordrecht, The Netherlands, 1995.

5. Kaufman, P. B., Dayanandan, P., Takeoka, Y., Bigelow, W. C., Jones, J. D., and Iler, R., Silica in shoots of higher plants, in *Silicon and Siliceous Structures in Biological Systems,* Simpson, T. and Volcani, B. E., Eds., Springer-Verlag, New York, 1983, pp. 409–499.
6. Martin, C. and Smith, A. M., Starch biosynthesis, *The Plant Cell*, 7, 971–985, 1995.
7. Suzuki, M. and Chatterton, N. J., *Science and Technology of Fructans,* CRC Press, Boca Raton, FL, 1993.
8. Edelman, J. and Jefford, T. G., The mechanism of fructosan metabolism in higher plants as exemplified in *Helianthus tuberosus* L., *New Phytolo.,* 67, 517, 1968.
9. Von Wettstein, D., Gough, S., and Kannangara, C. G., Chlorophyll biosynthesis, *The Plant Cell*, 7, 1039–1057, 1995.
10. Bartley, G. E. and Scolnik, P. A., Plant carotenoids: pigments for photoprotection, visual attraction, and human health, *The Plant Cell*, 7, 1027–1038, 1995.
11. McGarvey, D. J. and Crouteau, R., Terpenoid metabolism, *The Plant Cell*, 7, 1015–1026, 1995.
12. Holton, T. A. and Cornish, E. C., Genetics and biochemistry of anthocyanin biosynthesis, *The Plant Cell*, 7, 1071–1083, 1995.
13. Burbulis, I., E., Iacobucci, M., and Shirley, B. W., A null mutation in the first enzyme of flavonoid biosynthesis does not affect male fertility in Arabidopsis, *The Plant Cell*, 8, 1013–1025, 1996.
14. Croteau, R. and Winters, J. N., Demonstration of the intercellular compartmentation of *l*-menthone metabolism in peppermint (*Mentha piperita*) leaves, *Plant Physiol.,* 69, 975–977, 1982.
15. Dodson, C., Dressler, R., Hills, H., Adams, R., and Williams, N., Biologically active compounds in orchid fragrances, *Science*, 164, 1243–1249, 1969.
16. Galen, C., Regulation of seed set in *Polemonium viscosum*: floral scents, pollination and resources, *Ecology,* 66, 792–797, 1985.
17. Raguso, R. A., Light, D. M., and Pichersky, E., Electroantennogram responses of *Hyles lineata* (Sphingidae: Lepidoptera) to volatile compounds from *Clarkia breweri* (Onagraceae) and other moth-pollinated flowers, *J. Chem. Ecol.,* 22, 1735–1766, 1996.
18. Dodson, H. E. M., Floral volatiles in insect biology, in *Insect-Plant Interaction,* Vol. V, Bernays, E., Ed., CRC Press, Boca Raton, FL., pp. 47-81, 1993.
19. Galen, C. and Kevan, P., Bumblebee foraging and floral scent dimorphism: *Bombus kirbyellus* and *Polemonium viscosum, Can. J. Zool.,* 61, 1207–1213, 1983.
20. MacSwain, J., Raven, P., and Thorp, R., Comparative behavior of bees and Onagraceae. IV. *Clarkia* bees of the western United States, *University of California Publications in Entomology,* 70, 1–80, 1973.
21. Pellmyr, O., Three pollination morphs in Cimicifuga simplex: incipient speciation due to inferiority in competition, *Oecologia*, 78, 304–307, 1986.
22. Pichersky, E., Raguso, R. A., Lewinsohn, E., and Crouteau, R., Flower scent production in *Clarkia (Onagraceae)*. I. Localization and modulation of emission of monoterpenes and of linalool synthase activity, *Plant Physiol.,* 106, 1533–1540, 1994.
23. Picherski, E., Lewinsohn, E., and Crouteau, R., *S*-linalool synthase from *Clarkia* flowers: purification and characterization, *Arch. Biochem. Biophy.,* 316, 803–807, 1995.
24. Raguso, R. A. and Picherski, E., Floral volatiles from *Clarkia breweri* and *C. concinna* (Onagraceae): recent evolution of floral scent and moth pollination, *Plant System. Evol.,* 194, 55–67, 1995.

25. Croteau, R. and Karp, F., Origin of natural odorants, in *Perfume: Art, Science and Technology,* Muller, P. and Lamparsky, D., Ed., Elsevier Applied Sciences, New York, 1991.

26. Dudareva, N., Cseke, L., Blanc, M. B., and Pichersky, E., Evolution of floral scent in *Clarkia*: novel patterns of *S*-linalool synthase gene expression in the *C. breweri* flower, *The Plant Cell,* 8, 1137–1148, 1996.

27. Funk, C., Lewinsohn, E., Stofer Vogel, B., Steele, C. L., and Croteau, R., Regulation of oleoresinosis in grand fir (*Abies grandis*): coordinate induction of monoterpene and diterpene cyclases and two cytochrome P450-dependent diterpenoid hydroxylases by stem wounding, *Plant Physiol.,* 106, 999–1005, 1994.

28. Esau, K., *Plant Anatomy,* 2nd ed., John Wiley & Son, New York, 254, 1965.

29. Johnson, M. and Croteau, R., Biochemistry of conifer resistance to bark beetles and their fungal symbionts, in *Ecology and Metabolism of Plant Lipids,* Fuller, G. and Nes, W. D., Eds., ACS Symposium Series 325, American Chemical Society, Washington, D.C., p 76–91, 1987.

30. Sturm, A. and Crispeels, M. J., cDNA cloning of carrot extracellular β-fructosidase and its expression in response to wounding and bacterial infection, *The Plant Cell,* 2, 1107–1119, 1990.

31. Jones, R. A. and Kaufman, P. B., Multiple forms of invertase in developing oat internodes, *Plant Physiol.,* 55, 114–119, 1975.

32. Larcher, W., Physiological plant ecology, *Ecophysiology and Stress Physiology of Functional Groups,* 3rd ed., Springer-Verlag, New York, 19–31, 1995.

33. Zipf, A., Plant defenses and defensive compounds, *Plant Growth Regul. Soc. Am. Q.,* 24, 188–200, 1996.

34. Becerra, J. X., 1997, Insects on plants: macroevolutionary chemical trends in host use, *Science,* 276, 253–256, 1997.

35. Hebard, F. V. and Kaufman, P. B., Chestnut callus-cultures: tannin content and colonization by *Endothia parasitica*, *Proceedings of the American Chestnut Symposium*, Morgantown, WV, p 63–70, 1978.

36. Harborne, J. B., *Introduction to Ecological Biochemistry,* 3rd ed., Academic Press, New York, 1988.

37. Raguso, Rob, personal communication, 1997.

3 Regulation of Metabolite Synthesis in Plants

Leland J. Cseke and Peter B. Kaufman

CONTENTS

3.1 INTRODUCTION

In Chapter 2, we pointed out the importance of knowing the sequence of substrates and the respective enzymes that are involved in the biochemical pathways that lead to the synthesis of given metabolites in plants. Now, we need to consider a few environmental, biochemical, and molecular mechanisms which up-regulate or down-regulate the genes and/or enzymes that control the synthesis of these metabolites. Such information will help us to understand how plants respond to environmental and biotic stresses affecting their survival and how humans (1) can alter plant metabolism to favor the synthesis of a particular metabolite of medicinal or economic value or (2) can help plants of agronomic importance respond to adverse conditions. To get started, we shall explore how metabolite biosynthesis is increased or decreased by environmental and biotic stresses.

0-8493-3134-X/99/$0.00+$.50
© 1999 by CRC Press LLC

3.2   REGULATION BY ENVIRONMENTAL AND BIOTIC STRESSES

3.2.1   ENVIRONMENTAL STRESSES

A host of environmental factors are involved in the regulation of metabolite biosynthesis in plants. The need for this control of synthesis stems from the fact that plants must be able to adjust the production of metabolites according to changing factors if they are to survive. Light is obviously a key factor in the ultimate production of many compounds because it supplies the energy needed to fix carbon. It is also more directly necessary for the biosynthesis of compounds such as chlorophylls, as mentioned in Chapter 2. Here, photons trigger the enzymatic conversion of protochlorophyllide and phytol to chlorophylls a and b, and thence, to chlorophyll-protein complexes in chloroplasts.[1] Light also catalyzes the synthesis of anthocyanin pigment, via the plant pigment, phytochrome, in many tissues of many plants such as cotyledon (seed leaf) epidermal cells and hypocotyl (stem portion below the cotyledons) subepidermal cells in mustard seedlings.[1] Light intensity plays an important role in the biosynthesis of medicinally important metabolites. An excellent case in point is the tree of joy (*Camptotheca accuminata*) (Figure 3.1), where levels of the anti-prostate cancer drug, camptothecin (an alkaloid metabolite), significantly increase as the amount of light reaching the tops of the plants *decreases*. University

FIGURE 3.1   Author Peter Kaufman is shown standing in a plantation of tree of joy (*Camptotheca accuminata*) trees planted in southern Louisiana at the Citrus Experiment Station, located near Port Sulphur, LA, as part of a research project sponsored by the Agricultural Experiment Station of the Louisiana State University at Baton Rouge, LA and XyloMed Research, Inc. (Photo provided by Tracy Moore, President of XyloMed Research, Inc.).

of Michigan Biology students, Atul Rustgi and Ashish Goyal, provide the following essay on their Bachelor's research project on the effects of different light intensities on camptothecin levels in tree of joy plantlets.

Research Project

Objective

The objective of our experiment was to test the effects of light intensity on the production of camptothecin (CPT) in the tree of joy plants. It has been shown in previous experiments that a decrease in light intensity will increase the production of CPT. Our objective was to determine the effects of different light intensities on the biosynthesis of CPT.

Materials and Methods

Three trays containing seedlings of *Camptotheca accuminata* were grown in a greenhouse. Each tray contained plants of the same age and height. Each tray of plants was exposed to a particular light intensity different from that of the other two trays. In each tray, the seedlings were arranged in two rows. The first tray received no shading and had a light intensity at the top of the plants of 3000 $\mu Em^{-2}s^{-1}$. The second received 1x shading by means of a thin wire screen that was held above the plants by four posts at each corner of the tray. The light intensity measured at the top of this set of plants was 750 $\mu Em^{-2}s^{-1}$. The third tray received 2x shading by means of two wire screens. The light intensity measured at the top of this set of plants was 300 to 400 $\mu Em^{-2}s^{-1}$ (see Figures 3.2 and 3.3).

FIGURE 3.2 Students Ashish Goyal, Kathryn Timberlake, and Atul Rustgi measuring light intensity with a photo flux density meter (Ly-Cor, Inc.) in their shading experiment with seedlings of tree of joy, *Camptotheca accuminata*. (Photo courtesy of David Bay.)

FIGURE 3.3   Illustration of shading experiment with tree of joy (*Camptotheca accuminata*) seedlings grown at three different light intensities in the greenhouse at the University of Michigan. (Photo courtesy of David Bay.)

At the time of setup, a random sampling of the largest top leaves of the plants was taken. This was done in order to get a measurement of the initial concentration (T0) of CPT in these seedlings before any experimental variables were introduced. The following procedure was used in order to determine the concentration of CPT in this sample (T0) and in successive samples.

1. Freeze leaves in liquid nitrogen
2. Crush to a powder using a mortar and pestle
3. Add 1 g of crushed leaves to a beaker containing 50 ml of methanol (MeOH)
4. Cover beaker for 24 h
5. Vacuum filter
6. Transfer liquid portion to a clean beaker
7. Air dry
8. Add 0.01 g of dried filtrate to 400 μl of refrigerated MeOH in order to avoid evaporation
9. Cover using Parafilm™
10. Use a sonicator to fully dissolve
11. Analyze 10 μl with a high-pressure liquid chromatograph (HPLC)[3]

For the HPLC, 10 μl injections were used with a C18 column. The wavelength used was 347 nm, the temperature was 40°C and the flow rate was 1 ml·min^{-1}. The mobile phase used was run in acetonitrile (ACN). The 10 μl injection was run from a gradient of 20 to 80% ACN over the course of 60 min. The printer was set to an attenuation of 9 in order to get the best chromatograph.

In order to find out which peak on the graph from the HPLC represented CPT, a sample of T0 was spiked with extra CPT. For this run, 0.015 g of the air dried filtrate and 0.0004 g of CPT were added to 500 μl of MeOH. When the graph obtained for this run was compared to a run without the extra CPT (see Figure 3.4), one peak was noticeably larger (see Figure 3.5), thus indicating this peak to be the one representing CPT. For that peak, the given area under it represented the amount of CPT in a given injection.

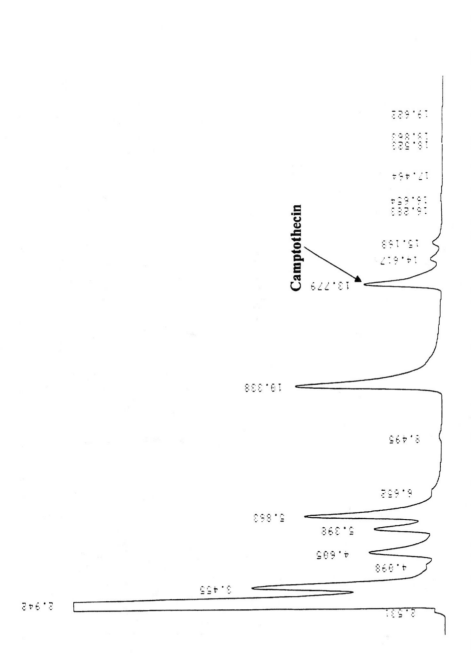

FIGURE 3.4 High pressure liquid chromatography trace illustrating camptothecin peak from nonspiked sample extract from tree of joy (*Camptotheca accuminata*) seedlings.

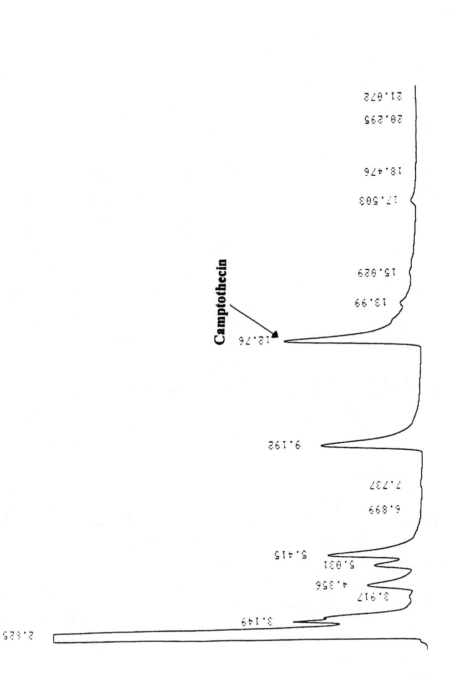

FIGURE 3.5 High pressure liquid chromatography trace illustrating camptothecin peak from a camptothecin-spiked sample extract from tree of joy (*Camptotheca accuminata*) seedlings.

At T1 (week one) six leaves were taken from each of the three trays. The six leaves were a collection of the largest top three leaves of two different plants in the same row of a particular tray. For the next 4 weeks, leaves were taken from the top of the plants of a new row so as to avoid getting young buds from a plant whose leaves had been removed the preceding week. The six leaves then used in the procedure described above in order to obtain data. The CPT peaks for the chromatographs of these successive trials could be identified by comparing these chromatographs to that of T0 and searching for similarities in the shape of and the time of elution of that CPT peak.

Results

With a standard curve, the amount of CPT in unknown samples can be determined. A sample chromatogram for 2.87E-03 *M* sample is shown in Figure 3.4. The standard curve results and a graphical representation are shown in Table 3.1 and Figure 3.6, respectively.

TABLE 3.1
Data on Areas under Curves for Respective
Camptothecin Concentrations

Area under the curve	Concentration of camptothecin (moles)
11,348	2.87E-06
183,077	2.87E-05
952,883	2.87E-04
6,196,572	2.87E-03

Note: Also used for the calculation of the standard curve for camptothecin in Figure 3.6.

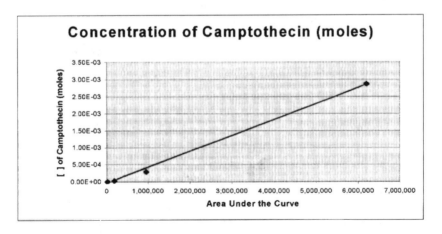

FIGURE 3.6   Standard curve for concentration of camptothecin plotted against areas under the high pressure liquid chromatography peaks.

The results for the three different amounts of shading are shown in Tables 3.2, 3.3, and 3.4.

TABLE 3.2
Time-Course Changes in Camptothecin Levels in Tree of Joy Seedlings Grown Without Artificial Shading

No Shading (Run 1)

Time (days)	Area under the curve	[ ] of Camptothecin (moles)
0 (T0)	2,469,311	1.10E-03
7 (T1)	2,384,273	1.06E-03
14 (T2)	2,101,377	9.22E-04
21 (T3)	2,311,930	1.02E-03
28 (T4)	3,031,237	1.36E-03
35 (T5)	4,062,633	1.85E-03

Note: Simulated full sunlight conditions

TABLE 3.3
Time-Course Changes in Camptothecin Levels in Tree of Joy Seedlings Grown Under 1x (Partial) Shading Conditions

1X Shading (Run 2)

Time (days)	Area under the curve	[ ] of Camptothecin (moles)
0 (T0)	2,469,311	1.10E-03
7 (T1)	1,887,101	8.21E-04
14 (T2)	2,530,378	1.12E-03
21 (T3)	5,339,954	2.45E-03
28 (T4)	5,370,632	2.47E-03
35 (T5)	7,592,100	3.51E-03

TABLE 3.4
Time-Course Changes in
Camptothecin Levels of Tree of Joy
Seedlings Grown Under 2x (Deep)
Shading Conditions

	2X Shading (Run 3)	
Time (days)	Area under the curve	[ ] of Camptothecin (moles)
0 (T0)	2,469,311	1.10E-03
7 (T1)	5,153,219	2.36E-03
14 (T2)	4,425,086	2.02E-03
21 (T3)	2,546,829	1.13E-03
28 (T4)	3,770,064	1.71E-03
35 (T5)	7,230,344	3.34E-03

A sample chromatograph of Run 1 (T1) is in Figure 3.5. A graphical representation of a comparison of all three runs is shown in Figure 3.7.

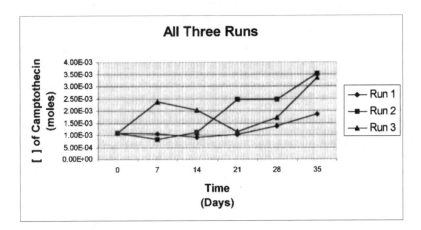

FIGURE 3.7   Graph of time-course changes in camptothecin levels in tree of joy seedlings (*Camptotheca accuminata*) grown under conditions of no shade (Run 1), 1x shading (Run 2), and 2x shading (Run 3). See text for the respective light intensities at the tops of the tree of joy seedlings for these three different light level regimes.

Conclusion

The data shows that Run 1, which had no shading, had a slow decrease in the amount of CPT production up to week 1. Thereafter there was a continuous slow rise in the production. The rise in production must be due to the effects of leaf growth. Run 2, which had 1x shading, showed an initial decrease in production of CPT, but then after week 1 showed a dramatic increase in production. The dramatic increase in production of CPT must be due to the shading effect. Run 3, which had 2x shading, showed continued increase in the production of CPT after the onset of the run, but started to decrease production after week 1 until week 3. This decrease must be due to poor leaf growth. After week 3, Run 3 showed an increase of CPT production, which was due to new leaf growth. Going into week 5, both Runs 2 and 3 were producing the same amount of CPT, but Run 2 showed a greater potential because its new growth was due to the shading effect.

A result that seemed to be surprising, was the fact that Run 3 had an initial rise in the production of CPT, while the other two runs showed an initial decrease in production. The results show that no shading produces the least amount of CPT. In the short run, it seems as through 2x shading produces the most amount of CPT. In the long run, it also seems as though both 1x and 2x shading produce the same amount of CPT, but 2x shading has more potential to produce greater amounts of CPT in the future due to its high growth time.

Social Benefits

CPT is known to be an anti-cancer agent and has shown activity against such cancers as ovarian tumors, leukemia, and lung cancer. CPT inhibits the growth of cancer by hampering DNA's ability to unwind and replicate.[4] It is, therefore, necessary for cancer research to be able to produce CPT in high amounts. One such way is to learn under what environmental factors CPT is produced in high amounts. CPT has apparent medicinal and anti-cancer value. Ongoing research can help eliminate cancer.

This experiment has shown that CPT is produced at higher levels in shading. It has also been shown that CPT is produced at different levels under varying degrees of shading. These results should be used to maximally produce CPT and help in the fight against cancer.

Temperature is another important factor that regulates plant metabolism. At reduced temperatures around 0°C, most enzymes are inactive but as the temperature increases the rate of enzyme activity increases up to about 40°C, after which time most plant enzymes become inactivated and even permanently damaged. Many enzymes are always present in plant cells at a certain level, but specific temperatures can trigger a dramatic change in these levels. For example, levels of heat shock proteins (HSP), constitutively present as chaperones, rapidly increase at temperatures of 40°C and above for most organisms. At this point HSP proteins act to help repair enzymes that may have been damaged due to the excess heat. Please note that not all organisms have enzymes that are only active between the temperatures of 0 and 40°C. Thermophilic bacteria, for example, thrive at high temperatures of 90 to 95°C.[4]

Carbon dioxide gas is the fundamental carbon source for all plant metabolites. Its levels can vary depending on the environment, and this variation causes changes in biosynthetic output. For example, elevated carbon dioxide levels in the Earth's atmosphere due to increased burning of fossil fuels and burning of tropical rainforests worldwide, together with elevated temperatures due to elevated levels of "greenhouse gases", are currently causing increases in total photosynthate produced in temperate zone plants.[4a] This is especially true for C-4 photosynthetic plants, which are adapted to higher temperature regimes and have little or no loss of carbon through photorespiration. However, the ultimate impact of such climatic perturbations on the biosynthesis of compounds other than photosynthetically produced sugars is unknown.

Flooding of plant root systems for variable periods of time is another kind of environmental stress. The stress imposed here is mainly due to oxygen deprivation to the roots. For terrestrial plants, too much water results in stunting of shoot growth, reduced chlorophyll biosynthesis in the leaves, and enhanced ethylene biosynthesis. However, aquatic plants such as rice (*Oryza sativa*) and cattail (*Typha* spp.) can tolerate continuous flooding because they have air passages in the root and shoot systems that allow atmospheric oxygen to permeate into the cells of their flooded roots. Where nonaquatic plants are periodically flooded by irrigation, after the soil has dried out plant growth and chlorophyll biosynthesis are not impaired, but rather, are stimulated. In the case of tree of joy, *Camptotheca accuminata*, such periodic flooding episodes result in greatly enhanced growth of new shoots which have significantly higher levels of CPT than the shoots of plants which have not been irrigated and only have old growth shoots.[4b]

3.2.2 BIOTIC STRESSES

Unlike environmental stresses which are the result of nonliving components of a plant's environment, biotic stresses are the result of living components of the environment. Herbivory (a process where herbivorous animals, insects, and mollusks eat plants as a food source) is one such biotic stress. According to Larcher,[5] biosynthesis of defense metabolites in plants is often induced or enhanced by herbivory. For example, intensively grazed grasses (members of the grass family, Poaceae) frequently contain more biogenic silica than grasses on nongrazed areas. Further, damage to plants elicited by herbivores causes an increase in amounts (per unit dry weight) of polyphenols, tannins, and terpenes within the tissues of many plant species such as birch (*Betula* spp.) and poplar (*Populus* spp.) trees which show increases in the levels of such metabolites after attack by insects and mollusks.[5] Larcher indicates that such up-regulation in the biosynthesis of these metabolites (defense compounds) occurs at the expense of biomass (dry matter) production in plants that are exposed to such stresses. So, defense against herbivory comes at a cost. It is best for the plant if it can down-regulate the production of defense compounds during times when the plant is not under attack.

Humans can make use of biotic stresses such as herbivory to increase yields of desired plant metabolites. As cited in Section 3.2.1, Dr. Zhijun Liu[6] has shown that bark tissue contains significantly higher amounts of the medicinal metabolite, CPT, than wood tissue by a factor of two in both roots and stems. As these trees grow larger in diameter, the proportion of bark tissue decreases substantially. Since bark tissue contains significantly more CPT per unit dry weight, Liu says that it is desirable to grow smaller diameter trees with many branches present because the ratio of bark to wood is much greater in such shoots. To achieve this condition, simulated herbivory, using coppicing (cutting of trees at ground level, to stimulate the development of new, vigorous shoot growth ["sucker" sprouts]), will induce the trees to regenerate plants with multiple, small shoots. These shoots can then be collected for the extraction of higher yields of CPT.

In Chapter 2 we mentioned that conifers secrete oleoresin (turpentine and rosin) in response to wounding and attack by insects (e.g., bark beetles) and fungal pathogens. This is well-documented in the classic work by Funk et al.[7] on the occurrence of oleoresinosis in grand fir (*Abies grandis*) elicited by physical wounding. The wounding treatments simulate the wounding that occurs after attack of stem bark tissues by bark beetles. This wounding is achieved by making a series of l-mm cuts ≈ 3 mm apart along the entire stem on opposite sides of 6-week-old saplings. The extent of up-regulation of oleoresin biosynthesis by these treatments is substantial. Over a 20-d period, one finds at the sites of wounding an accumulation of a viscous mass of resin acids and the loss of volatile monoterpenes. In response to an attack by the bark beetle itself (*Scolytus ventralis*), these oleoresins deter further attack by the beetles and act directly to kill eggs and larvae of the insect as well as to seal the wound.[7]

There are indeed some very interesting stories dealing with the action of volatile compounds produced in response to herbivory. These compounds do not always act directly on the attacking organism. For example, during the wounding caused by beet armyworm caterpillars feeding on plant leaves, the insect may produce an oral secretion of a newly discovered fatty acid-based elicitor/signal called volicitin [*N*-(17-hydroxylinolenoyl)-L-glutamine]. This elicitor, when applied to damaged leaves of corn (*Zea mays*) seedlings induces the seedlings to release a mixture of volatile compounds (octadecanoid-jasmonate signal complex) that attract females of parasitic or predatory wasps (natural enemies). These wasps then kill the feeding caterpillars thus removing the biotic stress from the plant.[8]

3.3   BIOCHEMICAL REGULATION

Apart from environmental and biotic factors which influence the synthesis of plant metabolites, there are also factors or conditions acting within the plant itself that influence the activity of biochemical pathways. An understanding of these factors and how they influence the individual steps of metabolic pathways holds significant benefit to humans. Some examples are given in the following sections.

3.3.1   METABOLITE FEEDS AND RADIOACTIVE PRECURSORS

One of the primary ways by which researchers study the pathways for synthesis of plant metabolites is to use ^{14}C-labeled metabolites, especially those which are known precursors in a given metabolic pathway. This not only helps one to identify intermediate substrates in a given pathway, but also helps one determine the rate-limiting step of that pathway. If one can discover the rate-limiting step in a pathway that produces a metabolite of interest, it is possible to up-regulate the synthesis of that metabolite (1) by up-regulating gene expression for the enzyme which catalyzes the rate-limiting step (see Section 3.4, Molecular Regulation), (2) by enhancing enzyme activity (in effect, lowering the K_m or affinity of the enzyme for its substrate) by feeding cells with the rate-limiting enzyme's preferred substrate or by increasing substrate concentration (see Section 3.3.2 on substrate activation), and (3) by removing the end-product of the rate limiting step (see Section 3.3.5 on end-product inhibition). A nice study illustrating the use of isotopes to help understand where and how up-regulation of metabolite biosynthesis occurs is that of Funk et al.[7] Here, they focused on the oleoresin biosynthetic pathway using *in vivo* [^{14}C] acetate feeding and analysis of intermediates produced by their respective enzyme activities. Two cytochrome P450-dependent disterpenoid hydroxylases involved in the synthesis of (–)-abietic acid (the principal resin acid in grand fir) increase in their activities 5- to 100-fold in wounded stems over the levels in nonwounded stems 10 d after wounding, after which time it declines. As it mentioned in Section 3.2.2, such resin acids are very effective in control of bark beetle attack.

3.3.2 SUBSTRATE ACTIVATION

In the biochemist's toolbox, one strategy used to enhance end-product biosynthesis is to elevate substrate concentration. This has the potential of enhancing the rate of a given enzyme's activity. A case in point is the hydrolysis of sucrose to D-glucose and D-fructose, mediated by the enzyme invertase (β-fructofuranosidase). When the photosynthetically produced sucrose level increases in source cells (green leaves and stems), this in effect causes invertase activity to increase in these cells so long as end-products are also being removed or metabolized. One of the consequences of this action, especially in the case of sucrose, is that other metabolic pathways are also up-regulated. The elevation in D-fructose levels leads to enhanced synthesis of the storage metabolite, fructan, found in cell vacuoles (see Chapter 2). Further, the parallel increases in the amount of the other product, D-glucose, leads to an enhancement in synthesis of cell-wall cellulose (a glucose polymer) and the storage polysaccharide, starch (also a glucose polymer), found in chloroplasts and colorless plastids called amyloplasts (as found in root-cap cells).

3.3.3 ENZYME ACTIVITY REGULATION BY PROTEIN PHOSPHORYLATION AND DEPHOSPHORYLATION AND BY CYTOSOLIC CALCIUM

Other biochemical factors that influence the production of metabolites act upon the structure of enzymes themselves. For example, in plant cells there are enzymes called protein kinases which act to phosphorylate (using ATP) other enzymes at particular amino acid residues. The additional phosphate group changes the conformation of the enzymes to which it is attached thus activating them. The phosphorylated enzyme can be rendered inactive again through the action of other enzymes called phosphatases, which releases the inorganic phosphates attached to the kinases. Such a mechanism is very important in carbon fixation through photosynthesis via RuP2-Case (ribulose *bis*phosphate carboxylase) in chloroplasts and dark fixation of carbon dioxide via PEP carboxylase (phosphoenolpyruvate carboxylase). Signal transduction cascades involving calmodulin-Ca^{2+} activation of protein kinases and phosphates, involved in protein phosphorylation/dephosphorylation reactions downstream in these cascades, are one of the primary mechanisms for enzyme activation.[9]

One of the key players here is cytosolic calcium. Once it is released from the endoplasmic reticulum (ER), it can bind to the calcium-binding protein, calmodulin, which in turn can activate specific protein kinases involved in protein phosphorylation reactions. It is of current interest to plant biologists that cytosolic calcium plays a key role in gravitropic response mechanisms in roots and shoots where one of the key metabolites is the plant protein, calmodulin.

3.3.4 ACTIVATION WITH FUNGAL ELICITORS AND PLANT GROWTH REGULATORS

During the course of evolution, plants have evolved intriguing defense strategies against attack by fungal pathogens that cause disease. When the fungus attacks the plant, it may synthesize and secrete into the plant's cells various fungal cell-wall polysaccharides (e.g., chitin) which we call *elicitors*. Such elicitors can act to up-regulate the synthesis of specific plant metabolites called *phytoalexins* (compounds that kill attacking fungal pathogens). Two such phytoalexins are the isoflavonoids, genistein and daidzein. In seedlings of soybeans and other members of the bean family (Fabaceae), the levels of these compounds increase dramatically when the plant is attacked by a fungal pathogen. They are toxic to the fungal pathogen and act to kill the fungus! This has an application with miso and tempeh, both fermented soybean food products. Miso is made by culturing soybean curd with the fungus, *Aspergillus oryzae*. The fungus secretes fungal elicitors that cause the soybean to synthesize significantly higher levels of the two isoflavonoids mentioned. This, in turn, produces the food's distinct flavor. It is also of interest that these two isoflavonoids are very important in preventing colon cancer and in treating patients suffering alcoholism.[10]

Naturally occurring or synthetic plant growth regulators have been used to up-regulate the biosynthesis of enzymes which produce useful metabolites either in intact plants or in plant cell cultures. Some of the classic examples are as follows:

- The induction of synthesis and rate of flow of latex (made up mostly of polyterpenes found in the latex of the stems) from wounds in the bark of Brasilian rubber trees (*Hevea brasiliensis*) by the naturally occurring plant hormone/growth regulator, ethylene[11,12]
- The induction of synthesis of invertase (β-fructofuranosidase) by the naturally occurring plant hormone/growth regulator, gibberellic acid (GA$_3$), in elongating stems of cereal grasses[13]
- The induction of synthesis of α-amylase in germinating seeds of cereal grains by the plant hormone, GA$_3$, which triggers the hydrolysis of starch to sugar (D-glucose) (This action by GA$_3$ on α-amylase activity is utilized in beer brewing, using modified barley (*Hordeum vulgare*) substrate (the D-glucose derived from starch stored in the grains)[14]
- The up-regulation of synthesis of shikonin (a red naphthoquinone pigment used as a medicine, dye, and cosmetic) in cell cultures of *Lithospermum erythrorhizon* in a two-stage bioreactor by kinetin (a synthetic cytokinin plant hormone) and by IAA (the naturally occurring auxin-type plant

hormone); and the down-regulation of skikonin biosynthesis by the synthetic auxin-type plant growth regulators, 2,4-D (2,4,-dichlorophenoxy-acetic acid) and α-NAA (α-naphthaleneacetic acid)[15]

- Induction of vanillic acid formation with the plant growth regulator, kinetin in cell suspension cultures of the vanilla orchid, *Vanilla planifolia*.[16] (The key to this up-regulation of vanillin biosynthesis is the enhancement in the activities of several enzymes in the phenylpropanoid biosynthetic pathway that leads to vanillin production, namely, phenylalanine ammonia lyase (PAL), 4-hydroxycinnamate:coenzyme A ligase and uridine 5'-diphosphate-glucose:*trans*cinnamic acid glucosyl transferase.)

3.3.5 END-PRODUCT INHIBITION

If end-products begin to accumulate in significant levels at the sites where metabolite synthesis is occurring, this can result in repression of enzyme activity for the last and preceding enzymes in a given biosynthetic pathway. In the example we cited earlier with invertase-mediated hydrolysis of sucrose, the accumulation of the end-products, D-glucose and D-fructose, can cause significant repression of invertase activity.[17] For the plant, this prevents the nonstop production of D-glucose and D-fructose which would use up the supply of sucrose needed for the production of many other metabolites. For humans, the strategy of feeding end-products to whole plants or cell cultures has been used to cause plants that have branched metabolic pathways to stop producing one type of metabolite at the end of one of the branches. This, in turn, causes the amount of end-product of the other branch to increase significantly.

3.3.6 DIRECT INHIBITION OF ENZYME ACTIVITY

Enzymes are inhibited by various molecules within the cell in two primary ways: (1) by competitive inhibition and (2) by noncompetitive inhibition.[9] In competitive inhibition, the inhibitor acts by binding to the active site of the enzyme, and in so doing, prevents the binding of normal substrate. To do this, the competitive inhibitor must resemble the enzyme's normal substrate. In noncompetitive inhibition, the inhibitor molecule binds to the enzyme, but it does not compete with the substrate for the active site. A good example of competitive inhibition is that between carbon dioxide and oxygen for the active site on the photosynthetic enzyme complex, ribulose-1,5-P2 carboxylase/oxygenase. If oxygen is occupying the active site, then CO_2 cannot be fixed. An example of noncompetitive inhibition is that manifested by the herbicide, glyphosate. It competes with phosphoenolpyruvate (PEP) for the PEP-binding site on the enzyme, 5-enolpyruvylshikimate-3P synthase (EPSP synthase), but it does not interfere with the actual active site on EPSP synthase.[9]

One relevant enzyme inhibition case involving medicinal natural products is that in which the glucoside of the isoflavonoid, daidzein, called daidzin [found in high levels in the seeds of Chinese Plum Flower (*Psoralea corylifolia*), soybean (*Glycine max*), and pinto bean (*Phaseolus vulgaris*)] inhibits the enzymes alcohol dehydrogenase and NAD-dependent alcohol aldehyde dehydrogenase. These enzymes work to catalyze the oxidation of acetaldehyde, the primary product of alcohol metabolism.[10,18] So, when daidzin is present, alcohol levels increase in the bloodstream and cannot be metabolized via alcohol dehydrogenase and alcohol aldehyde dehydrogenase. An important consequence of this is that alcoholics soon lose their appetite for alcohol. Another isoflavonoid produced in high amounts in soybeans (*Glycine max*), fermented soybean products, kudzu vine roots, and Indian breadroot (*Psoralea esculenta* and *P. corylifolia*) (Tables 1 and 2) is genistein. It acts as an anticancer agent in humans, in part, by inhibiting DNA topoisomerase which is functional in DNA synthesis and replication — especially in rapid growing tissues such as tumors.[19]

3.4 MOLECULAR REGULATION

Since the production of every enzyme along with its location and function within the cells of a given plant are ultimately controlled by the sequence of nucleotides on strands of DNA, one last category of factors that influence metabolite biosynthesis will be considered. These factors interact with the DNA molecules themselves to regulate the activity of the genes that govern the individual enzymes of each pathway.

3.4.1 Primary Steps in Gene Expression

The primary steps in gene expression are depicted in Figure 3.8. The steps of gene expression that take place in the nucleus to produce messenger RNA (DNA → Gene → Primary DNA Transcript → Mature mRNA) are known as *transcription*. The steps of gene expression which take place in the cytoplasm to form polypeptide chains from this mRNA (mRNA on the ribosomes → synthesis of polypeptides → formation of functional protein) are called *translation*. Note, in Figure 3.8, that degradation of DNA, mRNA, and functional proteins can also occur when the appropriate hydrolases are present (DNAases, RNAases, or proteases). Both synthesis and degradation of DNA, mRNA, and functional proteins are very important processes in gene regulation and are known as *turnover*. When the rate of synthesis exceeds the rate of degradation, there is a net synthesis of DNA, RNA, or protein; when the converse occurs, there is a net loss of DNA, RNA, or protein. This has a direct impact on the amount of production of a given enzyme within a given pathway. Synthesized mRNA and protein can also be stored within the cells for later times when changing environmental conditions trigger their activation (e.g., long-lived mRNA in seeds and animals egg; storage proteins in seeds). The absolute level of DNA, mRNA, or a given protein in a cell or tissue will then depend on the rates at which synthesis, degradation, and storage take place. We refer to this system of regulation of the steady-state level of such metabolites in cells as *homeostasis*.

3.4.2 How Plant Genes Are Turned On and Off

Regulation of gene expression can occur at the level of transcription (DNA to RNA), post transcription (initial RNA transcript to mRNA, translation of mRNA to polypeptide), or post translation (polypeptide to functional protein). These levels

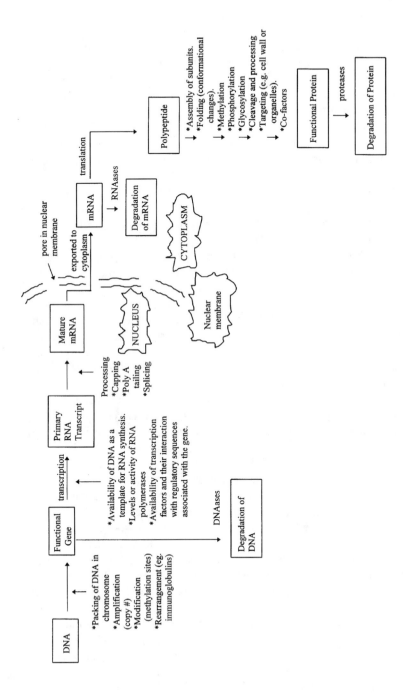

FIGURE 3.8 Diagram depicting the primary steps in gene expression and the control points that occur at steps leading from DNA to mRNA and protein synthesis in cells.

of regulation are controlled by a range of environmental or developmental signals. The various control points in gene expression at these different levels are designated in Figure 3.8.

What are some of the developmental and environmental signals which regulate gene expression in plants? One is light that may up-regulate the synthesis of mRNA for synthesis of the light-harvesting complex involved in photosynthesis. This is mediated by the phytochrome system involving red and far-red wavelengths of light. Other signals are stresses elicited by such factors as ultraviolet light, wounding, or pathogen attack, which can up-regulate, at the level of transcription, the synthesis of such enzymes as PAL that leads to synthesis of phenylpropanoid compounds. Still other signals can be attributed to plant hormones which are bound by protein transcription factors within the cell. For example, in germinating cereal grass seeds, gibberellins (GAs) can cause *de novo* synthesis of mRNAs for α-amylase that break down starch to sugar and of proteases that can break down stored proteins in seeds. In contrast, the plant hormone, abscisic acid (ABA) turns off such gene expression in germinating seeds and is thus partly responsible for the dormancy of these seeds as well as the dormancy of the buds of temperate zone trees.

The precise mechanisms by which environmental or developmental signals act to control gene expression are not yet completely understood. But, research so far has allowed several mechanisms to be promulgated including the following:

- The signal (such as GA or ABA) could stimulate the synthesis of a protein regulatory factor that binds to particular *trans*-acting (other proteins) or *cis*-acting (DNA sequence) elements located upstream in the promoter region of a gene to turn the gene on (as in the case of GA) or off (as in the case of ABA) gene expression.
- The signal (such as a cytokinin plant hormone, that acts to stimulate red light-induced synthesis of RuP2-Case and light-harvesting complex (LHCP) in greening tissue of duckweed, *Lemna gibba*) may act to stabilize particular mRNA species, i.e., retard the degradation of the initial RNA transcripts or mRNA produced from a given gene.[9]
- The signal may fail to act when plants are genetically engineered using constructs that have the gene of interest in anti-sense or reverse orientation. In this case the signal may be the plant hormone ethylene that causes ripening in fruits due to enhanced activities of pectinases (a class of cell wall-loosening enzymes, more properly known as polygalacturonases, which hydrolyze pectins or polygalacturonans, the cementing substances located mostly in the middle lamella between primary cell walls). This has the effect of producing RNA molecules that are complementary to the normal (correct orientation) RNA. Since mRNA is single stranded, when these two molecules bind together through their mutual affinity, the normal mRNA will not function in translation. The FLAVR SAVR™ tomato is one such genetically engineered product where the gene for pectinase was introduced into tomato plants in an anti-sense orientation to knock out gene expression of the plant's pectinase.[20]

To produce such transgenic plants as the FLAVR SAVR™ tomato, there is a specific order of questions and answers that must be elucidated. In natural products research, one of the first important biochemical questions to ask is: "How is the metabolite of interest synthesized?" Another is: "What are the enzymes for the respective steps in the pathway?" These are not easy questions to answer, but once these enzymes are isolated and purified, then the molecular biologist can potentially clone the genes that make these enzymes, determine their nucleotide sequences, and characterize their expression patterns within the various plant tissues. At this point the pathway for the metabolite of interest will be well understood and a new question arises. How can the expression of the gene(s) for the rate-limiting enzyme(s) in the biosynthetic pathway be up-regulated, or down-regulated, so as to make more, or less, of the metabolite through genetic engineering protocols? These protocols include the use of constitutive or super promoters attached upstream of the gene, the use of constructs to make anti-sense RNA, and the use of genetic transformation to express the gene of interest in organisms that normally do not express this gene. If all of the biochemistry is done properly, including (a) the purification of the proteins of interest, (b) the characterization of any isozymes for the particular enzyme being studied as well as their ultimate site(s) of action in the cell, and (c) the elucidation of the function of the enzymes in cell metabolism, then the above-outlined molecular biology work is not only feasible, but also allows one to turn specific genes on or off in a particular metabolic pathway thus changing the production of specific metabolites. In doing this kind of work, risk assessments are absolutely necessary to determine if a particular transgenic plant can have any detrimental effect on our environment. These are discussed in detail in References 20 through 22.

3.4.3 CASE STUDY: ISOLATION OF GENES IN THE ISOPRENOID BIOSYNTHETIC PATHWAY

Figure 3.9 illustrates the pathways of isoprenoid (also called terpenoid) biosynthesis in plants,[23] and Table 3.5 provides a key to the enzymes that operate at each of the respective numbered steps in these pathways.[23] The genes which have been cloned for many of the enzymes in the isoprenoid biosynthetic pathways are indicated in Table 3.6.[23] Often natural product researchers will search the Genebank database for these cloned genes to obtain the nucleotide sequences of such genes (see Reference 24 on how to access the Genebank database). The information obtained can then be used to tackle the problem of increasing or decreasing the production of a specific metabolite within the plant. Let us take as an example the synthesis of natural rubber, which comes from plants such as the Brasilian rubber tree (*Hevea brasiliensis*) and guayule (*Parthenium argentatum*). This example focuses on the following question: How can one increase the levels of natural rubber in these plants? According to Dr. Katrina Cornish[25] and C. Potena,[26] natural rubber is made up of isoprene units that are derived from isopentenyl pyrophosphate (see Figure 3.9). The polymerization step is catalyzed by the enzyme, rubber transferase, which requires allylic pyrophosphate to initiate the process. Dr. Cornish has focused her attention on identifying, isolating, and manipulating rubber transferase and its two substrates,

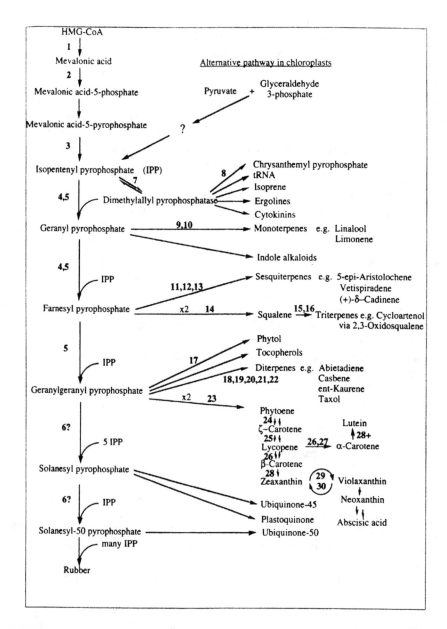

FIGURE 3.9 Pathways of isoprenoid biosynthesis. The enzymatic steps are numbered according to the key in Table 3.2. (From Scolnick, P. A. and Bartley, G. E., *Plant Mol. Bio. Rep.*, 14, 305, 1996. With permission.)

isopentenyl pyrophosphate and allylic pyrophosphate. Her results to date suggest that by raising the level of the initiator (through up-regulated gene expression of the enzyme which makes allylic pyrophosphate), she can enhance rubber production up to six times.[26]

TABLE 3.5
Key to Enzymatic Steps Shown in Figure 3.9

Step	Enzyme
1	3- Hydroxy-3-methyl glutaryl coenzyme A reductase
2	Mevalonic acid kinase
3	Mevalonate 5-pyrophosphate decarboxylase
4	Farnesyl pyrophosphate synthase
5	Geranylgeranyl pyrophosphate synthase
6	Hexaprenyl pyrophosphate synthase-related protein
7	Isopentenyl pyrophosphate isomerase
8	Chrysanthemyl pyrophosphate synthase
9	s- Linalool synthase
10	$4s$- Limonene synthase
11	5-epi- Aristolochene synthase
12	Vetispiradiene synthase
13	(+)-δ- Cadinene synthase
14	Squalene synthase
15	Squalene epoxidase
16	Oxidosqualene cyclase (cycloartenol synthase)
17	Geranylgeranyl pyrophosphate hydrogenase
18	Abietadiene synthase
19	Casbene synthase
20	ent- Copalyl pyrophosphate synthase A
21	ent- Kaurene synthase
22	Taxadiene synthase
23	Phytoene synthase
24	Phytoene desaturase
25	ζ- Carotene desaturase
26	Lycopene cyclase (β)
27	Lycopene cyclase (ε)
28	β- Carotene hydroxylase
29	Zeaxanthin expoxidase
30	Violaxanthin de-epoxidase

From Scolnik, P. A. and Bartley, G. E., *Plant Mol. Bio. Rep.*, 14, 305–319, 1996. With permission.

3.5 CONCLUSIONS

We have covered the primary ways by which metabolite biosynthesis is regulated by environmental, biotic, biochemical, and molecular signals. These mechanisms mostly impinge on regulation of rates of enzyme activity or on regulation of gene expression for particular enzymes. Also of importance in such regulation is the concept of DNA, RNA, and protein (enzyme) turnover, where one must consider rates of synthesis vs. rates of degradation. Metabolite homeostasis refers to all the inputs and outputs that affect the level of a given metabolite in plant cells. The inputs

TABLE 3.6
Some Cloned Plant Genes Involved in Isoprenoid Biosynthesis

Enzyme	Organism	Gene[a]	Clone type Genomic	Clone type cDNA	EST ID	Accession number	Ref.
Abietadiene synthase	*Abies grandis*	*Abs*1		+		U50768	Stoffer-Vogel et al., 1996
5-epi-Aristolo-chene synthase	*Nicotiana tabacum*	*Eas*1		+		?	Facchini & Chappell, 1992
				+		?	Facchini & Chappell, 1992
			+			L04680	Facchini & Chappell, 1992
			+			?	Facchini & Chappell, 1992
(+)-δ-Cadinene synthase	*Gossypium arboreum*	*Cdn*1		+		X96429	Chen et al., 1996
			+			X95323	Chen et al., 1996
				+		U23206	Chen et al., 1995b
				+		U23205	Chen et al., 1995b
				+		U27535	Chen et al., 1995a
ζ-Carotene desaturase	*Arabidopsis thaliana*	*Zds*1		+		U38550	Scolnik & Bartley, 1995c
				+	139J10T7	T46272	Newman et al., 1994
				+	158J7T7	T88256	Newman et al., 1994
	Capsicum annuum	*Zds*1		+		?	Albrecht et al., 1995
	Oryza sativa	*Zds*1		+	S14426_1A	D48291	Sasaki et al., 1994
β-Carotene hydroxylase	*Arabidopsis thaliana*	*Chyb*1		+		U58919	Sun et al., 1996
				+	VBVPH03	F13822	Höfte et al., 1993; Cooke et al., 1996
				+	VBVPH03	F13851	Höfte et al., 1993; Cooke et al., 1996
β-Carotene ketolase	*Haema-tococcus pluvialis*	*Bkt*1		+		D45881	Kajiwara et al., 1995
Chrysanthemyl pyrophosphate synthase	*Chrysanthe-mum cinerariaef olium*	*Cds*1		+		113995	Ellenberger et al., 1995
β-Cyclohexenyl carotenoid epoxidase	*Capsicum annuum*					X91491	Bouvier et al., 1966

TABLE 3.6 (CONTINUED)
Some Cloned Plant Genes Involved in Isoprenoid Biosynthesis

			Clone type				
Enzyme	Organism	Gene[a]	Genomic	cDNA	EST ID	Accession number	Ref.
Farnesyl pyrophosphate synthase	*Arabidopsis thaliana*	*Fps*1;1		+		X75789	Delourme et al., 1994
		*Fps*1;1	+			L46367	Cunillera et al., 1996
		*Fps*1;2		+		L46349	Cunillera et al., 1996
				+	240M1T7	N65905	Newman et al., 1994
			+			L46350	Cunillera et al., 1996
	Artemisia annua	*Fps*1		+		U36376	Matsushita et al., 1996
	Capsicum annuum	*Fps*1		+		X84695	Hugueney et al., 1996
	Hevea brasiliensis	*Fps*1		+		Z49786	Adiwilaga & Kush, 1996
	Lupinus albus	*Fps*1		+		U15777	Attucci et al., 1995a,b
				+		U20771	Attucci et al., 1995a
	Parthenium argentatum	*Fps*1;1		+		X82542	Pan et al., 1996
		*Fps*1;2		+		X82543	Pan et al., 1996
	Zea mays	*Fps*1		+		L39789	Li & Larkins, 1995
Geranylgeranyl pyrophosphate hydrogenase	*Arabidopsis thaliana*	*Gghy*1		+	41D1T7	T13808	Newman et al., 1994; Addlesee et al., 1996
				+	117D18T7	T43273	Newman et al., 1994; Addlesee et al., 1996
				+	FAFE38-1	Z26448	Höfte et al., 1992; Addlesee et al., 1996
				+	241A16T7	N65672	Newman et al., 1994; Addlesee et al., 1996
				+	F10G11T7	N96606	Newman et al., 1994; Addlesee et al., 1996

TABLE 3.6 (CONTINUED)
Some Cloned Plant Genes Involved in Isoprenoid Biosynthesis

			Clone type				
Enzyme	**Organism**	**Gene[a]**	**Genomic**	**cDNA**	**EST ID**	**Accession number**	**Ref.**
				+	150E22T7	T76292	Newman et al., 1994; Addlesee et al., 1996
				+	155D1T7	T76545	Newman et al., 1994; Addlesee et al., 1996
				+	OBO27	F14257	Höfte et al., 1992; Addlesee et al., 1996
				+	OBO27	F14220	Höfte et al., 1992; Addlesee et al., 1996
				+	4D9T7P	T04791	Newman et al., 1994; Addlesee et al., 1996
				+	194L8T7	H76435	Newman et al., 1994; Addlesee et al., 1996
				+	FA130	Z33939	Höfte et al., 1992; Addlesee et al., 1996
				+	175F3T7	H36592	Newman et al., 1994; Addlesee et al., 1996
				+	93N1T7	T21295	Newman et al., 1994; Addlesee et al., 1996
				+	F5G7T7	Nl95906	Newman et al., 1994; Addlesee et al., 1996
	Oryza sativa	Gghy1		+	S10735_4A	D46208	Sasaki et al., 1994
				+	S11188_1A	D46485	Sasaki et al., 1994
				+	S13017_4A	D47484	Sasaki et al., 1994
Geranylgeranyl pyrophosphate synthase	*Arabidopsis thaliana*	Ggps1;1		+		L25813	Scolnik & Bartley, 1994
		Ggps1;2		+		U44876	Scolnik & Bartley, 1996
		Ggps1;3		+		U44877	Scolnik & Bartley, 1996
		Ggps1;4	+			L22347	Bartley et al., 1994

TABLE 3.6 (CONTINUED)
Some Cloned Plant Genes Involved in Isoprenoid Biosynthesis

Enzyme	Organism	Gene[a]	Clone type Genomic	cDNA	EST ID	Accession number	Ref.
	Brassica campestris	*Ggps*1		+	F0958	L37477	Lim et al., 1996
	Capsicum annuum	*Ggps*1	+			X80267	Badillo et al., 1995
		*Ggps*1		+		P80042	Kuntz et al., 1992
	Catharan-thus roseus	*Ggps*1		+		X92893	Bantignies et al., 1996
	Lupinus albus	*Ggps*1		+		U15778	Aitken et al., 1995
	Sinapis alba	*Ggps*1		+		X98795	Welsch et al., 1996
Geranylgeranyl pyrophosphate synthase-related protein	*Arabidopsis thaliana*	*Ggr*1		+	42G9T7	L40577	Scolnik & Bartley, 1995a
				+	121A16T7	T43543	Newman et al., 1994
				+	145E2T7	T46450	Newman et al., 1994
				+	42G9T7	T04590	Newman et al., 1994
				+	221F20T7	N38654	Newman et al., 1994
Hexaprenyl-pyrophosphate synthase-related protein	*Arabidopsis thaliana*	*Hxsr*1		+	192H13T7	R90311	Newman et al., 1994
				+	193N19T7	H76026	Newman et al., 1994
				+	124L9T7	T44803	Newman et al., 1994
	Cyanophora paradoxa	PreA	+			M61174	Michalowski et al., 1991
Isopentenyl-pyrophosphate isomerase	*Arabidopsis thaliana*	*Ipi*1;1		+		U48961	Blanc et al., 1996
				+	1G2T7P	T04125	Newman et al., 1994
				+	168H5T7	R65272	Newman et al., 1994
		*Ipi*1;1		+		U47324	Leustek, 1996
		*Ipi*1;2		+		U49259	Leustek, 1996

TABLE 3.6 (CONTINUED)
Some Cloned Plant Genes Involved in Isoprenoid Biosynthesis

Enzyme	Organism	Gene[a]	Genomic	cDNA	EST ID	Accession number	Ref.
		*Ipi*1;2		+	11715T7	T43292	Newman et al., 1994
				+	173I16T7	H36484	Newman et al., 1994
				+	173J16T7	H36082	Newman et al., 1994
				+	VBVNC09	Z46571	Höfte et al., 1992; Cooke et al., 1996
				+	VBVNC09	Z46678	Höfte et al., 1992; Cooke et al., 1996
	Clarkia breweri	*Ipi*1;1		+		X82627	Blanc & Pichersky, 1995
		*Ipi*1;2	+			U48963	Blanc et al., 1996
	Clarkia xantiana	*Ipi*1;1	+			U48962	Blanc et al., 1996
	Oryza sativa	*Ipi*1;1		+	C1268_2A	D28222	Sasaki et al., 1994
ent-Kaurene synthase	*Cucurbita maxima*	*Eks*1		+		U43904	Yamaguchi et al., 1996
keto-Carotenoid synthase	*Capsicum annuum*	*Kcs*1		+		X77289	Deruère et al., 1994
				+		X76165	Bouvier et al., 1994
				+		X78030	Houlne et al., 1994
4s-Limonene synthase	*Mentha spicata*	*Lim*1		+		L13459	Colby et al., 1993
s-Linalool synthase	*Clarkia breweri*	*Lis*1		+		U58314	Dudareva et al., 1996
Lycopene cyclase (β)	*Arabidopsis thaliana*	*Lcyb*1		+		L40176	Scolnik & Bartley, 1995b
				+	VBVYH06	Z29210	Höfte et al., 1992; Cooke et al., 1996
				+	VBVYH06	Z29211	Höfte et al., 1992; Cooke et al., 1996
				+	OAO370	F13561	Höfte et al., 1992; Cooke et al., 1996
				+	OAO370	Z33983	Höfte et al., 1992; Cooke et al., 1996
				+	214C18T7	N37968	Newman et al., 1994
				+		U50739	Cunningham et al., 1996
	Capsicum annuum	*Lcyb*1		+		X86221	Hugueney et al., 1995

TABLE 3.6 (CONTINUED)
Some Cloned Plant Genes Involved in Isoprenoid Biosynthesis

Enzyme	Organism	Gene[a]	Clone type		EST ID	Accession number	Ref.
			Genomic	cDNA			
	Lycopersiscon esculentum	*Lcyb*1		+		X86452	Pecker et al., 1996
	Nicotiana tabacum	*Lcyb*1		+		X81787	Pecker et al., 1996
Lycopene cyclase (ε)	*Arabidopsis thaliana*	*Lcye*1		+		U50738	Cunningham et al., 1996
	Brassica campestris	*Lcye*1		+	F1516	L46452	Lim et al., 1996
	Brassica napus	*Lcye*1		+	khkh051-1	H07750	Sohn et al., 1995
	Solanum tuberosum	*Lcye*1		+	g16	R27545	Provart, 1996
Mevalonate 5-pyrophosphate decarboxylase	*Arabidopsis thaliana*	*Mpd*1		+	177G1T7	H36293	Newman et al., 1994
Mevalonate kinase	*Arabidopsis thaliana*	*Mki*1		+		X77793	Riou et al., 1994
				+	124O12T7	T44374	Newman et al., 1994
Oxidosqualene cyclase (cycloartenol synthase)	*Arabidopsis thaliana*	*Sqc*1		+		U02555	Corey et al., 1993
				+	103D18T7	T22249	Newman et al., 1994
	Oryza sativa			+	C21901_1A	D23492	Sasaki et al., 1994
				+	S14946_1A	D48622	Sasaki et al., 1994
	Zea mays			+	csuh00265	T18827	Keith et al., 1993
Phytoene desaturase	*Arabidopsis thaliana*	*Pds*1		+		L16237	Scolnik & Bartley, 1993
	Capsicum annuum	*Pds*1		+		X68058	Hugueney et al., 1992
	Glycine max	*Pds*1		+		M64704	Bartley et al., 1991
	Lycopersicon esculentum	*Pds*1		+		M88683	Giuliano et al., 1993
				+		X59948	Pecker et al., 1992
			+			X71023	Mann et al., 1994
			+			X78271	Aracri et al., 1994

TABLE 3.6 (CONTINUED)
Some Cloned Plant Genes Involved in Isoprenoid Biosynthesis

Enzyme	Organism	Gene[a]	Clone type Genomic	Clone type cDNA	EST ID	Accession number	Ref.
	Narcissus pseudonarc issus	Pds1		+		X78815	Albabili & Beyer, 1996
	Nicotiana benthamia na	Pds1		+		U19262	Kumagai et al., 1995
	Oryza sativa	Pds1		+	C2722_1A	D23378	Sasaki et al., 1994
	Zea mays	Pds1		+		L39266	Hable & Oishi, 1995
				+		U37285	Li et al., 1996
Phytoene synthase	Arabidopsis thaliana	Psy1;1		+		L25812	Bartley & Scolnik, 1994
	Capsicum annuum	Psy1;1		+		X68017	Romer et al., 1993
	Cucumis mela	Psy1;1		+		Z37543	Karvouni et al., 1995
	Lycoper- sicon esculentum	Psy1;1		+		A21360	Bird et al., 1991
				+		M84744	Bartley et al., 1992
				+		Y00521	Ray et al., 1987
			+			X60441	Ray et al., 1992
		Psy1;2		+		L23424	Bartley & Scolnik, 1993
			+			X60440	Ray et al., 1992
	Narcissus pseudo- narcissus	Psy1;1		+		X78814	Schledz & Beyer, 1996
	Oryza sativa	Psy1		+	S15075_1A	D48697	Sasaki et al., 1994
				+	S14375_1A	D48251	Sasaki et al., 1994
	Zea mays	Psy1		+		U32636	Buckner et al., 1995
Squalene epoxidase	Arabidopsis thaliana	Pqp1		+	250F2T7	W43353	Newman et al., 1994
				+	129F12T7	T44667	Newman et al., 1994
				+	191M9T7A	N64916	Newman et al., 1994
Squalene synthase	Arabidopsis thaliana	Sqs1		+		X86692	Kribii et al., 1995
				+		D29017	Nakashima et al., 1995

TABLE 3.6 (CONTINUED)
Some Cloned Plant Genes Involved in Isoprenoid Biosynthesis

Enzyme	Organism	Gene[a]	Clone type Genomic	Clone type cDNA	EST ID	Accession number	Ref.
				+	208D19T7	N37135	Newman et al., 1994
				+	229L21T7	N65466	Newman et al., 1994
	Nicotiana benthamiana	Sqs1		+		U46000	Hanley et al., 1996
Taxadiene synthase	Taxus brevifolia	Tsy1		+		U48796	Wildung & Croteau, 1994
Vetispiradiene synthase	Hyoscyamus muticus	Vsy1		+		U20188	Back & Chappell, 1995
				+		U20189	Back & Chappell, 1995
				+		U20190	Back & Chappell, 1995
Violaxanthin de-epoxidase	Arabidopsis thaliana	Vde1		+	207C23T7	N37612	Newman et al., 1994
			+			U39452	Lukowitz et al., 1996
				+		U44133	Bugos & Yamamoto, 1996a
	Lactuca sativa v. longifolia	Vde1		+		U31462	Bugos & Yamamoto, 1996b
Zeaxanthin epoxidase	Arabidopsis thaliana	Zep1		+	133D24T7	T45502	Newman et al., 1994
	Brassica campestris	Zep1		+	F0164	L37996	Lim et al., 1996
	Nicotiana tabacum	Zep1		+		X95732	Marin et al., 1996

[a] Designations for genes in isoprenoid metabolisms are currently under review by a working group of the Commission on Plant Gene Nomenclature.

From Scolnik, P. A. and Bartley, G. E., *Plant Mol. Bio. Rep.*, 14, 305–319, 1996. With permission.

refer to the rates of synthesis of a given metabolite. The outputs refer to the rates of degradation of the metabolite to other metabolites or oxidation products as well as to the rates of formation of conjugates of the metabolite (e.g., glucosyl, amide-linked conjugates, of myo-inositol ester conjugates, as seen with the plant hormone, indole-3-acetic acid). All of these inputs and outputs affect the level of a given metabolite. Once research provides an understanding of the above factors as well as an understanding of the enzymes controlling the biosynthetic pathways leading to the production of specific metabolites, steps can be taken to either increase or decrease the levels of these metabolites produced by the plants. These steps may include controlled environmental conditions, simulated herbivory, metabolite feeds, plant growth regulators, fungal elicitors, or transgenic technology. However, finding the best way to adjust the production of a given metabolite is by no means an easy task. Each individual biosynthetic pathway is regulated by a vast array of environmental, biotic, biochemical, and molecular factors. It is these factors that allow for the incredible variety of metabolite activities that control the overall growth, development, and environmental interactions of each plant.

REFERENCES

1. Mohr, H. and Schopfer, P., Biosynthetic metabolism, in *Plant Physiology*, Springer-Verlag, New York, 1995, chap. 18.
2. Pace, N. R., A molecular view of microbial diversity and the biosphere, *Science*, 276, 734–740, 1997.
3. Liu, Z., Carpenter, S. B., and Constantin, R. J., Camptothecin production in *Camptotheca accuminata* seedlings in response to shading and flooding, *Can. J. Bot.*, 75, 368–373, 1997.
4. O'Brien, J., The tree of life, *Your Health*, 6 August, 60–63, 1996.
4a. Teeri, J. A., personal communication 1997.
4b. Liu, Zhijun, personal communication 1997.
5. Larcher, W., Chemical interactions mediated by bioactive plant substances, in *Physiological Plant Ecology*, 3rd ed., Springer-Verlag, New York, 1995, sect. 1.1.4.2.
6. Liu, Z. and Adams, J., Camptothecin yield and distribution within *Camptotheca accuminata* trees cultivated in Louisiana, *Can. J. Bot.*, 74, 360–365, 1996.
7. Funk, C., Lewinsohn, E., Stofer Vogel, B., Steele, C. L., and Croteau, R., Regulation of oleoresinosis in grand fir (*Abies grandis*): coordinate induction of monoterpene and diterpene cyclases and two cytochrome P450-dependent diterpenoid hydroxylases by stem wounding, *Plant Physiol.*, 106, 999–1005, 1994.
8. Alborn, H. T., Turlings, T. C. J., Jones, T. H., Stenhagen, G., Loughrin, J. H., and Tumlinson, J. H., An elicitor of plant volatiles from beet armyworm oral secretion, *Science*, 276, 945–949, 1997.
9. Anderson, J. W. and Beardall, J., Enzymes and post-translational enzyme regulation, in *Molecular Activities of Plant Cells, An Introduction to Plant Biochemistry*, Blackwell Scientific Publications, Oxford, U.K., 1991, chap. 5.
10. Duke, J., Commentary — novel psychotherapeutic drugs: a role for ethnobotany, *Psychopharm. Bull.*, 31, 177–182, 1995.
11. Schery, R. W., Latex products, in *Plants for Man*, 2nd ed., Prentice-Hall, Englewood Cliffs, NJ, 1972, chap. 8.
12. Weaver, R. J., Size control and related phenomena, in *Plant Growth Substances in Agriculture*, W. H. Freeman and Company, San Francisco, 1972, chap. 11.
13. Kaufman, P. B. and Dayanandan, P., Gibberellin-induced growth in *Avena* internodes, in *The Biochemistry and Physiology of Gibberellins*, Vol. I, Crosier, A., Ed., Praeger Publishers, New York, 1983, p 129–157.

14. Jacobsen, J. V., Gubler, F., and Chandler, P. M., Gibberellin action in germinated cereal grains, in *Plant Hormones, Physiology, Biochemistry and Molecular Biology,* Davies, P. J., Ed., Kluwer Academic Publishers, Dordrecht, The Netherlands, 1995, chap. D-3.

15. Tabata, M. and Fujita, Y., Production of shikonin by plant cell cultures, in *Biotechnology in Plant Science: Relevance to Agriculture in the Eighties,* Zaitlin, M., Day, P., and Hollaender, A., Eds., Academic Press, New York, 1985, 207–218.

16. Funk, C. and Brodelius, P. E., Phenylpropanoid metabolism in suspension cultures of *Vanilla planifolia* Andr. IV. Induction of vanillic acid formation, *Plant Physiol.,* 99, 256–262, 1992.

17. Kaufman, P. B., Ghosheh, N. S., LaCroix, J. D., Soni, S. L., and Ikuma, H., Regulation of invertase levels in *Avena* stem segments by gibberellic acid, sucrose, glucose, and fructose, *Plant Physiol.,* 52, 221–228, 1973.

18. Kaufman, P. B., Duke, J. A., Brielmann, H., Boik, J., and Hoyt, J. E., A comparative survey of leguminous plants as sources of the isoflavones, genistein, and daidzein: implications for human nutrition and health, *J. Altern. Complem. Med.,* 3, 7–12, 1997.

19. Boik, J., Dietary non-nutrient factors and their effects on cancer, in *Cancer and Natural Medicine. A Textbook of Basic Science and Clinical Research,* Oregon Medical Press, Princeton, MN, 1996, chap. 14.

20. Redenbaugh, K., Hiatt, W., Martineau, B., Kramer, M., Sheehy, R., Sanders, R., Houck, C., and Emlay, D., *Safety Assessment of Genetically Engineered Fruits and Vegetables. A Case Study of the FLAVR SAVR™ Tomato,* CRC Press, Boca Raton, FL, 1992.

21. Rissler, J. and Mellon, M., *The Ecological Risks of Engineered Crops,* The Massachusetts Institute of Technology Press, Cambridge, MA, 1996.

22. Krimsky, S. and Wrubel, R., *Agricultural Biotechnology and the Environment. Science, Policy, and Social Issues,* University of Illinois Press, Urbana, IL, 1996.

23. Scolnik, P. A. and Bartley, G. E., A table of some cloned plant genes involved in isoprenoid biosynthesis, *Plant Mol. Biol. Rep.,* 14, 305–319, 1996.

24. Wu, W., Welsh, M. J., Kaufman, P. B., and Zhang, H. H., *Methods in Gene Biotechnology,* CRC Press, Boca Raton, FL., 1997.

25. Cornish, K. and Siler, D. J., Characterization of *cis*-prenyl transferase activity localized in a buoyant fraction of rubber particles from *Ficus elastica* latex, *Plant Physiol. Biochem.,* 34, 337–384, 1996.

26. Potera, C., Genetic engineering improves natural rubber, *Gene. Eng. News,* 16, 1, 1996.

4 Good and Bad Uses of These Compounds by Humans

Peter B. Kaufman, Lynn Pennacchini,
Maureen McKenzie, and James E. Hoyt

CONTENTS

0-8493-3134-X/99/$0.00+$.50
© 1999 by CRC Press LLC

4.1   INTRODUCTION

Throughout human history, we have seen both good and bad uses of natural products from plants. We know of notable cases of poisons from plants that caused the demise of Socrates (coniine, an alkaloid from Poison Hemlock, *Conium maculatum*)[1,2] and St. Anthony's Fire disease in humans due to eating ergot fungus-infected (*Claviceps purpurea*) rye grains used to make rye bread flour.[3] *C. purpurea* produces toxic alkaloids that are derivatives of lysergic acid and the hallucinogenic d-lysergic acid amide. Good uses of natural products from plants by humans are countless; they include natural pesticides, flavorings, fragrances, medicinal compounds, fibers, beverages, and food metabolites. For the sake of consistency, we have grouped the examples of good and bad uses of natural products from plants in this chapter under the same major classes of compounds cited in Chapter 1. After reading this chapter, we challenge you to think of both good and bad uses of natural products from plants from your own experience. Almost everyone has gotten a skin rash from plants like poison ivy (*Toxicodendron radicans*) or has an allergy due to pollen from grasses, walnut trees (*Juglans* spp.), or ragweed (*Ambrosia* spp.) and can recall vividly the adverse effects. And for most of us, flower scents can be a delight or really a bad experience, depending on what plant's flowers you sniff. Even some of the foods you eat may taste bad but still be good for you nutritionally. So, let's explore some of the most notable case examples that humans classify as either good or bad. One of the surprises you will encounter is that some plants can be both good or bad for human health depending on the kinds and amounts of the metabolite(s) involved and on the uses to which they are put.

4.2   THE HISTORY OF HERBAL MEDICINE: A CAPSULE SUMMARY

Penelope Ody, in her book, *The Complete Medicinal Herbal*,[1] has compiled a very nice summary on the origins of western herbalism, starting with ancient civilizations of Egypt, the Greeks, and Romans. She continues with Islamic influences from the Arab world, Ayurvedic medicine from ancient India (*ajur* meaning life and *veda* meaning knowledge, refers to "knowledge of how to live" and places emphasis on good health being the responsibility of the individual), and Tibetan herbalism. The legacy of Chinese herbal medicine and its basic principles of five elements (wood, water, metal, earth, and fire) is cited in great detail together with beautiful illustrations in color of Chinese herbs. She has a marvelous treatise on herbs and herbal medicine that came out in magnificently illustrated herbals during the dark ages in Europe. She ends her synopsis with North American Indian herbal traditions and practices and a nice summary on the use of herbs in modern medicine.

The use of herbs in modern medicine is well expressed in terms of home accessible ***homeopathic medicine*** (the use of herbal drugs to treat human diseases). Don DeSander has prepared the following account based on his own research and personal experience.

Research Project

It has long been known that Native Americans used the native plants surrounding them to treat everything from relieving headaches to curing snakebites. After the general population's 75-year hiatus from folk healing, medicinal herbs are once again gaining popularity in the treatment of disease and illness. However, because of our society's aesthetic influences, there is now more emphasis on the appearance of medicinal plants than before. By this I mean that homeopathy is being used in conjunction with home gardening and landscape design. Therefore, this research project focuses not only on the medicinal values of six herbs (*Echinacea purpurea, Achillea millefolium, Verbascum thapsus, Nepeta cataria, Monarda didyma*, and *Geranium maculatum*), but also, on creating a visually stimulating medicinal garden to display these native plants (Figures 4.1 through 4.4).

FIGURE 4.1   Making and planting a medicinal plants garden in Saginaw County, MI. Shown are purple coneflower (*Echinacea purpurea*) plants in the foreground. Planters are Donald and David DeSander (May, 1997). (Photo by Don DeSander.)

Because my father owns a landscaping business, I felt this was the perfect project for me in order to broaden my knowledge of medicinal plants. Although I was being introduced to a field in which I had no previous knowledge, I always had a good sense of landscape design and experience with decorative plants on which to fall back. My plant selection was based on herbs used by Native Americans because their usage reflects my own feelings towards homeopathy, namely, using one's natural surroundings to treat oneself rather than introducing foreign techniques or materials into the body.

FIGURE 4.2 A lakeside view of same medicinal plants garden shown in Figure 4.1. (Photo by Don DeSander.)

FIGURE 4.3 Flowers of the prairie coneflower, *Echinacea purpurea,* growing in the Gateway Garden and University of Michigan Matthaei Botanical Gardens (October, 1996). (Photo by Don DeSander.)

FIGURE 4.4 Close-up view of a single flower of the prairie coneflower, *Echinacea purpurea*. (Photo by Don DeSander.)

All of my selected medicinal herbs have been used by Native Americans and are still being used by them today. They have used *E. purpurea* (purple coneflower) for everything from sore throats to snake bites.[2] Being one of the most popular herbs in homeopathy, it is an immune system booster which has antiviral, antifungal, and antibacterial properties.[3] Although bitter tasting (as I can attest), *Echinacea* root is excellent for treating headaches, fevers, bladder infections, rheumatism, hemorrhoids, and numerous other wounds and illnesses.[4] *V. thapsus* (mullein) was used to treat coughs, congestion, and tuberculosis. In addition, Native Americans smoked the flowers to counteract pulmonary diseases.[5] The flowers of *A. millefolium* (yarrow) were used by the Great Lakes Ojibwa Amerinds to break fevers and a root decoction was used on skin "eruptions". Various other parts of this plant also are used for wounds, toothaches, diarrhea, gas, intestinal ailments, and even chapped hands.[6] The Otsego Indians were noted for their use of *M. didyma* (Bergamot or Oswego tea). This pleasant tasting herb is still widely used in teas today. It is also recommended for coughs, sore throats, nausea, flatulence, and menstrual cramps.[7] Although *N. cataria* (catnip) came from Europe, the Great Lakes Ojibwa used a decoction from this plant to cure fevers.[8] The tops are also used in a tea as a sedative and to heal symptoms of colds, headaches, and indigestion.[9] An infusion made from the *G. maculatum* (storksbill or wild geranium) roots was used by the Great Lakes Ojibwa to treat diarrhea and to help relieve mouth soreness.[10]

The three most common ways that I encountered for the extraction of these herbs' medicinal properties and administering them were through tinctures, teas, and decoctions. The recipes I used were a combination of those found in *Rodale's Illustrated Encyclopedia of Herbs* and interviews with people who have had first-hand experience with these herbs.[11]

Of the three, tinctures are the most potent. I combined 4 oz (114.3 g) of powdered or finely cut *Echinacea* herb with 1 pt (0.48 l) of 100 proof (50% ethyl alcohol) spirits (vodka, brandy, or gin). I poured 1 pt (0.48 l) of vodka over 4 oz (114.3 g) of *Echinacea* root which I had powdered using liquid nitrogen. I then sealed the container and turned it on its side, rotating it daily for a 2-week period. After 2 weeks passed, I allowed the plant material to settle and then strained off the liquid into another clean container using a paper filter circle (32 cm diameter). I used this as a preventative measure for illnesses. I mixed 2 to 5 drops (2 to 5 ml) of the tincture with a cup (230 ml) of water both in the morning and at night throughout the winter months. To disguise the taste of *Echinacea*, I sometimes combined the dosage with orange juice. However, dealing with the taste was apparently worth it. The treatment succeeded as an immune system booster. I also avoided the flu that seemingly all my friends had developed.

Herbal teas are a fun and much better tasting way for me to extract the plants' medicinal constituents. The general recipe for teas is to use 1 oz (28.57 g) of dried tops (leaves, flowers, and stems) per 1 pt (0.48 l) of water or 2 oz (57.2 g) of fresh herbs per 1 pt (0.48 l) of water. Keep in mind that the herb will soak up approximately one quarter of the total water you use. Also, be sure to use distilled water (you don't want chlorine or other chemicals in your tea). I poured boiled hot water over the herb in a large pot, and then allowed the herb to steep for 15 min, but one can steep it for longer periods. The longer the infusion is under the hot water, the stronger the tea's potency. Meanwhile, I kept a tight lid on the pot because I didn't want any of the medicinal products to escape. If you can smell the aroma of the tea, then your tea is losing its potency. Finally, I strained the infusion into another container once again using a paper filter. The general dosage is a half-cup (115 ml) three times daily.

I found the process of making decoctions to be very similar to teas. In fact, the only difference between the two is that decoctions are used for tougher woody parts of herbs such as roots, bark, stems, and leaves too heavy for teas. I used the same ratio of plant material to water as I did for teas: 1 oz (28.5 g) of dried plant material for every pint (0.48 l) of distilled water. Using the same method as teas, I allow the herbs to simmer in hot water that is just below boiling (100°C or 212°F) for 30 min. Then, I separate the plant matter from the liquid and store it for use. Even the dosage is similar to that of teas: a half-cup (115 ml) three times daily. Once again, I found this form of treatment much more tasteful than tinctures as long as I didn't use any *Echinacea* in my decoction.

One of the primary goals of this project is simply to provide a general introduction into the realm of medicinal plants in preparation for further studies on the subject. But with homeopathy's resurgence, I feel this project is also important because it aims to show that healing with plants can be accessible and attractive to everyday gardeners. I particularly liked this project because of its combination of history, design, and the introduction to medicine that it offered. Hopefully, in the end, this research will offer a link between the past human knowledge of plants and contemporary society's stress on beauty to attract weekend gardeners to experiment with something old yet new.

4.3   PROTEIN METABOLITES

4.3.1   MEDICAL USES OF THE ANTI-AIDS DRUG, TRICHOSANTHIN

Trichosanthin is a protein produced primarily in the tuberous roots of the Chinese Cucumber plant (*Trichosanthes kirilowii*). The gene that is responsible for the synthesis of trichosanthin has been cloned and sequenced by Clontech Company in California. The Chinese cucumber has been used in oriental medicine for several thousand years to promote abortions. More recently, it has been demonstrated that trichosanthin selectively inhibits replication of HIV virus *in vitro* by inhibiting ribosomal protein synthesis and cellular reproduction (mitotic activity). However, we need to point out that extracts of Chinese cucumber are very toxic and should never be ingested unless under the supervision of a physician. Further, the use of this plant is contraindicated in pregnant women or those of child-bearing age. Trichosanthin can cause severe reactions in humans, including pulmonary and cerebral edema, cerebral hemorrhage, myocardial damage, seizures, fever, damage to blood cells, and even death in one AIDS patient.[15,16]

4.3.2 Uses of Seed Storage Proteins in Human Nutrition

During grain-fill in cereals, proteins [albumin (water-soluble), prolamin (alcohol-soluble) globulin (salt-soluble), and glutelin (alkali-soluble)] are stored in the outermost layer of the starch-storing endosperm tissue.[17] This is called the ***aleurone layer***; it is located just inside the seed-coat or ***pericarp***. Such proteins are stored in protein body organelles in the cells of this layer. In cereal grains such as rice (*Oryza sativa*), wheat (*Triticum aestivum*), barley (*Hordeum vulgare*), oats (*Avena sativa*), and rye (*Secale cereale*), the percent dry weight of the grain that is made up of protein varies between 6 and 12%. If one eats white rice, one gets little or no protein except for a very small amount in the embryos (or germs). Why is this so? It is because white rice, during milling, has been polished to remove the brown aleurone layer that contains the bulk of the protein in the grain. If white rice is eaten, and one is a vegetarian, other protein sources must be found, such as whole grains of oats, wheat, rye, or barley; or of high protein seeds of soybean (*Glycine max*), teff (*Eragrostis abyssinica*), millets (*Panicum* spp., *Pennisetum glaucum, Eleusine coracana*), wild rice (*Zizania aquatica*), and grain amaranth (*Amaranthus* spp.).[18] Some of these contain much higher amounts of protein, varying between 20% (millets, grain amaranth, and teff) and 45 to 48% (soybean).

4.3.3 Secretion of Enzymes that Digest Insects in the Leaves of Pitcher Plants

Pitcher plants are insectivorous plants that live in wetlands, such as bogs [*Sarracenia* spp., *Darlingtonia californica* (Figures 4.5, 4.6)] or on limbs of trees in tropical rain forests as ***epiphytes*** (literally, on plants) (*Nepenthes* spp., Figures 4.7, 4.8). These environments are nitrogen-poor, and none of these plants form symbiotic relationships with bacteria or blue-green algae which fix atmospheric nitrogen. So, to obtain nitrogen, these plants have evolved various mechanisms to trap insects and to secrete hydrolytic enzymes (e.g., chitinases, proteases) that digest soft tissues of the insects, thereby releasing nitrogen-containing compounds in the form of amino acids that the pitcher plants can assimilate.

The insect-trapping mechanisms of the leaves ("pitchers") developed by these plants are ingenious.[19,20] They include the following:

FIGURE 4.5   A native population of the cobra plant, *Darlingtonia californica*, near Coos Bay, Oregon Coast. (Photo by Dr. Larry Mellichamp, University of North Carolina at Charlotte.)

- Emission of distinctive odors from the leaves that attract insects to the leaves
- Occurrence of an attractive color (red, due to anthocyanin pigments) that serves to lure the insect to the pitchers (as seen with *Nepenthes,* and even more so, with *Sarracenia* leaves)
- Provision of "ladders" along the outside of the pitcher that facilitate access of the insect to the mouth of the pitcher (as seen in *Nepenthes* spp.)[19a]
- Possession of translucent window-like areas in the hood over leaf pitchers in *Darlingtonia californica* that confuse and disorient the insect once it is inside the hood portion of the leaf
- Secretion of sugary nectar from glands located near the rim of the pitcher that provide a food source lure/reward for the insect that visits the leaf
- The occurrence of a slippery, thick waxy cuticle on the inside of the pitcher; "as the insect struggles up the surface, their feet become coated with wax, which builds up until the victims seem to have acquired heavy clodlike boots"[20]
- The occurrence of downward-pointing hairs near the rim of the pitcher that prevent the insect from getting out of the pitcher, once trapped inside
- The occurrence of a pool of liquid (up to 1 l in some *Nepenthes* spp.) containing digestive enzymes that not only drowns the insect, but also results in their being partially digested

How many insects can a single pitcher trap per day? The answer, according to ecologists who have examined this question for *Darlingtonia californica* in the state

FIGURE 4.6 Close-up view of the cobra plant, *Darlingtonia californica*, photographed at Gasquet, CA. (Photo by Dr. Larry Mellichamp, University of North Carolina at Charlotte.)

of Oregon, is as many as 50 per day. That many could provide a lot of nitrogen to the plant!

Of what concern and use are pitcher plants to humans? Are they good? Are they bad? What is their relevance to protein metabolites? They are indicator plant species of habitats (wetlands like fens, tropical ecosystems) that are basically nitrogen-poor. Because of their rarity in such habitats, they are placed on rare and endangered plant species lists. They are wonderful subjects to study and to photograph by people of all ages who visit such ecosytems or see them on display in conservatories.[19a] They have relevance to protein metabolites in that these plants get most of their nitrogen for synthesis of proteins from the insects that they digest in their leaf-pitchers!

4.4 CARBOHYDRATE METABOLITES

4.4.1 HUMAN USES OF STARCH FROM PLANT STORAGE ORGANS

Starch is a primary energy source for humans. We obtain it in abundance from cereal grains such as wheat (*Triticum aestivum*), rice (*Oryza sativa*), barley (*Hordeum*

FIGURE 4.7A   The pitcher plant, *Nepenthes,* growing in the conservatory at the University of Michigan Matthaei Botanical Gardens. (Photo by Peter Kaufman.)

vulgare), oats (*Avena sativa*), rye (*Secale cereale*), and maize (*Zea mays*). It also occurs in abundance in tubers of potato (*Solanum tuberosum*), oxalis (*Oxalis* spp.) and tuberous roots of yam (*Dioscorea* spp.), yautia (*Xanthosoma brasiliense*), sweet potato (*Ipomoea batatus*) (Figure 4.8), cassava (*Manihot esculenta*), hog peanut (*Apios americana*), and Indian turnip (*Psoralea esculenta*). Jerusalem artichoke (*Helianthus tuberosus*) tubers (Figure 4.9) do not contain starch, but rather, fructan as the primary storage carbohydrate (see Chapter 2). Starch is used to make breads, tortillas, enchiladas, cakes, pies, pancakes, waffles, and crackers.

Regarding the potato (Figure 4.10), it has been in cultivation since 500 B.C.[5] In the 17th and 18th centuries, potatoes formed a major staple item in the diets of people from Ireland (an intake of almost 2.5 kg/d per person). However, during the mid-1840s, the late blight of potato fungal pathogen (*Phytophthora infestans*) destroyed more than 80% of the crop 2 years in a row. This resulted in the starvation of more than 3 million Irish people and the mass emigration of Irish citizens to New England in the U.S.[15,18]

FIGURE 4.7B Close-up view of one of the pitchers (a modified leaf). (Photo by Peter Kaufman.)

The following is an essay on starch from the tuberous roots of Indian breadroot (*Psoralea esculenta,* in the legume family, Fabaceae) (Figure 4.11A and B) prepared by University of Michigan SROP (Summer Research Opportunity Program) student Lynn Pennacchini (Figure 4.12A), based on her own personal experiences with the uses of this plant by Amerinds in South Dakota.*

* Resource information for this essay on Indian breadroot was obtained from the following sources: (1) Ansel Woodenknife, Interior, SD 57750; (2) **Plants For a Future,** a Resource and Information Centre for Edible and Other Useful Plants: Species Database (Homepage: http://www.npsc.nbs.gov); (3) Northern Prairie Science Center, Northern Prairie Resources (Homepage: http://www.npsc.gov); (4) Daniel Moerman's two volume paperback text, *Medicinal Plants of Native America*, University of Michigan Museum of Anthropology Publication, Ann Arbor, Michigan, 1986; (5) Heart of the Earth Market, WoodenKnife Indian Fry Bread Mix (Homepage: http://www.blackhills.com/~hrtearth/); and (6) Soule, J.P. and Piper, J.K., *Farming in Nature's Image*, Island Press, Washington, D.C. and Covelo, CA, 1990.
Special thanks go to Dan Marcus, who provided transportation to Lynn Pennacchini for this study in South Dakota, and to the Summer Research Opportunity Program at the University of Michigan for their financial support.

FIGURE 4.8   Tuberous roots of the sweet potato, *Ipomoea batatus*. (Photo by Peter Kaufman.)

Research Essay

P. esculenta (Indian breadroot) has historically been found in Zone 4 hardiness regions of North America, south to Texas and north to Canada. It is most prevalent in the Great Plains of the U.S. and Canada. Indian breadroot is an indicator of a healthy prairie and is now most commonly found in the prairies of North and South Dakota. Human development has caused the plant's disappearance in many regions, mainly because it is very intolerant of root disturbance. Cultivation has not been successful because modern farming practices impose undue stress on the plant. Therefore, it would be most beneficial to cultivate the plant in a natural prairie setting, using alternative farming techniques, such as prescribed burns to prevent succession to a shrub/woodland ecosystem.

Indian breadroot is very hardy in that it tolerates very warm summer and very cold winter temperatures and can withstand drought. It will grow in a variety of soils, but prefers gravely soils, particularly those that have their origin from limestone. In such locations, it usually grows in patches of four or more plants. The plants are most prevalent on the southeast sides of hills, growing in association with grama (*Bouteloua* spp.) grass. They are characterized by having purple flowers and masses of silvery hairs on the stems and leaves. Like many other herbaceous legumes (e.g., peas, beans, clovers, alfalfa), it has nitrogen-fixing bacteria in association with root nodules. This provides considerable benefit to neighboring plants.

Indian breadroot was probably the most important food plant gathered by the Indians who lived on the prairies of the U.S. and Canada. It was so important and sacred to these people that it influenced the selection of their animal hunting grounds. It is known to the Lakota Indians as Tinpsila. Tinpsila has been a source of food and commerce in the Great Plains for centuries. In the Lakota language, the month of June is called "tinpsila i Kwaca wi", meaning the moon when the breadroot is ripe. This is the time

FIGURE 4.9   Tubers of the Jerusalem artichoke, *Helianthus tuberosus*. (Photo by Peter Kaufman.)

when the flowers die down and the tuberous roots are ready to harvest. The tops of the plant are left in place to dry so that they can be blown across the prairie by wind to re-seed itself.

The starchy tuberous root (Figures 4.11A and B and 4.12B) has traditionally been the food source of the plant (see Table 4.1 on the amounts of phytochemicals found in *P. esculenta* roots). Women and children used to gather the tuberous roots to eat raw, to put in stews, or to grind up into a powder for different cooking uses. Today, Ansel Woodenknife, a Lakota Indian of South Dakota, carries on traditional uses of tinpsila by marketing Woodenknife Indian Fry Bread Mix, which uses tinpsila powder as an ingredient. Other ingredients include: wheat flour, sugar (sucrose), corn starch, bicarbonate of soda ($NaHCO_3$), sodium aluminum sulfate, and acid phosphate of calcium.

Indian breadroot has also been used by Great Plains Indians for its medicinal properties. The Cheyenne Indians used it as a burn dressing and as a diuretic. The Blackfeet Indians brewed a tea from Indian breadroot to treat sore throats, chest problems, gastroenteritis, and to treat earaches. The chewed root was applied to sprains and fractures and would be sprayed into a baby's rectum to treat gas pains. The Arapaho Indians used it as a moisturizer, a tea for headaches, and to clear a throat.

Currently, there is a very small market for this plant. An increase in demand for the root, due to current evidence of its medicinal value, may have health benefits,

FIGURE 4.10 Elementary school children and teacher collecting tubers of the potato plant, *Solanum tuberosum*. (Photo by Peter Kaufman.)

A

FIGURE 4.11A Xerox image of plant of Indian breadroot, *Psoralea esculenta*, vegetative plant obtained from a prairie ecosystem in South Dakota. (Courtesy of Lynn Pennacchini.)

FIGURE 4.11B Xerox image of an older plant of *P. esculenta* at a later, flowering stage of development. (Courtesy of Lynn Pennacchini.)

but could potentially have damaging ecological and cultural effects. The plant requires several years to flower and seed itself. Therefore, it is important to wait until the plant is fully mature before harvesting it. Because of the current socioeconomic issues of Native Americans, an increase in demand may trigger over-harvesting of the plant in its natural prairie habitat and lead to its decline in the wild. In addition, financial rewards due to an increased market may attract outsiders to begin harvesting the plant as well.

To successfully harvest Indian breadroot in the wild, *people must be respectful, knowledgeable, and sensitive to the plant and its needs*! Cultivation may be the answer. Modern farming techniques to date have been unsuccessful. Alternative farming techniques which rely on the natural dynamics of a prairie are being investigated at the Land Institute in Kansas. Using such alternative approaches to farming, Indian breadroot has the potential to be high-yielding and successful in commercial cultivation if it is domesticated and grown in a natural prairie setting.

FIGURE 4.12A   Author of Indian breadroot essay, Lynn Pennacchini, standing in South Dakota prairie where this plant is native, with Ansel Woodenknife, a Lakota Indian of South Dakota. (Photo courtesy of Lynn Pennacchini.)

FIGURE 4.12B   A plant of Indian breadroot showing the tuberous root; top of plant, the shoot, is shown at the bottom of the photo. About one fourth the actual size. (Photo courtesy of Lynn Pennacchini.)

TABLE 4.1
Amounts of Phytochemicals Found in
Roots of *Psoraea esculenta,* Indian
Breadroot

Chemical	Amount (ppm)	Low (ppm)	High (ppm)
Alanine	1,139		
Arginine	7,976		
Ascorbic Acid		4	171
Aspartic Acid	8,989		
β-Carotene		0	1
Calcium	5,100		
Cystine-(Half)	802		
Fat	36,800		
Fiber	160,300		
Glutamic Acid	12,319		
Glycine	1,097		
Histidine	1,266		
Iron	39		
Isoleucine	1,308		
Leucine	2,448		
Lysine	2,743		
Magnesium	1,400		
Methionine	253		
Phenylalanine	1,224		
Phosphorus	500		
Potassium	2,800		
Proline	1,266		
Protein		70,000	422,000
Serine	1,857		
Starch	698,400		
Sugars	56,000		
Threonine	1,477		
Tryptophan	633		
Tyrosine	1,899		
Valine	1,730		
Water	569,800		

From Internet, Hot-Bot Program, Phytochemical Data Base.

4.4.2 SECONDARY FERMENTATION FROM MODIFIED BARLEY SUBSTRATE (STARCH)

Here, we return to the grains of barley (*Hordeum vulgare*), the rich source of starch as a carbohydrate food source we mentioned in Section 4.4.1. This same starch is used for alcoholic fermentation in the brewing of beer. First, the starch must be hydrolyzed by α- and β-amlyases, ultimately yielding the sugar, D-glucose. This happens during seed germination during which time the plant hormone, gibberellic acid (produced in the embryo or germ and the scutellum tissue of the grain and secreted to the aleurone layer), stimulates the *de novo* synthesis of α-amylase by transcriptional up-regulation in the aleurone layer of the grain. Under anaerobic conditions (in the absence of oxygen), beer yeasts (*Saccharomyces cerevisiae*), convert the sugar released from the hydrolysis of starch to carbon dioxide and ethyl alcohol (≈ 4% ethanol). Some of the flavor of beer comes from hops (*Humulus lupulus*) (Figure 4.13) which produces the primary flavoring constituent, lupulin, in glandular hairs located on bracts of the flowers.

FIGURE 4.13   A hops plant vine (*Humulus lupulus*). (Photo courtesy of Peter Kaufman.)

4.5   TERPENOIDS AND PHENOLICS

4.5.1   POKEWEED (*PHYTOLACCA AMERICANA*) CAN BE BOTH GOOD AND BAD

Pokeweed (Figures 4.14, 4.15, and 4.16) is a common perennial garden weed with lush green shoots and bright red stems (due to presence of anthocyanin pigment).

Leaves of its green shoots are harvested in the spring in Appalachia, boiled several times, each time pouring off the hot water. All parts of the plant except the above-ground leaves that grow in the spring contain toxic triterpene saponins (phytolac-cigenin, jaligonic acid, phytolaccinic acid, esculenic acid, and pokeberrygenin).[15] The ritual of boiling several times is to make sure that the green leaves are rendered edible by removing any low amounts of toxic saponins. Only green leaves are used for this purpose; older leaves, which become reddish in color, are *not* safe to eat. The toxic saponins are in highest amounts in the roots, less in the mature leaves and stems, and least in the fruits. If these parts are ingested, they cause severe stomach cramping, nausea, diarrhea and vomiting, slow and difficult breathing, spasms, hypotension, severe convulsions, and death.[15]

FIGURE 4.14   Young vegetative shoot of pokeweed, *Phytolacca americana*. (Photo courtesy of Dr. Larry Mellichamp, University of North Carolina at Charlotte.)

It is essential to know the identification of this plant when it is being collected. Pokeweed poisonings were commonplace in the 1800s in Eastern North America. The berries and roots of pokeweed were eaten by mistake, the collectors thinking the parts collected were parsnip (*Pastinaca sativa*), Jerusalem artichoke (*Helianthus tuberosus*), or horseradish (*Amoracia lapathifolia*).[15]

FIGURE 4.15 Flowering and fruiting shoot of pokeweed, *Phytolacca americana*. (Photo courtesy of Dr. Larry Mellichamp, University of North Carolina at Charlotte.)

4.5.2 *CANNABIS SATIVA* (MARIJUANA OR HEMP): HALLUCINOGENIC AND MEDICINAL USES

Marijuana has been used for more than 4000 years for the treatment of catarrh, leprosy, fever, dandruff, hemorrhoids, obesity, asthma, urinary tract infections, loss of appetite, inflammatory conditions, and cough.[15] More recently, it has been used in medicine to treat patients who have glaucoma (an eye disease) and to alleviate the adverse side effects of chemotherapy used to treat patients who have had cancer. These are the only legal medical uses of marijuana at the present time.

The fibers from this plant also have been used to make fibers for ropes and clothing. This plant was brought to the New World by European settlers to grow primarily for this purpose, not for medical or psychoactive drug purposes.

Marijuana contains 20 sedative-type drugs. These are called tetrahydrocannabinoids. The most active cannabinoid metabolite is tetrahydrocannabinol (THC). This and the other cannabinoids are produced in glandular hairs on the shoots of the plants (Figure 4.17).[21] The plant's sedative effects were recognized by the Chinese, but its widespread use as a psychoactive drug plant has only been fairly recent.[15]

FIGURE 4.16   Fruits of pokeweed, *Phytolacca americana*. (Photo by Dr. Larry Mellichamp, University of North Carolina at Charlotte.)

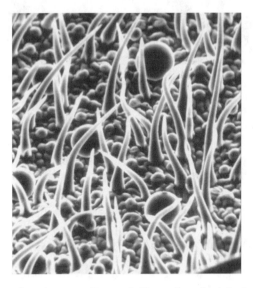

FIGURE 4.17   Scanning electron micrograph illustrating glandular hairs (Trichomes) on leaves of the hemp plant (*Cannabis sativa*). (Photo by Dr. P. Dayanandan, Department of Botany, Madras Christian College, Tambaram, Madras, India.)

4.5.3   MAYAPPLE: SOURCE OF A PHENOLIC COMPOUND

Mayapple (*Podophyllum peltatum*) is an understory (beneath the forest canopy) plant found in deciduous forests native to Eastern North America and Canada (Figure 4.18).[22]

FIGURE 4.18   Illustration of a mayapple (*Podophyllum peltatum*) plant in flower. About 1/20 normal size. (Photo by Peter Kaufman.)

All parts of the plant except the ripe fruits are toxic to humans and other animals. The leaves and roots contain the toxic, phenolic compound, podophyllotoxin.[14] The ripe fruits, in contrast, are edible. They are large, yellow fleshy berries which are used to make mayapple jelly, marmelade, or to eat fresh. In the southern U.S., a drink is prepared from the juice of the fruit and combined with Madeira and sugar or squeezing the juice from the fruit into lemonade or other fruit-drinks.[23]

4.6   NITROGEN-CONTAINING COMPOUNDS

4.6.1   Alkaloids from Plants that Are Used to Treat Malaria: Quinine (*Chinchona* spp.)

The bark of Chinchona trees is the source of the drug, quinine, which has been used to treat patients afflicted with malaria. Malaria is a disease in humans caused by the parasite, *Plasmodium falciparum*. The parasite is transmitted by the mosquito, *Anopheles* spp. Quinine, or more generally, the drug, chloroquine, since 1986 in

FIGURE 4.19   A vegetative plant of sweet annie, *Artemisia annua,* source of the antimalarial drug, artemisinin. (Photo by David Bay.)

Brasil and Africa, has been shown to be no longer effective in treating malaria due to resistance by *P. falciparum* to the drug.[24] The consequences of this are dire. In Africa, 80% of the world's malaria cases were recorded in 1990 with over 95% of the deaths occurring on this continent, mainly due to the ineffectiveness of chloroquine in treating patients having malaria.[24]

Are there any solutions? One possible solution is the use of the drug artemisinin or quinghaosu (Chinese name for the drug) (a sesquiterpene lactone compound) from the common "weed", *Artemisia annua* (Sweet Annie) (Figure 4.19). This compound is very effective in treating patients with malaria at the present time because it kills chloroquine-resistant *P. falciparum* cells[25] (see also Chapter 6, Section 6.2.5 on antimalarial compounds and Table 6.2). It also has fewer adverse side effects. Artemisinin has now been synthesized and derivatives of artemisinin have been synthesized which have an improved action in curtailing malaria. Artemisinin is an unusual drug with a 6-member lactone ring which is termed α-methyl-γ-lactone. The compound is also a peroxide, which is required for its activity.[25] One of the primary questions raised here is whether or not *P. falciparum* will soon develop resistance to artemisinin or its derivatives as it did with chloroquine.

4.6.2   THE INDOLE ALKALOIDS, PSILOCYBIN AND PSILOCIN
(*PSILOCYBE* MUSHROOMS), THAT ACT AS HALLUCINOGENS

Psilocybe mushrooms (Figure 4.20) are the source of two metabolites, psilcybin (4-phosphoryloxy-*N,N*-dimethyltryptamine) and psilocin (4-hydroxy-*N,N*-dimethyl-tryptamine).[14] These are both psychoactive compounds that produce hallucinations in humans. *Psilocybe* mushrooms are called Teonanacatl by Mexican Amerinds, a term which means "flesh of the gods".

FIGURE 4.20   The Liberty Cap mushroom, *Psilocybe semilaceata*, common in Northern Europe and the western coast of the U.S. north to British Columbia, Canada. The Mexican species are the hallucinogenic species collected by Mexican Amerinds. (Photo by David Bay.)

In Mexican Indian religious ceremonies, the *Psilocybe* mushrooms, are collected and dried by women, then served to tribal members in pairs in all night ceremonies. The ingested mushrooms have a vision-inducing action that is typified by the following symptoms: muscular relaxation, flaccidity, and pupil dilation in the early stages followed by emotional disturbances and visual and auditory hallucinations, and eventually causing lassitude and mental and physical depression. The subject feels isolated from the world around him/herself without loss of consciousness and becomes indifferent to the environment.[14] Richard Evans Schultes, the famous ethnobotanist (now retired) from Harvard University, described these hallucinogenic effects of *Psilocybe* mushrooms and the religious ceremonies surrounding their use by Mexican Amerinds when he presented a seminar on this subject at the University of Michigan in the 1960s.[25a]

4.6.3   The Protoalkaloid, Mescaline (*Lophophora williamsii*), that Acts as a Hallucinogen

Mescal button or peyote (*Lophophora williamsii*) (Figure 4.21) is native from South Texas to central Mexico. In the Aztec Empire of Mexico, it was referred to as "Peyote". The hallucinogenic compound of interest in the "buttons" (top parts) of this plant is the protoalkaloid, mescaline. It is used in Amerind religious ceremonies and eaten as mescal buttons or the dried, brown pieces of the above-ground parts of the cactus (which has no spines), or occasionally, as fresh green pieces. Sometimes, it is brewed and drunk as a tea. Lewis and Elvin-Lewis[14] describe how humans respond to this plant when it is ingested as dry or fresh tissue or as a tea. It produces nausea, chills, and vomiting often accompanied by terror, anxiety, and a dislocation of visual perspective. These symptoms then subside and are followed by mental stimulation expressed in terms of clarity and intensity of thought, brilliant colored visions, and exaggerated sensitivity to sounds. Dr. Elzada U. Clover, former Botanist at the University of Michigan, while collecting cacti along the Colorado River in the Grand Canyon of Arizona in Havasupi Indian territory, ingested some of the peyote cactus and described her visions as a technicolor display of sheep jumping fences.

FIGURE 4.21   Potted plants of the hallucinogenic cactus, mescal button, *Lophophora williamsii,* growing in a cage at the University of Michigan Matthaei Botanical Gardens. About 1/10 natural size. (Photo by Peter Kaufman.)

4.6.4 THE STEROID (*ACONITUM* SPP.), USED FOR ACONITE POISON WHALING IN ASIA AND ALASKA*

Aconitine is a toxic alkaloid derived from all parts of *Aconitum* spp. plants. The roots are the most dangerous part, but the leaves are greatest in toxicity just before flowering commences.[14] Symptoms of poisoning include numbness followed by paralysis of upper and lower extremities. A weak pulse develops accompanied by respiratory paralysis. Convulsions typically occur and death follows after about 2 h.[14]

Kodiak and Aleutian Indians as well as natives to the Pacific littoral zone of northeast Asia (Kurile Islands and the Kamchatkan coast) use aconite poison obtained from the roots of monkshood plants on their whaling spears to kill whales. The procedure used is relatively simple. The Indians throw a lance into the whale, the stone point or dart (typically made of ground slate) becomes detached when it enters the whale. The tips of the stone points are smeared with aconite poison.[26] The spear heads were also greased with human fat, grease obtained from corpses found in burial caves. The purpose of this ritual is to deceive others not familiar with the whaling techniques by leading them to believe that the human fat is the source of the poison! In effect, they do this so as not to divulge the real source of the poison.[26]

4.7 SULFUR-CONTAINING COMPOUNDS**

4.7.1 FUNCTIONS OF SULFUR-CONTAINING COMPOUNDS IN GARLIC IN MEDICINE, NUTRITION, AND PEST CONTROL

4.7.1.1 Natural History and Association with Humanity[15,27,28]

Garlic, *Allium sativum* (Liliaceae or lily family) has an ancient association with humans. It is so ancient that garlic is not found in nature and its exact origins are unknown. The earliest documented occurrence of garlic's use by humans is the fine clay sculptures of garlic cloves found at El Mahasna, Egypt which have been dated to 3750 BC, almost 6000 years ago. Its use by humans certainly predates even this artifact. While the exact origin of garlic is unknown, the best determination for its origin is the Eurasian continent north of Afghanistan and northwest of Tibet. This is an area rich in *Allium* species, most of which are found growing in narrow gorges. *A. longicuspis* bears a close similarity to *A. sativum* and may be its progenitor.

Alliums are usually classed as members of the lily family (Liliaceae). The genus contains about 450 species including several cultivated species such as *A. cepa* (onion), *A. porrum* (leeks), *A. schoenoprasum* (chives), and at least one species cultivated as an ornamental, *A. gigantium*. A bulb of garlic is derived from leaf tissue, with each bulb divided into sections or "cloves". Roots develop below the bulb. Bulbs are the food storage organs for the plant and contain rich reserves of

---

* The primary source material for this section was provided by Maureen McKenzie, CEO, Naniquah Corp., Girdwood, AK.

** This section of Chapter 4 on garlic has been prepared by James E. Hoyt, Department of Biology, University of Michigan.

carbohydrates (including fructans, see Chapter 3) and proteins. As the food storage organs of the plant, the bulbs would be ideal targets for predation by numerous animal species. Alliums, garlic in particular, have evolved powerful antiherbivory compounds as an adaptation for protection against predation. It is these compounds which yield garlic's second great benefit to humans — effective medical drugs.

As a valuable food and medical species, garlic has been introduced by humans to all temperate climates throughout the world, although it prefers hot, dry climates. It is unable to thrive in moist tropical and arctic habitats. As mentioned above, the known use of garlic by humanity extends back for 6000 years, and its use before that time can reasonably be inferred. In addition to ancient Egypt, we have visual images of garlic being dispensed in Sumeria. Written records are found in Egypt, with 22 garlic preparations cited as being used for the treatment of heart problems, headache, animal bites, worms, and tumors. People of Greece and Rome used garlic extensively. Pliny the Elder gives 61 remedies using garlic for rheumatism, hemorrhoids, ulcers, and loss of appetite. Rome gave garlic to its legions to boost endurance and to be used as a prophylactic against disease. In China the 5th century text, Ch'i-min-yao-shu (essential arts for people) lists garlic as a prominent medicinal. The Indian text, Charaka-Samhita, dates from the 1st century but is based on an older text. In it, garlic is offered as a diuretic, gastrointestinal tonic, for the treatment of eye ailments, as a heart stimulant, and as an antiheumatic. Ayurvedic, Unani, and Tibbi medical systems list garlic as a prophylactic and as a cure for colic, cholera, dyspepsia, typhoid, dysentery, arterio-sclerosis, gastric and laryngeal tuberculosis, lupus, duodenal ulcers, lung gangrene, whooping cough, pulmonary phthisis, and bronchestasis. Garlic is constantly mentioned in western medicine. Hildegard von Bingen writes of its use, as does Culpepper. Louis Pasteur in 1858 provides the first modern scientific report of garlic's action as an antibacterial while tons of garlic were used in World Wars I and II as field dressings for wounds. Current literature searches on "garlic" yield almost 1000 papers since 1985 on all aspects of garlic agriculture and biological actions.

4.7.1.2   Garlic as Food

Garlic bulbs are a rich source of carbohydrates and proteins. Analysis of garlic indicates it contains 61 to 64% moisture, 31% carbohydrate, 5 to 6% protein, and only 0.2% fat. Significant levels of phosphorous (3.9 to 4.6 mg·g^{-1}), potassium (1.0 to 1.2 mg·g^{-1}) and calcium (0.5 to 0.9 mg·g^{-1}) are present.

4.7.1.3   Antiherbivory and Insect Defenses

Garlic produces a variety of sulfur-based compounds which are effective as insect repellents and insecticides, antiherbivory compounds and probable antifungal agents. All *Allium* species produce volatile chemicals which act as repellents to many insects. Planting alliums, especially garlic, with other agricultural or orna-mental species will often protect those species against insect predation. Garlic has additional protection for those insects and animals not deterred by its volatile metabolites. Sequestered in vacuoles within the plant's cells is an odorless, sulfur-based compound, (+)S-allyl-l cysteine sulfoxide or alliin, (Figure 4.22). Alliin

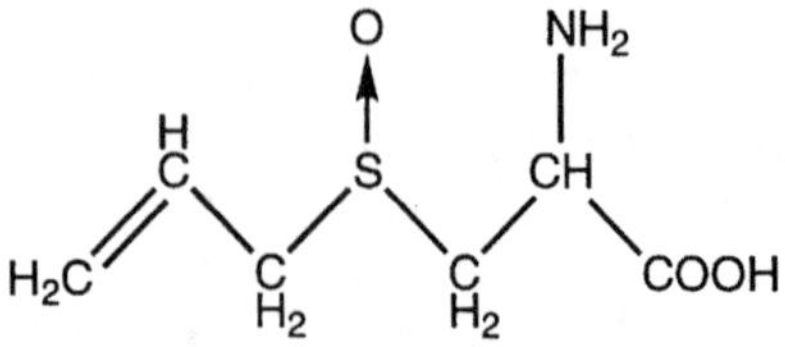

FIGURE 4.22 Chemical structure of alliin, (+)-allyl-l-cysteine sulfoxide from garlic, *Allium sativum*. (Chemical structure by James E. Hoyt.)

comprises approximately 0.24% of the fresh weight of the bulb, or about 3.7 $mg \cdot g^{-1}$. In the cytosol, normally separate from alliin, is an enzyme, allinase. Allinase is found in all alliums and is capable of acting on several different substrates. When the cells are disrupted, e.g., an animal eating the bulb, the cell's vacuoles are broken and alliin (substrate) mixes with allinase (enzyme). The result is a chemical change in which alliin is converted to sulfenic acid (Figure 4.23)

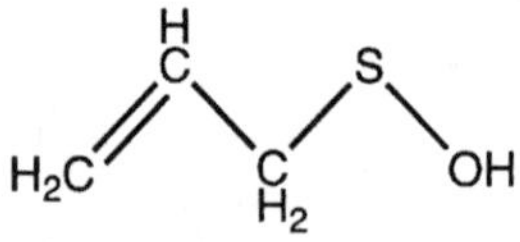

FIGURE 4.23 Chemical structure of sulfenic acid from garlic, *Allium sativum*. (Chemical structure by James E. Hoyt.)

and ammonia. Two molecules of sulfenic acid then combine to form one molecule of water and one of allyl-2-propenethiosulfinate or allicin[15] (Figure 4.24), a strong smelling and fiery tasting chemical that repels almost every animal except most humans. Usually one taste is sufficient to deter further predation and the plant survives. Allicin soon breaks down into diallyl disulfide (Figure 4.25) which is another strong smelling copound and has been shown to be a powerful insecticide. Further transformations occur with time resulting in a variety of sulfite compounds. Commercial insecticide/repellent preparations are now available to farmers and home users (Garlic Barrier AG, EPA #66352-2 from Allium Associates) and are certified for use against mites, nematodes, and mosquito larvae on a wide variety of crops. It is further used as a fungicide and as an antibiotic for poultry, cattle, and aquaculture shellfish.

4.7.1.4 Garlic as Medicine

4.7.1.4.1 Antibiotic Action (Bactericide, Fungicide, Parisiticide)

The significant antibiotic compound of garlic is allicin. It has been shown to be effective against a broad range of bacteria species at dilutions of 1:10.[5] Bacteria shown to be susceptible include some of the most dangerous to human health, such

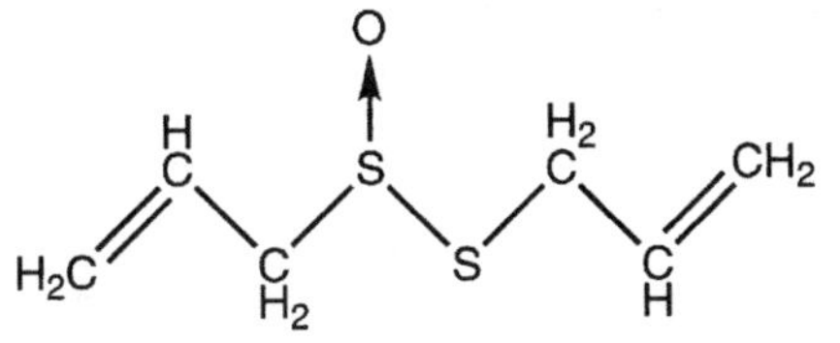

FIGURE 4.24 Chemical structure of allicin, allyl-2-propenethiosulfinate from garlic, *Allium sativum*. (Chemical structure by James E. Hoyt.)

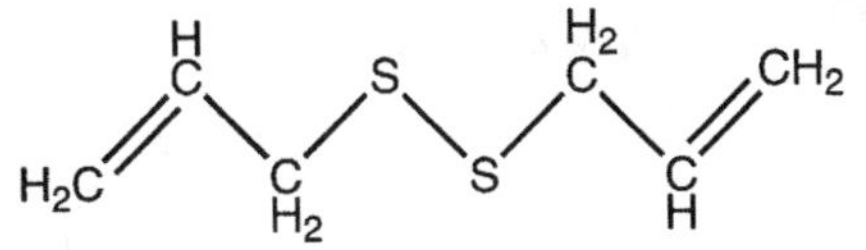

FIGURE 4.25 Chemical structure of diallyl disulfide from garlic, *Allium sativum*. (Chemical structure by James E. Hoyt.)

as tuberculosis (*Mybacterium tuberculosis*), staph (*Staphylococcus aureus* and *S. faecalis*), and salmonella (*Salmonella typhimurium*). Antibiotics such as penicillin are generally stronger then allicin. However, drug resistance is an increasing danger to human health. For instance, there are now strains of tuberculosis which are fully or partially resistant to every class of antibiotic currently available. Allicin could prove to be a life-saving alternative therapy and a model for the development of drug families.

One underlying mechanism of action for garlic's antibiotic (and anticancer) action is the stimulation of the immune system, both systemically and locally. Lymphocytes and macrophages are attracted to the sites where garlic is injected.[29] Therefore, injections near the sites of localized disease can mobilize the body's immune system to that site. Ingestion of garlic stimulates a general systemic increase in the immune system.

Fresh garlic extracts have been found to inhibit many fungal species and have been used to protect plants and stored foods as well as in medicine. Another pathogen that has been developing resistance to conventional antibiotics is *Candida albicans*. Garlic has been shown to be effective in suppressing candida infections. Indeed, it is as effective as conventional drugs such as amphotericin and nystatin, acting faster but without the difficulties in administration or the side effects those drugs produce. A similar result was found for ringworm.

Fresh garlic has been found to be effective against amebic dysentery, hookworm, and pinworm. Garlic has been a traditional remedy for tapeworm, lice, and other parasites.

4.7.1.4.2 *Antitoxin (Heavy Metals and Biotic Toxins)*

Garlic has been shown to be a strong antioxidant. It reduces damage to liver cells by inhibiting the formation of free radicals and preventing the oxidation of lipid peroxides. This can protect the liver from chemicals such as carbon tetrachloride.

It has also been shown to prevent lysis of blood cells *in vitro* due to toxic levels of metals such as lead, aluminum, mercury, and copper.

4.7.1.4.3   Lipoprotein and Cholesterol Regulator

Garlic can lower blood lipid levels. Evidence indicates three possible means by which this can occur. First, garlic reduces or inhibits lipogenesis (the formation of fats in the liver and adipose tissue). Second, garlic accelerates the breakdown and excretion of lipids. Finally, garlic enhances the transfer of lipids from storage in adipose tissue to the bloodstream.[29] This last is an interesting effect. People who are on a garlic regime will show an increase in blood lipid levels during the first few months as fat is recruited from storage and transferred to the blood stream. As fat reserves are reduced, however, the other two activities of garlic, reduced lipogenesis and enhanced breakdown and excretion, act to reduce serum lipid levels. In addition to lowering overall serum lipid levels, garlic shifts the ratio of low- and high-density lipoproteins, increasing the amount of high-density lipoproteins, the so-called "good" lipoproteins.

Garlic is used as a treatment for atherosclerosis in Germany where it has been shown to reduce serum cholesterol levels by 10%. When combined with reduced cholesterol diet, serum cholesterol levels decline 20%.

4.7.1.4.4   Garlic and Cancer

Cancer can be seen as a failure of the cell's regulatory mechanism caused by mutation in specific regulatory genes called protooncogenes. Mutations can occur from exposure to chemical mutagens, radiation, and viruses. As noted above, garlic contains compounds which are antioxidants and can protect the liver from mutagenic chemicals such as carbon tetrachloride. Some fractions from garlic have been shown to protect cells grown *in vitro* from radiation. However, fresh garlic proved fatal.[29] Garlic also appears to have some antiviral properties. Whether this is caused by direct action on the virus or through stimulation of the immune system is not known. Taken together, the indications are that garlic should be effective in preventing cancer. When cancer is already present, studies indicate that garlic stimulates the immune system, especially when applied directly to the tumor. Another possible means by which garlic compounds may act is by directly inhibiting guanylate cyclase activity[31] which would interfere with tumor metabolism.

4.7.1.4.5   Other Effects

Garlic has been shown to reduce platelet aggregation. Together with its effects on sero-lipids and cholesterol, this can have a tremendous impact on heart disease and is an official treatment for hypertension in Japan. It has also been shown to reduce blood glucose levels and to increase blood insulin levels. Traditionally, garlic has been used as a treatment for diabetes though its effects are small. It appears to lower the levels of "stress hormones" and to increase stamina, two effects with which the Roman legions were apparently familiar.

4.7.1.5   Problems with Garlic Use

The most apparent problem with using garlic in human medicine is its strong odor. This is a problem of social esthetics. A problem of clinical importance is that some people are allergic to sulfur-based compounds. In most such cases, garlic would be unsuitable as either a food or a medical preparation. Garlic appears to suppress spermatogenesis. This is related to its effect on cholesterol and glucose metabolism. Perhaps the most difficult problem with widespread use of garlic compounds is the development of resistant strains of pathogens. The dimensions of this problem are unknown; on the one hand garlic has been used successfully against pathogens for millennia, apparently without the evolution of resistance. On the other, we have ample general evidence of the evolution of drug resistance by pathogens (bear in mind that most of these drugs are derived from naturally occurring compounds themselves). Alliums themselves are subject to pathogenic attack; *Sclerotium cepivorum* is a fungus specialized to infect alliums. So the evidence is clear that pathogens can evolve defenses to these drugs.

It is critically important for us to understand the mechanisms of action of garlic compounds on the cellular and molecular level, both to develop new antibiotic and immune-boosting agents, and to protect the efficacy of what we currently have available.[30,31]

4.8   CONCLUSIONS

Plants, as we have learned from the many examples cited in this chapter, have metabolites that may be either good or bad for humans. Many are very toxic compounds that should be avoided. Even in the field, one should avoid poisonous "look-alike" plants! If by chance you have ingested or come in contact with a poisonous plant, it is best to call a Poison Control Center at a local hospital immediately. Plants also contain slow-acting poisons that are toxic due to long-term use and addiction. The case in point is nicotine from tobacco (*Nicotiana tabacum*). It is a prime cause of lung cancer in tobacco smokers and those who experience second-hand smoke. If, however, you are savvy about the plants you use for food, teas, or medicine, they can be very helpful to you for your health and well-being and may help to prevent such maladies as osteoporosis, cancer, heart disease, and hopefully in the future, different forms of dementia, including Alzheimer's disease.

Plants Good and Bad

Plants can be good; plants can be bad,
Tis very important to know what you have had!
Beware of the aconite,
Tis a poison you cannot fight.
But try edible mayapple fruits,
But not this plants' roots,
Because they contain podophyllotoxin that is bad,
Tis not to be had!
On the other hand, Echinacea is OK,
Your immune system will obey,
Have no fear,
Your cold may disappear!

REFERENCES

1. Ody, P., *The Complete Medicinal Herbal*, Dorling Kindersley, Limited, London, 1993.
2. Moerman, D. E., *Medicinal Plants of Native America*, University of Michigan Museum of Anthropology Technical Reports, Vol. 19, The Regents of the University of Michigan, The Museum of Anthropology, Ann Arbor, 1986, 156–157.
3. Ody, P., *The Complete Medicinal Herbal*, Dorling Kindersley, London, 1993, 55.
4. Kowalchik, C. and Hylton, W. H., Eds., *Rodale's Illustrated Encyclopedia of Herbs*, Rodale Press, Emmaus, PA, 1987, 176.
5. Kowalchik, C. and Hylton, W. H., Eds., *Rodale's Illustrated Encyclopedia of Herbs*, Rodale Press, Emmaus, PA, 1987, 204.
6. Meeker, J. E., Elias, J. E., and Heim, J. A., *Plants Used by the Great Lakes Ojibwa*, Great Lakes Indian Fish and Wildlife Commission, Odanah, WI, 1993, 93.
7. Kowalchik, C. and Hylton, W. H., Eds., *Rodale's Illustrated Encyclopedia of Herbs*, Rodale Press, Emmaus, PA, 1987, 39.
8. Meeker, J. E., Elias, J. E., and Heim, J. A., *Plants Used by the Great Lakes Ojibwa*, Great Lakes Indian Fish and Wildlife Commission, Odanah, Wisconsin, 1993, 115.
9. Kowalchik, C. and Hylton, W. H., Eds., *Rodale's Illustrated Encyclopedia of Herbs*, Rodale Press, Emmaus, PA, 1987, 72.
10. Meeker, J. E., Elias, J. E., and Heim, J. A., *Plants Used by the Great Lakes Ojibwa*, Great Lakes Indian Fish and Wildlife Commission, Odanah, WI, 1993, 290.
11. Stark, P., owner of Renaissance Acres Organic Herb Farm of Ann Arbor, MI, interviewed by author, 23 November 1996, Ann Arbor, MI.
12. Muenscher, W. C., *Weeds*, 2nd ed., Cornell University Press, Ithaca, NY, 1955, 325, 326.
13. Robinson, T., *The Organic Constituents of Higher Plants*, 6th ed., Cordus Press, North Amherst, MA, 1991, 272, 278.
14. Lewis, W. H. and Elvin-Lewis, M. P. F., *Medical Botany. Plants Affecting Man's Health*, John Wiley & Sons, New York, 1977.
15. Burnham, T. H., *The Review of Natural Products*, Facts and Comparisons, a Wolters Kluwer Company, St. Louis, MO, 1997.
16. Duke, J. A. and Foster, S., *Trichosanthes kirilowii*: a new hope in the AIDS-relief search?, *HerbalGram*, 20, 20–47, 1989.
17. Juliano, B. O., Polysaccharides, proteins, and lipids of rice, in *Rice: Chemistry and Technology*, 2nd ed., Juliano, B. Ed., American Association of Cereal Chemists, St. Paul, MN, 1985, chap. 3.
18. Schery, R. W., The cereals and other food feeds and forages, in *Plants for Man*, 2nd ed., Prentice Hall, Publishers, Englewood Cliffs, NJ, 1952, chap. 16, 17.
19. Wilkins, M., Mineral nutrition, in *Plant Watching. How Plants Remember, Tell Time, Form Partnerships, and More*, Facts on File Publications, New York, 1988, chap. 11.
19a. Keller, R., personal communication, University of Michigan Matthaei Botanical Gardens.
20. Stern, K. R., Leaves, in *Introductory Plant Biology*, 7th ed., William C. Brown, Dubuque, IA, 1997, chap. 7.
21. Dayanandan, P. and Kaufman, P. B., Trichomes of *Cannabis sativa* (*Cannabinaceae*), *Am. J. Bot.*, 63, 578–591, 1976.
22. Fernald, M. L., *Gray's Manual of Botany*, 8th ed., D. Van Nostrand, New York, 1970.
23. Fernald, M. L., Kinsey, A. L., and Rollins, R. C., *Edible Wild Plants of Eastern North America*, Harper and Row, New York, 1958.

24. Garrett, L., The revenge of the germs or just keep inventing new drugs, in *The Coming Plague, Newly Emerging Diseases in a World Out of Balance,* Farrar, Straus, and Giroux, New York, 1994, chap. 13.

25. Bisset, N. G., Houghton, P. J., and Hylands, P. J., Some current trends in medicinal plant research, in *The Medicinal Plant Industry,* Wijesekera, R. O. B., Ed., CRC Press, Boca Raton, FL, 1991, chap. 10.

25a. Schultes, R. E., personal communication.

26. Heizer, R. F., Aconite poison whaling in Asia and America: an Aleutian transfer to the new world, Anthropological papers, No. 24, *Bur. Am. Ethol. Bull.,* 133, 415–468, 1943.

27. Huang, K. C., *Pharmacology of Chinese Herbs,* CRC Press, Boca Raton, FL, 1993.

28. Duke, J. A., *Handbook of Biologically Active Phytochemicals and Their Activities,* CRC Press, Boca Raton, FL, 1992.

29. Lau, B., *Garlic For Health,* Lotus Light Publications, Wilmot, WI, 1988.

30. Fulder, S. and Blackwood, J., *Garlic Nature's Original Remedy,* Healing Arts Press, Rochester, VT, 1991.

31. Fenwick, G. R. and Hanley, A. B., The genus *Allium.* 1. *Critical Review in Food Science and Nutrition,* 22, 199, 1985; The genus *Allium.* 2. *Critical Review in Food Science and Nutrition,* 22, 273, 1985; The genus *Allium.* 3. *Critical Review in Food Science and Nutrition,* 23, 1, 1985.

5   Modes of Action at Target Sites

Sara Warber

CONTENTS

5.1   INTRODUCTION

Plants and humans have sustained one another for eons. All cultures have some
definable plant knowledge that includes appropriate edible plants, medicinal plants,
and ceremonial plants. Even in the Western tradition the first botanists were physicians who kept their own herb gardens for treating the sick. Modern allopathic
medicine is derived predominantly from the alchemical practice, but even here some
well-known plants have become part of common practice. Cardiac glycosides, i.e.,
digitalis, from the foxglove is a well known example. Much of what we know about
the nervous system function has been defined through the use of plant alkaloids,
i.e., muscarine, nicotine, atropine, ephedrine. As has already been demonstrated in
this book, plants produce a wide variety of chemicals. Our knowledge and appreciation of how these botanical chemicals interact with the human body is growing
with each year. The mechanisms of action of phytochemicals are far more complex
than previously suspected. Plants like *Echinacea* are found to modulate the immune
system through such unlikely candidates as polysaccharides. Plant-based medicines
are becoming an important part of cancer chemotherapy regimes. The public, disil-
lusioned with allopathic medicine, has an intense interest in herbal preparations
which will further stimulate research into the mechanisms of action of phytochemicals.

This chapter will consider some known mechanisms of action of specific plant
preparations. We will consider how phytochemicals participate in cell-cycle interactions, signaling across cell membranes, immunomodulation, and toxic reactions.

5.2   CELL LIFE CYCLE AND CANCER TREATMENT

5.2.1   INTRODUCTION

Cancer is one of the predominant killers in the western world today. Despite much
advancement in cancer therapy, many cancers are still ineffectively treated or become
resistant or recur. In addition, the methods of treating cancer are often difficult for
patients to tolerate due to the side effects. Thus, there continues to be great interest
in the search for new and better treatments. Plant-based medicines have definitely
found a role in this type of treatment and the mechanism of interaction between
many phytochemicals and cancer cells has been studied extensively.

5.2.2   CELL LIFE CYCLE

In order to understand phytochemical-cell interactions it is first important to understand a little about the life cycle of human cells, including proliferation, differentiation, and cell death. The cell reproductive life cycle has four phases: (see Figure
5.1) G_1, S, G_2, M. G_0 is a stage of quiescence which can be of variable length.
During this time the cell is carrying out its ordinary role for the organism. If there
is a commitment to proliferation, then purines and pyrimidines, the building blocks
for DNA synthesis, must be produced. The cell then enters the G_1 state in which
nucleotides and enzymes are synthesized. In the S phase DNA synthesis occurs.
Many enzymes must work together to reproduce an accurate replication of DNA for

Cell Life Cycle

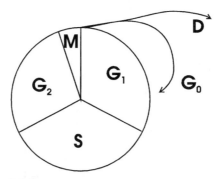

G₀ = Resting
G₁ = Synthesis of purines and pyrimidines
S = Synthesis of DNA
G₂ = Synthesis of components for mitosis
M = Mitosis
D = Differentiation

FIGURE 5.1   Cell life cycle.

the new cell. One enzyme of this system that seems to be particularly vulnerable to exogenous plant chemicals is topoisomerase. Its job is to separate the daughter DNA strands. The next phase is G_2 when the cell prepares other structures needed for mitosis. The M phase is mitosis itself and the production of two daughter cells which will then enter the cycle themselves.

In most cell systems there is a period of normal growth which is a time of proliferation of cells. With more maturity of the tissue, the cells differentiate into the various specialized subsets required for tissue function. These differentiated cells no longer proliferate, instead they synthesize the proteins, steroids, and other chemicals required for maintenance or function of the organism. Within the tissue there remain stem cells capable of proliferation. In some areas such as bone marrow (where blood cells form), skin, and the lining of the gastrointestinal tract, there is a high turnover of cells. This requires a high density of stem cells and constant proliferation.

Cancer cells can be thought of as cells that become capable of proliferation. Much work has been done to identify oncogenes and tumor suppressor genes which are thought to control this abnormal proliferative state. One recent approach to therapy has been to try to induce cells to differentiate into more specialized cells and, therefore, stop proliferating.

Although stem cells and cancer cells may be nearly immortal due to their proliferative capacity, cell death does occur. Necrosis is the process of cell death due to external events such as hypoxia, chemical exposure, radiation injury, and many others. Cells are observed to swell, become vacuolized, and finally be digested by either their own enzymes or the enzymes of neutrophils. The critical insult is to the cell membranes, through lipid peroxidation. This causes permeability changes and allows massive influx of calcium ions. Excess calcium ions inactivate mitochondria and denature proteins and enzymes. Necrosis generally occurs in contiguous cells and is accompanied by an inflammatory response. Many currently available cancer treatments induce necrosis.

In contrast to necrosis, apoptosis is programmed death, whereby physiologic signals, such as hormones or growth factors, trigger rapid DNA damage, condensation of chromatin, and fragmentation of DNA. The cell, too, becomes fragmented and is phagocytized by nearby macrophages or neutrophils without causing inflammation. Several chemotherapeutic agents that cause DNA damage also lead to apoptosis. Recent research has looked more seriously at apoptosis as a goal of chemotherapy. Some natural agents may have more application in this area.

In the section which follows we will highlight some of the plant chemicals which are currently in use as anticancer agents or which are being studied for their potential application. We will not attempt an exhaustive coverage of this field but rather a representative one. In turn, these examples will illustrate some of the ways phytochemicals interact with mammalian or human cells.

5.2.3 GENISTEIN

Epidemiological studies have shown that populations that have a high soy intake have a lower incidence of breast and prostate, as well as other carcinomas. Genistein is an isoflavone (Figure 5.2) found in high quantities in soybean products. Genistein-containing soy diets have been shown to decrease incidence and number of tumors, and to increase latency in animal models of cancer.[1] Much work has been done in cell-culture models which demonstrate that genistein inhibits proliferation of some types of cancer cells.[2] Cell culture and other *in vitro* techniques have been used to elucidate the mechanism by which genistein might alter cancer cell kinetics. There is evidence to support several hypotheses of the target site and mechanisms of action of genistein. Some of these are inhibition of angiogenesis,[3] interaction with steroid hormone receptors, inhibition of tyrosine kinase, inhibition of radical oxygen species formation, and interaction with topoisomerase.[2,4] In this section, we will focus on the interaction with topoisomerase which appears to be one of the more important mechanisms in regulating cellular proliferation.

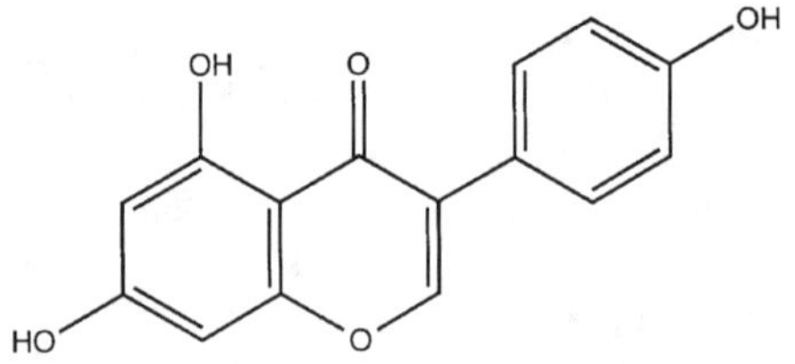

Genistein

FIGURE 5.2 Chemical structure of genistein, an isoflavonoid, commonly found in members of the legume family, Fabaceae.

DNA in its resting state (does it ever really rest?) is highly twisted to conserve intracellular space. In order for transcription to occur, the DNA must be relaxed. The topoisomerase enzymes relax the DNA by nicking single strands. This allows normal gene expression to occur and cells to proliferate. Genistein is postulated to stabilize the enzyme/DNA complex in such a way that both strands are nicked and DNA breaks occur. Hypothetically this leads to altered gene expression and cell differentiation and a concomitant decrease in cell proliferation. Experiments have shown that at genistein concentrations high enough to induce cell differentiation, all types of cells tested had extensive DNA breakage. In a cell-free system containing supercoiled plasmid DNA and genistein, linear DNA (i.e., broken DNA) was produced only when topoisomerase II was present. This supports topoisomerase as the active site for genistein.[5] Further support comes from other experiments where cell lines were developed that were resistant to the effects of genistein. Resistant cells showed altered activity of topoisomerase II[6] or markedly reduced expression of the topoisomerase II β isoform.[7] Because of genistein's site of activity, it will be further tested as an anticancer agent. Soy products, in general, are an important part of a diet to promote wellness.

5.2.4 TAXOIDS AND *VINCA* ALKALOIDS

Several anticancer agents create their effects by interrupting cell division. Since cancer cells are dividing at a more rapid rate than the normal cells around them, the chemotherapeutic agents have a proportionally greater impact on the tumor cells. The target site for the taxoids and the well-known *Vinca* alkaloids is microtubule formation. Microtubules are critical to spindle and aster formation in all cells as they prepare for mitosis. Microtubules also have other cellular functions, such as maintenance of cell shape, cellular motility and attachment, and intracellular transport. Tubulin dimers polymerize to form microtubules. This is in dynamic equilibrium controlled according to the cell's needs by intracellular messengers, such as calcium and guanosine triphosphate (GTP).[8]

The *Vinca* alkaloids, vinblastine and vincristine (Figure 5.3), are derived from the periwinkle (*Catharanthus roseus*). They have been used for many years in treating lymphomas and acute childhood leukemia, respectively. Vincristine and vinblastine inhibit cancer cell reproduction by promoting microtubule disassembly. They bind to the tubulin dimers. When the tubulin-alkaloid complex attaches to the microtubule, polymerization is terminated and depolymerization begins. Mitosis is arrested at metaphase.[9]

The taxoids, paclitaxel and the related semisynthetic docetaxel, are examples of novel new anticancer agents provided by plants. Paclitaxel is extracted from the bark of the Pacific yew (*Taxus brevifolia*), as well as needles and stems of other yews (*Taxus* spp). Docetaxel is derived from a precursor, baccatin III, found in the needles of the English yew (*Taxus baccata* L.).

In contrast to the *Vinca* alkaloids, paclitaxel and docetaxel (Figure 5.4) induce assembly of microtubules and stabilize microtubule networks. Cells treated *in vitro* with paclitaxel form disorganized bundles of microtubules in all phases of the cell cycle. During cell division, paclitaxel induces the formation of many abnormal

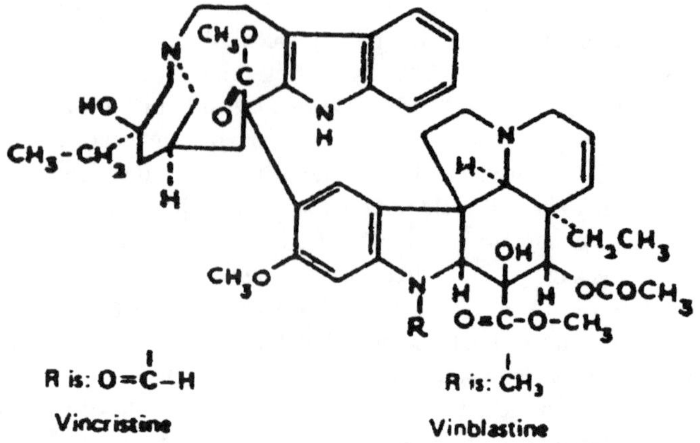

FIGURE 5.3 Chemical structures of vincristine and vinblastine, two alkaloids from the medicinal plant, Madagascar pink, *Catharanthus roseus*.

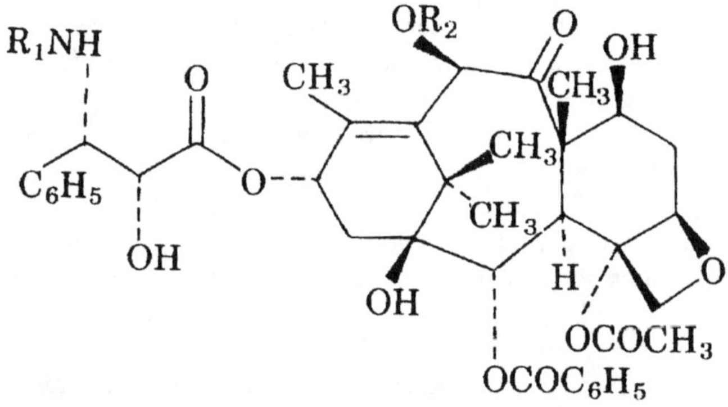

Paclitaxel $R_1 = COC_6H_5$, $R_2 = CH_3CO$
Docetaxel $R_1 = COOC(CH_3)_3$, $R_2 = H$

FIGURE 5.4 Chemical structures of paclitaxel and docetaxel, two taxoids from yews, *Taxus* spp., that are used for treatment of ovarian and breast cancer.

spindle asters. Cells are either arrested in mitosis or in G or S phases. Docetaxel has twice the potency of paclitaxel in inducing microtubule polymerization. Treated cells accumulate in the mitotic phase of the cell cycle.[10]

The taxoids are being used successfully in refractory ovarian cancer,[11] breast cancer, and non-small-cell lung cancer. Their side effect profile is largely predictable from the mechanism of action. Normal body cells with a high turnover or with processes dependent on microtubule formation, such as white blood cells, gastrointestinal mucosa, neurons, and secretory cells are preferentially incapacitated to some degree by paclitaxel and docetaxel. These effects are generally reversible and dose schedules have been developed to maximize tumor response and minimize side effects. Overall cancer response rates have varied from 30 to 70%. These taxene compounds are and will continue to be important anticancer agents, particularly if supply problems are solved (see other chapters in this work).[8,10,11]

5.2.5 HOMOHARRINGTONINE AND PROTEIN SYNTHESIS

Chinese traditional medicine has been preserved, respected, and incorporated into the modern approach in that country. Many of the plants used in that system have potential anticancer efficacy. The bark of the Chinese evergreen, *Cephalotaxus harringtonia,* is used for several indications, including treatment of malignancy.[12] The alkaloids extracted from the seeds of this tree were tested in the National Cancer Institute (NCI) screening program of the 1960s and shown to have cytotoxic activity. There are several related active substances, all of which are esters of the alkaloid, cephalotaxine.

Homoharringtonine (HHT) (Figure 5.5) is the most active of the alkaloids. Further testing in animal models confirmed its ability to prolong the life of animals bearing implanted tumors. HHT is now in phase II and phase III trials in humans for treatment of acute nonlymphoblastic leukemias and chronic myelogenous leukemia. The initial results are promising.[13]

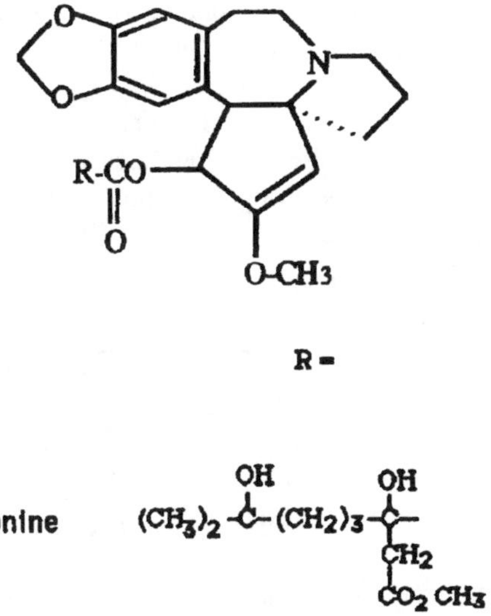

FIGURE 5.5 Chemical structure of homoharringtonine, an anticancer drug obtained from the bark of Chinese evergreen, *Cephalotaxus harringtonia.*

HHT has its cytotoxic effects in the G_1 and G_2 phases of the cell cycle.[14] These are the times of intense protein synthesis. Protein synthesis involves two major steps: initiation and elongation. During initiation the messenger ribonucleic acid (mRNA), bearing the code for the new protein, associates itself with the ribosome. The first transfer RNA (tRNA) then attaches to the mRNA, bringing the initial amino acid building block for the protein. Elongation is the process by which subsequent tRNA's attach to the mRNA and bonds are formed between the amino acids to produce the polypeptide protein. HHT inhibits the elongation step, most likely not from inhibiting the bonding of tRNA to mRNA, but by competitively inhibiting the enzyme, peptidyl transferase, which catalyses the formation of the polypeptide bond.[13] There is evidence that HHT also disrupts protein synthesis in other ways, such as detaching ribosomes from endoplasmic reticulum, degrading ribosomes, inhibiting release of completed proteins from ribosomes, and inhibiting glycosylation of completed proteins.[13] Through these mechanisms HHT may induce both apoptosis and differentiation of cancer cells, making it an important new anticancer agent.

5.2.6 RHEIN AND NECROSIS

Rhein is an anthraquinone found in rhubarb (*Rheum* spp.) and other purgatives (Figure 5.6). Rhein is also antineoplastic. Several hypotheses exist as to the mechanism of action by which rhein exerts its antitumor effects. Studies show that it exerts an effect at the membrane level. In electron microscopic evaluation, rhein appears to distort and disrupt the membranes of both mitochondria and cells. Membrane disruption appears to be mediated through altered actin microfilaments, which collapse into ring-like structures in the cell cytoplasm. In addition, the christae of mitochondria are disrupted. This may lead to impairment of energy metabolism, variations in cellular permeability, and altered receptor molecule activity.[15] Others have hypothesized that rhein alters the fluidity of membranes and hence the uptake of glucose.[16] The net result is decreased energy available for vital cellular functions and eventual cellular necrosis. Because of rhein's proposed mechanisms of action, it is a phytochemical that may warrant further examination as an antineoplastic agent.

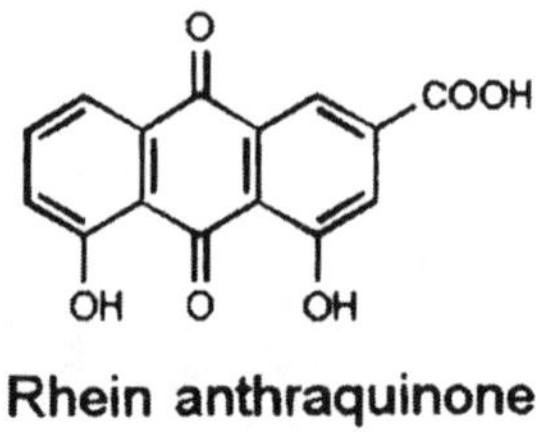

Rhein anthraquinone

FIGURE 5.6 Chemical structure of rhein anthraquinone, an anticancer drug found in rhubarb (*Rheum* spp.).

5.2.7 MISTLETOE AND APOPTOSIS

Mistletoe, well known for its amorous seasonal effects, is also well known in Europe as an adjuvant cancer therapy. Aqueous extracts of *Viscum album* L. are used for their combined effect as immunostimulatory and cytotoxic agents. The polysaccharide portion of the extract is thought to be responsible for the immunostimulatory effects, much in the same manner as *Echinacea* polysaccharides (see Section 5.4.2). Recent work has focused on the lectin portion of mistletoe extract. Lectins are proteins that cause agglutination and mitoses of mammalian cells. Studies with tumor cell lines *in vitro* show that mistletoe lectins inhibit tumor growth. Further analysis indicates that the DNA in these cells is fragmented as would be expected in apoptosis.[17] Other researchers found evidence of both membrane damage leading to necrosis and DNA damage indicative of apoptosis.[18] It may be that mistletoe extracts or purified mistletoe lectins will be validated with further studies as an effective means of treating some cancers.

5.2.8 SUMMARY

In this section we have seen how phytochemicals interact with various parts of the human cell life cycle (see Figure 5.7). These mechanisms can be employed to target rapidly proliferating tumor cells and to induce differentiation, apoptosis, or necrosis. The *Vinca* alkaloids and the taxoids are currently used in mainstream cancer treatment. Homoharringtonine is in human trials to determine dosage schedules and effects on a broad population. Genistein, mistletoe, and rhein are promising in their mechanisms, but work remains to be done before they will be approved for use in the U.S. Opportunities for research abound in these important applications of phytochemicals to the cancer epidemic of our current times.

5.3 TRANSMEMBRANE SIGNALING

5.3.1 INTRODUCTION

We have examined ways in which plant molecules effect the synthetic capacity of cells and their ability to proliferate or complete their life cycle. Another important way that exogenous molecules interact with cells and their functions is by various types of transmembrane signaling. Two types of signaling, ligand gated ion channels and G-protein/second messenger, are particularly relevant to the function of nerves and muscles. We will discuss these in detail and look at examples of how phytochemicals interact with them.

5.3.2 LIGAND GATED ION CHANNELS

Signaling of nerve cells and contraction of muscle cells are controlled in part by ion channels. Ion channels regulate the flow of sodium, potassium, and calcium across the cell membrane. Depending on the relative polarity on either side of the membrane, the cell will either be resting, activated (depolarized), or in a recovering state (hyperpolarized) (Figure 5.8). Ion channel opening and closing can be regulated by purely electrical forces as in the heart muscle. Cardiac cells depolarize and contract in unison via current flow at gap junctions along the membrane. Most ion channels, however, are opened or closed by the binding of chemicals, i.e., ligands. Binding causes conformational changes in the ion channel, allowing or inhibiting ion flow. As ions shift, the electrical potential across the membrane changes and the cell depolarizes. Depending on cell type, depolarization results in neurotransmission or muscle contraction. A hallmark of this kind of interaction is the extremely rapid reaction induced. Phytochemicals have historically played an important role in

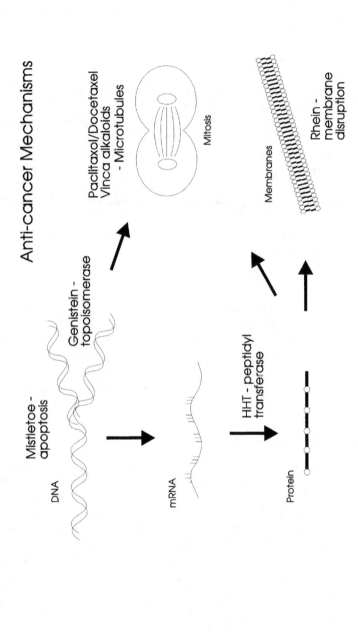

FIGURE 5.7   Anticancer mechanisms.

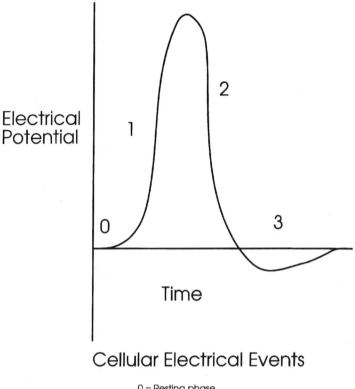

Cellular Electrical Events

0 = Resting phase
1 = Depolarization
2 = Repolarization
3 = Hyperpolarization

FIGURE 5.8 Cellular electrical events. Depending on the polarity on either side of the membrane, the cell will be in one of the four phases.

elucidating the nature of ligand gated ion channels. Nicotinic receptors at the neuromuscular junction on skeletal muscle are so named because the alkaloid nicotine causes depolarization of the muscle cells.

Plant-based medicines continue to have therapeutic value based on their ability to modify the actions of ion channels. In Ghana *Desmodium adscendens* is used to treat asthma. The symptoms of asthma can be modified by inhibiting the contraction of smooth muscles lining the airways. *D. adscendens* extracts can inhibit contractions in guinea pig intestinal smooth muscle. Three triterpenoid glycosides (Figure 5.9) have recently been isolated from *D. adscendens*. These glycosides increase the probability that calcium-dependent potassium channels of bovine tracheal smooth muscle will be open.[19] If potassium channels are open, the cell will hyperpolarize. It is then much more difficult to depolarize the cell and cause contraction. The traditional use of this herbal medicine in treating asthma is validated by understanding its mechanism of inhibiting smooth muscle contraction.

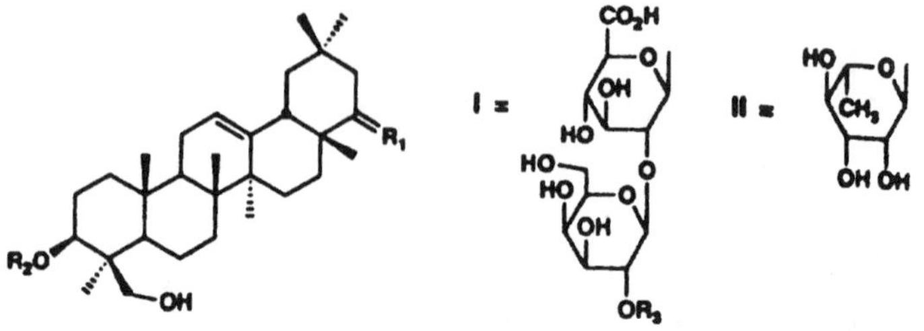

Compound	R₁	R₂	R₃
Soyasaponin I	H, OH	I	II
Dehydrosoyasaponin I	O	I	II
Soyasaponin III	H, OH	I	H
Soyasapogenol B	H, OH	H	-

FIGURE 5.9 Chemical structures of three triterpenoid glycosides from *Desmodium adscendens* from Ghana. These compounds modify the actions of ion channels.

5.3.3 G-Proteins and Second Messengers

Transmembrane signaling via G-proteins and second messengers is far more complicated than ligand gated ion channel signals and therefore has potential for many interactions with exogenous molecules. A G-protein sits within the membrane and is bound to guanosine diphosphate (GDP). In this mechanism, ligand binding to the receptor causes a change in the G-protein. GDP is phosphorylated to GTP. This activates a cascade of enzymatic reactions which are the second messengers. Within this process there is amplification of the signal. There are two different series of second messenger reactions that can be stimulated. One is set in motion by formation of cyclic adenosine monophosphate (cAMP) which activates protein kinases. These enzymes in turn catalyze the phosphorylation of regulatory enzymes. Cells processes are turned on or off based on the phosphorylation state of the regulatory enzymes. The other second messenger reaction series begins with formation of inositol triphosphate which triggers release of intracellular stores of calcium ions. Calcium in conjunction with calmodulin activates or deactivates regulatory cellular enzymes. Protein kinase C is also activated and causes phosphorylation of other enzymes. No matter which second messenger pathway is activated, the net result is a change in the products or function of the effected cell based on the enzymes that are turned on or off. This produces the cellular response to the original message-bearing ligand.

Catacholamines, of which there are many analogs found in natural products, act on the sympathetic nervous system effector organs through two basic types of receptors, alpha and beta. The α-receptor reactions are mediated through the calcium/inositol system. β-receptors are connected to the cAMP pathway. The overall reaction of cells and organs to catacholamine stimulation will be based on the relative number and type of receptor on the individual cells.

Ma Huang (*Ephedra sinica* or *E. equisetina*) has been used for thousands of years in China. It is said to facilitate the circulation of lung Qi and control wheezing.[20] It is also used to promote sweating and urination. *Ephedra* spp. are often found in cold and flu remedies, "energy" formulas, and weight loss formulas.[21] These myriad effects might seem unreal until one realizes that all are related to stimulation of the sympathetic nervous system through α- and β-receptors. Pharmacological studies done at the turn of the century isolated ephedrine (Figure 5.10) and pseudoephedrine

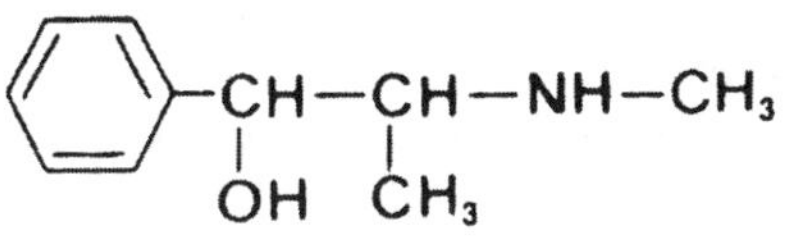

Ephedrine

FIGURE 5.10 Chemical structure of ephedrine from the stems of the plant, Ma Huang or *Ephedra sinica*. This drug acts to decongest the nose, relieving the symptoms of the common cold.

from the stems of *E. sinica*.[22] Ephedrine directly stimulates β-receptors to dilate bronchioles in the lung thus decreasing wheezing. Because of its lipid solubility, ephedrine crosses the blood-brain barrier and causes central nervous system stimulation and appetite suppression. Through indirect effects on other β-receptors, ephedrine and pseudoephedrine increase heart rate and the force of heart contractions. This leads to increased blood flow to the kidneys and increased urine formation. Actions on α-receptors cause increased sweating and the constriction of blood vessels in the nasal mucosa. The later effect decongests the nose, relieving the symptoms of the common cold. Over-the-counter cold preparations often contain pseudoephedrine for this purpose. All these helpful effects have made *Ephedra* spp. popular ingredients in modern herbal preparations. However, a plant with all these powerful effects may also cause harm. Heart attacks, seizures, psychotic episodes, and deaths have been associated with the use of ephedrine containing herbal supplements. The FDA is currently considering regulation of these products.[23] Persons with heart problems and high blood pressure should be especially careful when using these supplements.

5.3.4 SUMMARY

Phytochemicals can have potent effects when they stimulate cells through the body's transmembrane signaling mechanisms. We have seen how *Desmodium* glycosides inhibit smooth-muscle contraction consistent with its traditional use in asthma. The ephedrine in Ma Huang has its multitude of actions mediated through G-proteins and second messengers. Another way phytochemicals can influence signal transmission is by increasing the signal itself, for example, increasing neurotransmitters (see essay on St. John's wort). There are many forms of cell-to-cell communication in the body. Phytochemicals have an important place in the modulation of that communication.

Essay on St. John's Wort: Increasing the signal

St. John's wort, *Hypericum perforatum*, has long been used in folk medicine. It is currently licensed in Germany for treatment of anxiety, depression, and sleep disorders. A recent meta-analysis of 23 randomized trials with data from 1757 outpatients shows that St. John's wort preparations are consistently superior to placebo for the relief of mild to moderately severe depression. The same study reports that when compared to standard pharmaceutical treatment, St. John's wort is as effective and appears to have fewer significant side effects.[24]

The exact mechanism of action of St. John's wort is unknown. There are several groups of compounds that may mediate its effects or act synergistically to produce enhanced mood. Lipid soluble hypericins (Figure 5.11) (0.06 to 0.75%), flavonoids (2 to 4%), xanthones (0.0004%), procyanidines (8%), hyperforin (2.8%), and ethereal oil (0.1 to 1%) are possible candidates because they can cross the blood-brain barrier. Of these, hypericins, xanthones, and hyperforin are characteristic of St. John's wort, while the other substances are ubiquitous in the plant kingdom.[25] St. John's wort preparations are currently standardized by their hypericin content.

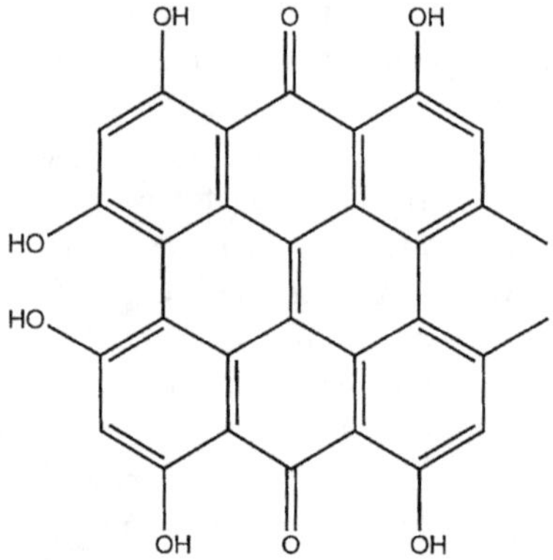

Hypericin

FIGURE 5.11   Chemical structure of hypericin, from St. John's wort, *Hypericum perforatum*. It is a drug used for the treatment of anxiety, depression, and sleep disorders.

Early *in vitro* work showed hypericin to be an irreversible monoamine oxidase (MAO) inhibitor.[26] Inhibition of this catabolic enzyme increases the amount of neurotransmitters in the synapse between neurons and leads to enhanced mood. Recent studies in the same lab and others have questioned the viability of this mechanism *in vivo*.[27] Another group has shown that the flavonoids and xanthones in *Hypericum* extracts inhibit catechol-*o*-methyltransferase, another enzyme that catabolizes neurotransmitters.[28] Still others have shown that *Hypericum* extract decreases the uptake of the neurotransmitter, serotonin, in rat synaptosomes.[29] The same lab also showed decreased expression of serotonin receptors in a neuroblastoma cell line after exposure to *Hypericum*.[30] Decreased catabolism, decreased uptake, or decreased numbers of receptors all result in a relative increase in the amount of neurotransmitter signals the receiving cell experiences. St. John's wort extract appears to have many potential mechanisms which may, in fact, be acting synergistically to increase the neurotransmitter signal.

5.4 IMMUNOMODULATION

5.4.1 INTRODUCTION

The mammalian immune system consists of many cells and signal molecules which act in concert to protect the organism from that which is "nonself". The chief cellular effectors are macrophages ("big eaters") and white blood cells. Neutrophils and lymphocytes are the most important of the white blood cells. Some of the signal molecules are interleukin-1 (IL-1), tumor necrosis factor-α (TNF-α), and γ-interferon (IF-γ, also known as IL-6). Cells that come in contact with foreign organisms begin to secrete signal molecules to call other effector cells into the area and to activate them. The invaders are immobilized or killed by numerous strategies including phagocytosis, antibody production, and radical oxygen species production. There is increased blood flow and the area becomes swollen, red, warm, and painful. This next section will first examine two plants, *Echinacea* and *Aloe vera* that may act to boost the response of the immune system in fighting disease. Then we will discuss how plants stimulate the immune system in ways that cause the organism discomfort, i.e., allergic or hypersensitivity reactions.

5.4.2 ECHINACEA

Echinacea has long been known in the Native American materia medica.[31] It was also known in Europe for its immune stimulating effects and skin repairing properties as early as 1831.[32] Today *Echinacea* products are widely used in Europe as an aide to boost the immune system in its struggle with the viruses that cause colds and flu. Recent clinical trials in Germany have supported this usage.[33] In addition, research with extracts of *Echinacea* have begun to elucidate the interactions between this herb and the mammalian immune system.

Initially, echinacoside, a caffeic-acid glycoside that showed weak antibacterial activity, was thought to be the active ingredient. Further work showed this was not the case.[33] In a series of elegant experiments spanning more than a decade, M. L. Lohmann-Matthes, H. Wagner, and colleagues have steadfastly expanded our knowledge of how *Echinacea* works.[34-38] Early on, this group pinpointed polysaccharides from the aqueous extracts of *E. purpurea* as the active fraction. Further work showed that the effective polysaccharides were cell wall-derived arabinogalactan and two fucogalactoxyloglucans. They developed a plant-cell-culture system whose supernatant provided them with a solution of the polysaccharides that could be standardized. Then they applied this purified extract in a host of carefully executed experiments. They have shown that this polysaccharide fraction stimulates macrophages to produce signal molecules, TNF-α, IL-1, and IL-6 (interferon). These signals activate other parts of the immune system and promote the migration of other effector cells, such as neutrophils, from the bone marrow to the blood. The activated macrophages produce more oxygen radicals, phagocytize more and are more cytotoxic to tumor cells. Overall, there is a higher rate of killing of *Listeria monocytogenes* bacteria and *Candida albicans* yeast, such that a lethal dose of either can be withstood by both immunocompetent and immunosuppressed mice that have been treated with the polysaccharides. Similar results have been obtained in humans. Although the polysaccharides stimulate the immune system, much as an invading organism would, they are completely nontoxic. Another group has done preliminary work using *E. purpurea* extracts in combination with cyclophosphamide and thymostimulin to stimulate the immune system of patients with hepatocellular and advanced colorectal cancer. Their results are encouraging.[39,40]

These experiments give new credence to the herbalists' claims of the immune-enhancing effects of *Echinacea* spp. Soon it may be an integral part of accepted therapy for withstanding cancer and other infectious diseases.

5.4.3 ALOE VERA

Humans have used the aloe plant since the ancient times of Egypt and Greece for skin infections and wound healing.[41] The leaf contains three medically important and distinct parts: the leaf exudate, the leaf epidermis, and the leaf pulp. Much of the medical literature on aloe use is confusing because the part and formulation is not specified clearly. This may account for the often widely divergent results obtained. While researchers divide and extract these different parts to find the active ingredients, many others advocate for study of the use of whole leaf preparations, since that is the way it has been used throughout history.

The leaf exudate, a bitter yellow liquid, is produced by pericyclic cells.[42] This can be heated, concentrated, and dried to a black powder. This is the source of drug aloes, also known as Cape Aloes USP, which is used as a purgative.[43] *Aloe ferox* is grown commercially for this purpose. Of the dried exudate, 70 to 97% is made up of aloeresin, aloesin, and aloin in a ratio of 4:3:2.[44] The exudate also contains aloe-emodin, and anthraquinone which is a gastrointestinal irritant, hence the purgative effects.[42]

Recent studies have centered on a lectin purified from the leaf epidermis of *Aloe arborescens* Miller.[45] An aloe lectin has been reported to inhibit the growth of a fibrosarcoma in mice through a host-mediated effect.[46] A possible mechanism may be activation of the immune system as purified aloe lectin has been shown to increase mitogenic activity in mouse lymphocytes.[47] This will undoubtedly be an area for further research.

The aloe leaf pulp or gel is a clear mucilaginous substance which is 98.5% water.[48] The mucilage is predominantly made up of polysaccharides which are partially acetylated glucomannans.[49] Recently, an acetylated mannan, acemannan, extracted from *Aloe vera*, has been shown to have immune-system modulating effects. This appears to be mediated through macrophages which synthesize and release nitric oxide, IL-1, and TNF-α when activated by acemannan.[50,51] The activated macrophages and other immune cells are then able to respond to viral or cancer cells. This product of aloe plants will be studied more thoroughly in the future.

The whole leaf of *Aloe vera*, or products extracted from whole leaf, have been used directly on radiation burns, thermal burns, partial thickness wounds, stasis ulcers, and diabetic ulcers. Most researchers report an initial increase in necrosis and then more rapid healing when compared to other treatments or no treatment.[41-43] This may be a reflection of the above-identified immune modulating effects. *Aloe vera* has enjoyed a great popularity in household remedies and cosmetics. Research is just beginning to unravel the reasons why this botanical has been highly regarded by healers and the healed alike.

5.4.4 PLANT CONTACT DERMATITIS

There are several different ways in which plants can affect the skin of humans, some beneficial and some causing discomfort. Many plants, like *Aloe vera*, promote healing of wounds. Other plants, such as poison ivy (*Toxicodendron* spp), are well known for their toxicity to the skin. Plant contact dermatitis is subdivided based on causative mechanisms. One such division is: (1) irritant contact, (2) immediate contact, (3) phytophotosensitivity, and (4) allergic contact.[52,53] As more is learned about these mechanisms, it is clear that there is some overlap. The divisions are useful, however, in determining appropriate treatment. In each of the following sections, we will define and describe the clinical picture of each type of dermatitis. Each will be illustrated with one or two examples, along with more detail about the mechanism of interaction when known.

Irritant contact dermatitis occurs when humans encounter thorns, spines, irritant hairs, and chemical substances which primarily protect plants from herbivores. In the human, these plant defenses usually cause some kind of persistent skin reaction which may be due to physical trauma or chemical interaction with skin or nerves.[54] Stinging nettles, *Urtica diocia* and *U. ureus*, are commonly known for the intense burning and stinging that begins just a few minutes after brushing up against the plant. The skin turns red and warm and itchy. There may be persistent itching or tingling for about 12 h. These *Urtica* spp. have glandular hairs which inject four chemicals into the skin, namely, histamine, acetylcholine, serotonin, and a fourth unidentified compound. The histamine causes immediate vasodilatation and edema, producing redness and swelling. The serotonin is responsible for the pain and itch.[55] Another irritant contact dermatitis is caused by capsaicin, an alkaloid in red peppers, chili peppers, and paprika of the genus *Capsicum*. It produces redness and intense burning. Capsaicin stimulates a specific receptor on cutaneous sensory neurons which in turn probably increases intracellular calcium ions. This causes massive release of neuropeptides, including substance P. These molecules are responsible for both pain signal transmission to the brain via depolarization of unmyelinated type C and thin myelinated A delta sensory neurons and modulation of the local inflammatory response. Repeated application depletes the neuropeptides, and therefore, pain signals can no longer be transmitted. This is the basis for the use of capsaicin in products used to treat diabetic neuropathy, post herpetic neuralgia, and arthritis.[56,57]

Immediate contact dermatitis occurs when skin previously sensitized is re-exposed to the offending agent. In some people, strawberries, kiwifruit, tomato, castor bean, and others trigger a type I hypersensitivity response typified by redness, swelling, and itching.[52] On first exposure, the plant antigens stimulate B lymphocytes to produce immunoglobulin E (IgE) antibodies which then bind mast cells. No reaction is apparent. At the second exposure, when the antigen cross-links the antibodies on the mast cell, there is an influx of calcium ions into the cell. This causes release of preformed mediators, such as histamine, heparin, enzymes, chemo-tactic, and activating factors, and stimulates formation of longer acting mediators, such as prostaglandins and leukotrienes. These mediators, among other things, cause vessel dilatation fluid leakage and recruitment of other blood cells to the area. These changes cause the observed skin reactions.[58]

A third type of dermatitis associated with plants is *phytophotodermatitis*. This occurs when there is direct or airborne contact or ingestion of plant furocoumarins and then exposure to sunlight. The result is a painful, red, itchy rash with watery blister formation which lasts 1 to 2 weeks. Hyperpigmentation follows which can last for months. This type of reaction can be caused by rue (*Ruta* spp), gas plant (*Dictamnus albus*), citruses (*Citrus* spp), Apiaceae (angelica, parsley, parsnip) and others.[52] The best studied of the furocoumarins are psoralens. They cross-link DNA in the cells and, when exposed to ultraviolet light, cause cell death, inhibit normal mitosis, or cause mutations. Dermatologists use ingested psoralens (Figure 5.12) and ultraviolet-A light in the treatment of psoriasis.[53]

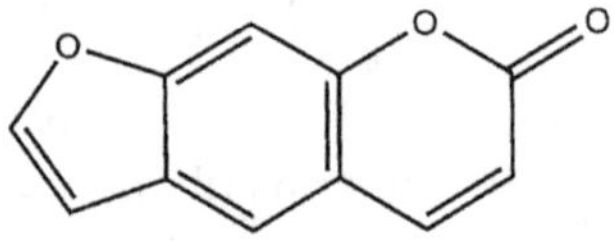

Psoralen

FIGURE 5.12   Chemical structure of Psoralen, which is used in combination with ultraviolet light for the treatment of psoriasis.

The most well-known plant-skin interaction in North America is that caused by poison ivy, poison oak, and poison sumac (*Taxicodendron* spp.). These plants cause *allergic contact dermatitis* typified by red, itchy skin with weeping blisters, scabs, and crusts which peaks about 48 h after exposure. Affected areas may appear in a linear distribution because of the mechanism of contact or early scratching. The lesions may erupt over 3 weeks, which is the time it takes the plant resin to evaporate. It is not spread through leakage of the blisters. Delayed eruption is due to re-exposure from resin on clothes, tools, or pet fur. There is usually no long-term scarring or hyperpigmentation.[59] Similar type IV or delayed-hypersensitivity reactions can be caused by sesquiterpene lactones in the Asteraceae (thistle) family and quinones in toxic woods.[52,60] In the *Toxicodendron* spp. The allergen is urushiol, a catechol nucleus with a 15-carbon lipophilic tail containing 2 to 3 unsaturated bonds. Urushiol binds to epidermal cells (keratinocytes, Langerhans cells, and endothelial cells) stimulating release of mediators (ICAM-1, ELAM-1, VCAM-1) which form adhe-sive networks and promote migration (via IL-8) of T-cell lymphocytes to the area.

Pathology is then T-cell mediated through lymphokine production, antigen-specific and nonspecific cytotoxicity, and recruitment of other effector cells.[61,62]

5.4.5 SUMMARY

We have seen how plant and human interaction can have significant immunomodulatory affects. In the case of *Echinacea* and *Aloe*, plant polysaccharides stimulate the immune system in a beneficial way, promoting healing and increased defensive capacity. When human and plant defense systems clash, the interaction can leave humans with painful, red, swollen, itchy, and blistered skin through a variety of mechanisms. Sometimes these very mechanisms can be used to lessen symptoms of other diseases like psoriasis and neuropathy.

5.5 TOXIC EFFECTS

5.5.1 INTRODUCTION

The last section on plant contact dermatitis serves as a good bridge to this portion on the harmful affects of plants. We have already seen that plants have powerful potential in their interactions with humans. This can benefit or harm. Some significant aspects of the negative interactions will be covered with respect to congenital anomalies (teratogenesis), carcinogenesis, and toxicity.

5.5.2 TERATOGENESIS

Teratogenesis (literally "monster formation") occurs when cell proliferation, cell migration, or cell differentiation in a developing human embryo is altered. Human embryos are most vulnerable to the effects of teratogens during the third through the ninth week of pregnancy during a time when women may not be aware they are pregnant. About one quarter of all birth defects are genetic aberrations, and 65 to 70% are from unknown causes. Drugs and chemicals account for only about 1% of birth defects.[63] There are several plant-derived compounds that are known teratogens, notably some alkaloids from angiosperms such as: colchicine, reserpine, tubocurarine, caffeine, nicotine, and quinine.[64]

Ethyl alcohol derived from fermentation of grapes or grains is a commonly ingested plant product with recognized teratogenic effects. The fetal alcohol syndrome is diagnosed by its constellation of growth retardation, microcephaly, atrial septal defects, short palpebral fissures, maxillary hypoplasia, and other minor anomalies. The mechanism behind these effects is multifactorial. Fetal hypoxia and nutrient deficiencies may be involved. At the cellular level, enzyme activities, cell division, and maintenance of membrane integrity are altered by exposure to ethanol.[65]

In general, it is very difficult to establish causality in a situation where multiple factors may play a role. The high proportion of unknown causes of birth defects indicates that much that we are exposed to may be less benign than we think. Accordingly, most drugs should be avoided in pregnancy, including plant-based remedies and beverages, unless the benefit to be obtained far outweighs the often unknown risk to the developing offspring.

5.5.3 CARCINOGENESIS

In Section 5.2 we discussed various phytochemicals and their role in treating cancer. Natural products or their metabolites can also be implicated in causing cancer, although far more synthetic chemicals are known culprits at this time. Viruses and irradiation are also responsible for much neoplastic transformation. Chemical carcinogenesis is proposed to occur via a two-step process of initiation and promotion. Initiation is accomplished when damaged DNA is passed on to daughter cells unrepaired. Particular portions of DNA known as proto-oncogenes may be transformed through mutation to become active oncogenes. Other genes known as tumor suppresser genes may be inhibited. These genes would normally control cell growth and differentiation. Once damaged, the state is set for uncontrolled proliferation. This will not occur, however, unless there is a second type of stimulus called promotion. One well-studied promoter exerts its action through multiple effects including activation of protein kinase C. This in turn causes a host of protein phosphorylations which regulate multiple cellular functions including membrane receptor, ion channel, and enzymatic activity. The result is altered proliferation and differentiation and neoplasia.[66,67]

The most well-known plant carcinogen is tobacco, the leaf of *Nicotiana tabacum*. It contains many compounds that may be volatilized during burning. More importantly, several aromatic hydrocarbons are known to be formed during combustion. Wherever these are applied experimentally they cause local cancer formation. They are metabolized to dihydrodiol epoxides, which are strong electrophilic reactants. They exert their cancer initiating effects by combining with nucleophilic sites on DNA, RNA, and proteins. Tobacco aromatic hydrocarbons may be complete carcinogenic agents in that they are sufficient to cause tumors without a promoter. On the other hand, tobacco acts synergistically with betel nut juice (*Areca catechu*) chewed in south Asia. The betel nut alone causes tumors in 38% of hamster cheek pouches, but when combined with tobacco the number rises to 78%. In this study tobacco alone did not induce malignancy; however, it caused leukoplakia, which may enhance the susceptibility to cancer.[66,68]

5.5.4 TOXICITY

Many plant-based medicines and herbal remedies have side effects just as prescribed synthetic medicines do. Gastrointestinal effects, such as nausea and diarrhea, and skin reactions are common to many ingested products. There are a few plant-based products with well-known toxicities to the liver and the central nervous system. The next section will explore the mechanism of toxicity of comfrey root and jimsonweed seed.

Comfrey (*Symphytum officinale*) has been used for the treatment of stomach ulcers and as a blood purifier among other things. The roots are the part most often used. They contain pyrrolizidine alkaloids (Figure 5.13) which can cause liver toxicity, as well as carcinogenesis and teratogenesis. These alkaloids have a 1,2 double bond and esterified hydroxymethyl groups (see Figure 5.13). In the liver they are dehydrogenated to pyrrole derivatives, which then act as potent alkylating agents.

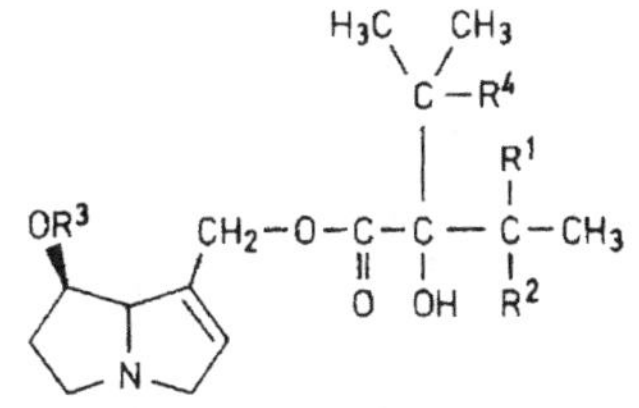

	R^1	R^2	R^3	R^4
Intermedine	OH	H	H	H
Acetylintermedine:	OH	H	Acetyl	H
Lycopsamine:	H	OH	H	H
Acetyllycopsamine:	H	OH	Acetyl	H
Symphytine:	H	OH	Tigloyl	H
Echimidine:	H	OH	Angeloyl	OH

FIGURE 5.13 Chemical structures of pyrrolizidine alkaloids, which can cause liver toxicity, as well as carcinogenesis and teratogenesis.

They react with bases in the DNA strand, cross-linking strands and causing strand breakage. Studies in rats have supported the hepatotoxic, carcinogenic, and teratogenic role of comfrey root.[69] In humans, a form of Budd-Chiari syndrome known as veno-occlusive disease has been the primary concern. Clinical manifestations are hepatomegally and refractory ascites, often progressing to hepatic failure. Untreated, there is a high mortality rate. Pathologically, the liver shows tissue necrosis in the center of lobules as well as dilation of the central vein. The small venules of the liver have fibrous deposition in and around them, which leads to obstruction of blood outflow and the resultant ascites. Many cases have been reported in the world literature, attributable to *Symphytum*, as well as pyrrolizidine alkaloid-containing species of *Heliotropium*, *Senecio*, or *Crotalaria*.[70,71] Internal consumption of comfrey is officially banned or discouraged in Australia, New Zealand, the United Kingdom, and Germany.[69]

Neurotoxicity is another common result of ingestion of plant products. In another section, we have discussed interaction of phytochemicals with various cellular membrane signaling mechanisms. Neurotoxicity can occur when the plant molecule acts as a blocker to neurotransmission. Jimson weed (*Datura stramonium*) has been used as a tea for treatment of asthma. The atropine-like substances, hyoscyamine and scopolamine (Figure 5.14), are in all portions of the plant. They

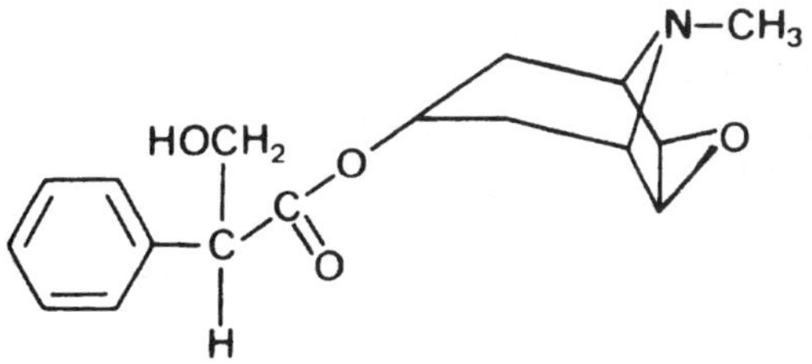

FIGURE 5.14 Chemical structures of scopolamine, a drug which can be neurotoxic to humans and other animals.

act to block neurotransmission by acetylcholine which is the predominant neurotransmitter of the parasympathetic nervous system. The signs and symptoms of *Datura* toxicity pervade many organ systems including: dry mouth, dry skin, blurred vision, disorientation, excitability, aggressiveness, tachycardia, tachypnea, and hyperpyrexia. Death can occur from cardiac arrest.[72]

5.5.5 SUMMARY

In this final section we have discussed some of the problems associated with ingestion of certain toxic, teratogenic, and carcinogenic plant products. Medical literature has often focused on these negative effects alone. As we have seen in other portions of the chapter, plants can have a host of salutary effects as well. While usage of plants medicinally may be steeped in tradition, scientific investigation often uncovers the mechanisms by which phytochemicals interact with the human body. This area of investigation currently offers more questions than answers. There is a growing need for well-trained and thoughtful ethnobotanists, basic scientists, and clinicians to carry forward the work of understanding how best to join with the plants to create good health.

REFERENCES

1. Barnes, S., Effect of genistein on *in vitro* and *in vivo* models of cancer, *J. Nutr.*, 125, 777S–783S, 1995.
2. Peterson, G., Evaluation of the biochemical targets of genistein in tumor cells, *J. Nutr.*, 125, 784S–789S, 1995.
3. Fotis, T., Pepper, M. et al., Genistein, a dietary ingested isoflavonoid, inhibits cell proliferation and *in vitro* angiogenesis, *J. Nutr.*, 125, 790S–797S, 1995.
4. Barnes, S. and Peterson, T. G., Biochemical targets of the isoflavone genistein in tumor cell lines, *Proc. Exp. Biol. Med.*, 208, 109–115, 1995.
5. Conctantinou, A. and Huberman, E., Genistein as an inducer of tumor cell differentiation: possible mechanisms of action, *Pro. Soc. Exp. Bio. Med.*, 208, 109–115, 1995.
6. Markovits, J., Linassier, C. et al., Inhibitory effects of the tyrosine kinase inhibitor genistein on mammalian DNA topoisomerase II., *Canc. Res.*, 49, 5111–5117, 1989.
7. Markovits, J., Junqua, S. et al., Genistein resistance in human leukaemic CCRF-CEM cells: selection of a diploid cell line with reduced DNS topoisomerase II B isoform, *Biochem. Pharm.*, 50, 177–186, 1995.
8. Rowinsky, E. K., Cazenave, L. A., and Donehower, R. C., Taxol: A novel investigational antimicrotubule agent, *J. Nat. Canc. Inst.*, 82, 1247–1259, 1990.
9. Salmon, S. E. and Sartorelli, A. C., Cancer chemotherapy, in *Basic and Clinical Pharmacology*, Katzung, B. G., Ed., Appleton and Lange, Norwalk, CT, 1989, 694.
10. Pazdur, R., Kudelka, A. P. et al., The taxoids: paclitaxel (Taxol®) and docetaxel (Taxotere®), *Canc. Treat. Rev.*, 19, 351–386, 1993.
11. Runowicz, C. D., Wiernik, P. H. et al., Taxol in ovarian cancer, *Canc. (Suppl.)* 71, 1591–1596, 1993.
12. Ohnuma, T. and Holland, J. F., Homoharringtonine as a new antileukemic agent, *J. Clin. Oncol.*, 73, 604–606, 1985.

13. Zhou, D. C., Zittoun, R., and Marie, J. P., Homoharringtonine: an effective new natural product in cancer chemotherapy, *Bull. Canc.,* 82, 987–995, 1995.

14. Dwyer, P. J., King, S. A. et al., Homoharringtonine — perspectives on an active new natural product, *J. Clin. Oncol.,* 4, 1563–1568, 1986.

15. Iosi, F., Santini, M. T., and Malorni, W., Membrane and cytoskeleton are intracellular targets of rhein in A431 cells, *Anticanc. Res.,* 13, 545–554, 1993.

16. Castiglioni, S., Maurizio, F. et al., Rhein inhibits glucose uptake in ehrlich ascites tumor cells by alteration of membrane-associated functions, *Anti-Cancer,* 4, 407–414, 1993.

17. Janssen, O., Scheffler, A., and Kabelitz, D., In vitro effects of mistletoe extracts and mistletoe lectins, *Arzneim.-Forsch./Drug Res.,* 43, 1221–1227, 1993.

18. Bussing, A., Suzart, K. et al., Induction of apoptosis in human lymphocytes treated with *Viscum album* L. is mediated by the mistletoe lectins, *Canc. Let.,* 99, 59–72, 1996.

19. McManus, O. B., Harris, G. H. et al., An activator of calcium-dependent potassium channels isolated from a medicinal herb, *Biochemistry,* 32, 6128–6133, 1993.

20. Bensky, D. and Foster, S., *Chinese Herbal Medicine, Materia Medica,* Eastland Press, Seattle, WA., 1986, 32–34.

21. Leung, A. Y. and Foster, S., *Encyclopedia of Common Natural Ingredients,* 2nd ed., John Wiley & Sons, New York, 1996. 227–229.

22. Olin, B. R., Ed., The Ephedras, *The Lawrence Review of Natural Products,* Facts and Comparisons, a Wolters Kluwer Company, St. Louis, MO, 1995.

23. Zwillich, T., FDA considers regulation of ephedrine as drug, *Fam. Prac. News,* Oct. 1, 30, 1996.

24. Linde, K., Ramirez, G., Mulrow, C. .D. et al., St. John's wort for depression — an overview and meta-analysis of randomized clinical trials, *Brit. Med. J.,* 313, 253–258, 1996.

25. Wagner, H. and Bladt, S., Pharmaceutical quality of *Hypericum* extracts, *J. Geriatr. Psych. Neurol.,* 7(suppl.), S65–S68, 1994.

26. Suzuki, O., Katsumata, Y., Oya, M. et al., Inhibition of monoamine oxidase by hypericin, *Planta Medica,* 50, 272–274, 1984.

27. Bladt, S. and Wagner, H., Inhibition of MAO by fractions and constituents of *Hypericum* extract, *J. Geriatr. Psych. Neurol.,* 7(suppl.), S57–S59, 1994.

28. Thiede, H. M. and Walper, A., Inhibition of MAO and COMT by *Hypericum* extracts and hypericin, *J. Geriatr. Psych. Neurol.,* 7(suppl. 1), S54–S56, 1994.

29. Perovic, S. and Muller, W. E. G., Pharmacological profile of *Hypericum* extract, *Arzneim.-Forsch./Drug Res.,* 45, 1145–1148, 1995.

30. Muller, W. E. G. and Rossol, R., Effects of Hypericum extract on the expression of serotonin receptors, *J. Geriatr. Psych. Neurol.,* 7(suppl 1), S63–S64, 1994.

31. Gilmore, M. R., Thirty-third annual report of the Bureau of American Ethnology, U.S. Government Printing Office, Washington, D.C., 1919, 1–145.

32. Dierbach, H. (1831), cited in Tragni, E., Galli, C. L., et al., Antiinflammatory activity of *Echinacea angustifolia* fractions separated on the basis of molecular weight, *Pharm. Res. Comm.,* 20, Supple. 5, 87–90, 1988.

33. Foster, S., Echinacea: the cold and flu remedy, *Altern. Compl. Ther.,* June/July, 254–257, 1995.

34. Stimpel, M., Proksch, A. et al., Macrophage activation and induction of macrophage cytotoxicity by purified polysaccharide fractions from the plant, *Echinacea purpurea, Infec. Imm.,* 46, 845–849, 1984.

35. Luettig, B., Steinmuller, C. et al., Macrophage activation by the polysaccharide arabinogalactan isolated from plant cell cultures of *Echinacea purpurea, J. Nat. Canc. Inst.*, 81, 669–675, 1989.

36. Roesler, J., Steinmuller, C. et al., Application of purified polysaccharides from cell cultures of the plant *Echinacea purpurea* to mice mediates protection against systemic infections with *Listeria monocytogenes* and *Candida albicans, Int. J. Immunopharm.*, 13, 27–37, 1991.

37. Roesler, J., Emmendorffer, A. et al., Application of purified polysaccharides from cell cultures of the plant *Echinacea purpurea* to test subjects mediates activation of the phagocyte system, *Int. J. Immunopharm.*, 13, 931–941, 1991.

38. Steinmuller, C., Roesler, J. et al., Polysaccharides isolated from plant cell cultures of *Echinacea purpurea* enhance the resistance of immunosuppressed mice against systemic infections with *Candida albicans*, and *Listeria monocytogenes, Int. J. Immunopharm.*, 15, 605–614, 1993.

39. Lersch, C., Zeuner, M. et al., Stimulation of the immune response in outpatients with hapatocellular carcinomas by low doses of cyclophosphamide (LDCY), *Echinacea purpurea* extracts (echinacin) and thymostimulin, *Arch. Geschwulstforsch,* 60, 379–383, 1990.

40. Lersch, C., Zeuner, M. et al., Nonspecific immunostimulation with low doses of cyclophosphamide (LDCY), thymostimulin, and *Echinacea purpurea* extracts (echinacin) in patients with far advanced colorectal cancers: preliminary results, *Canc. Invest.*, 10, 343–348, 1992.

41. Shelton, R. M., *Aloe vera*: its chemical and therapeutic properties, *Int. J. Dermatol.*, 30, 679–683, 1991.

42. Klien, A. D. and Penneys, N. S., *Aloe vera, J. Am. Acad. Dermatol.*, 18, 714–720, 1988.

43. Grindlay, D. and Reynolds, T., The *Aloe vera* phenomenon: a review of the properties and modern uses of the leaf parenchyma gel, *J. Ethnopharm.*, 16, 117–151, 1986.

44. van Wyk, B. E. et al., Geographical variation in the major compounds of *Aloe ferox* leaf exudate, *Plant Med.*, 61, 250–253, 1995.

45. Koike, T., Titani, K. et al., The complete amino acid sequence of a mannose-binding lectin from "Kidachi Aloe" (*Aloe arborescens* Miller var. *natalensis* Berger), *Biochem. Biophys. Res. Comm.*, 214, 163–170, 1995.

46. Imanishi, K., Ishiguro, T. et al., Pharmacological studies on a plant lectin, aloctin A. I. Growth inhibition of mouse methylcholanthrene-induced fibrosarcoma (meth A) in ascites form by aloctin A, *Experienta*, 37, 1186–1187, 1981.

47. Koike, T., Beppu, H. et al., A 35 kDa mannose-binding lectin with hemagglutinating and mitogenic activities from "Kidachi Aloe" (*Aloe arborescens* Miller var. *natalensis* Berger), *J. Biochem.* (Tokyo), 118, 1205–1210, 1995.

48. Rowe, T. D. and Parks, L. M., Phytochemical study of *Aloe vera* leaf, *J. Am. Pharm. Assoc.*, 30, 262–266, 1941.

49. Gowda, D. C., Neelisiddaiah, B., and Anjaneyalu, Y. V., Structural studies of polysaccharides from *Aloe vera, Carbohyd. Res.*, 72, 201–205, 1979.

50. Peng, S. Y., Normal, J. et al., Decreased mortality of Norman murine sarcoma in mice treated with the immunomodulator, acemannan, *Mol. Biother.*, 3, 79–87, 1991.

51. Karaca, K., Sharma, J. M., and Nordgren, R., Nitric oxide production by chicken macrophages activated by acemannan, a complex carbohydrate extracted from *Aloe vera, Int. J. Immunopharm.*, 17, 183–188, 1995.

52. Juckett, G., Plant dermatitis, *Prostgrad. Med.*, 100, 159–171, 1996.

53. Epstein, W. L., Plant-induced dermatitis, *Annals Emer. Med.*, 16, 950–955, 1987.

54. Southcott, R. V. and Haegi, A. R., Plant hair dermatitis, *Med. J. Aust.,* 156, 623–632, 1992.

55. Oliver, F., Amon, E. U. et al., Contact urticaria due to common stinging nettle (*Urtica dioica*) — histological, untrastructural and pharmacological studies, *Clin. Exp. Dermatol.,* 16, 1–7, 1991.

56. Williams, S. R., Clark, R. F., and Dunford, J. V., Contact dermatitis associated with capsaicin: Hunan hand syndrome, *Annals Emer. Med.,* 25, 713–715, 1995.

57. Girolomoni, G. and Tigelaar, R. E., Capsaicin-sensitive primary sensory neurons are potent modulators of murine delayed-type hypersensitivity reaction, *J. Immunol.,* 145, 1105–1112, 1990.

58. Roitt, J. M., Brostoff, J., and Mala, D. K., Hypersensitivity-Type I, in *Immunology,* C. V. Mosby, St. Louis, MO, 1985, 19.2–19.11.

59. Quick, G., Scratching below the surface of poison ivy rash, *Consultant,* 34, 545–549, 1995.

60. Woods, B. and Calnan, C. D., Toxic woods, *Br. J. Derma.,* 95 (suppl.), 1–95, 1976.

61. Griffiths, C. E. M., Barker, J. N. W. N., Kunkel, S., and Nickoloff, B. J., Modulation of leucocyte adhesion molecules, a T-cell chemotoxin (IL-8) and a regulatory cytokine (TNF-ol) in allergic contact dermatitis (rhus dermatitis), *Brit. J. Dermat.,* 124, 519–526, 1991.

62. Kalish, R. S., The use of human T-lymphocyte clones to study T-cell function in allergic contact dermatitis to urushiol, *J. Invest. Dermat.,* 94, 108S–111S, 1990.

63. Cotran, R. S., Kumar, V., and Robbins, S. L., *Robbins Pathologic Basic of Disease,* W.B. Saunders, Philadelphia, 1989, 520–522.

64. Lewis, W. H. and Elvin-Lewis, M. P. F., Cell modifiers: mutagens, teratogens, and lectins, in *Medical Botany, Plants Affecting Man's Health,* John Wiley & Sons, New York, 1977, 90–96.

65. Zajac, C. S. and Abel, E. L., Animal models of prenatal alcohol exposure, *Int. J. Epidem.,* 21 (suppl. 1), S24–S32, 1992.

66. Csotran, R. S., Kumar, V., and Robbins, S. L., *Robbins Pathologic Basis of Disease,* W.B. Saunders, Philadelphia, 1989, 267–272.

67. Boik, J., *Cancer and Natural Medicine,* Oregon Medical Press, Princeton, MN, 1996, 5–7.

68. Lewis, W. H. and Elvin-Lewis, M. P. F., Cancer, in *Medical Botany, Plants Affecting Man's Health,* John Wiley & Sons, New York, 1977, 120–121.

69. Bisset, N. G., Ed., *Herbal Drugs and Phytopharmaceuticals,* Medpharm, Stuttgart, Germany and CRC Press, Boca Raton, FL, 1994, 461–484.

70. McDermott, W. V. and Ridker, P. M., The Budd-Chiari syndrome and hepatic venoocclusive disease, *Arch. Surg.,* 125, 525–527, 1990.

71. Olin, B. R., Ed., Comfrey, *The Lawrence Review of Natural Products,* Facts and Comparisons, a Wolters Kluwer Company, St. Louis, MO, 1995.

72. Combs, S. P., Acute anticholinergic syndrome: Jimson week strikes again, *Res. Staff Phys.,* 43, 54–57, 1997.

6 The Synergy Principle at Work in Plants, Pathogens, Insects, Herbivores, and Humans

James A. Duke and Mary Jo Bogenschutz-Godwin

CONTENTS

0-8493-3134-X/99/$0.00+$.50
© 1999 by CRC Press LLC

6.1   INTRODUCTION

One verbal battle between herbalists, pharmacists, and physicians, if not chemists, often centers around the word *synergy*, a word difficult to define, and in biological systems and mixtures, at least, a process a bit difficult to prove. To us, synergy is best defined as the interaction of two or more agents such that the combined effect is greater than the expected sum of the individual effects. This is perhaps an antonym of *antagonism*, the interaction of two or more agents such that the combined effect is less than the sum of the expected individual effects. Intermediate is *additive*, the interaction of two or more agents where the combined effect is exactly the sum of their individual effects.

For the sake of argument, let us pretend that we stumbled onto a nice new mutation, a new plant species with a succulent edible leaf with a well-balanced array of amino-acids, fatty acids, minerals, vitamins, and other important nutrients, with absolutely no distasteful antinutrient or harmful compounds in it. There are no such leaves, but some bland head lettuce comes close! We will call it Sweet Leaf. We like it. So do a lot of bacteria, insects, fungi, and viruses. If we all eat it before it has had a chance to reproduce itself, Sweet Leaf becomes extinct. In an evolutionary sense, it does not have a chance, because it has been devoured without reproducing itself. But suppose one or two individuals of Sweet Leaf, through mutation (a rearrangement of a fragment of its genetic and hereditary mechanisms) modifies one of its heretofore tasty and salubrious compounds to produce a new and bitter compound, distasteful to us, the bacteria, insects, fungi, herbivorous mammals, and/or the viruses. We have a distasteful new incipient species which we will call Bitter Leaf. One of Bitter Leaf's edible amino acids has become a distasteful bitter chemical which we will call Alkaloid-1. The herbivores much prefer Sweet Leaf to Bitter Leaf, eating more and more of Sweet Leaf, and less and less of Bitter Leaf. With no more leaves, individuals of the Sweet Leaf species cannot make a living (photosynthesize) and literally starve to death before they have progeny. Bitter Leaf, on the other hand, thanks to its bitter Alkaloid-1, is rejected as food by the grazers (herbivores) and continues to make a living. The uninitiated grazer may take a bite, taste the alkaloid, spit it out and move on to Sweet Leaf. The grazers prefer Sweet Leaf and eat more of it than of Bitter Leaf. With more leaves left intact than Sweet Leaf, Bitter Leaf prospers and has a lot more progeny, all bearing the bitter Alkaloid-1, which makes them distasteful to grazers. The Bitter Leaf progeny prosper and have a lot of offspring. Gradually there are more great "grandchildren" of Bitter Leaf as Sweet Leaf hovers on the verge of extinction.

Among the progeny a second mutation occurs to Alkaloid-1 producing a second distasteful Alkaloid-2, which is chemically very similar to Alkaloid-1, but is twice as bitter. The interactions of this pair of alkaloids may be synergistic, additive, or antagonistic in their distaste or bitterness to grazers. Roughly speaking, if the two are synergistic, the combination of 1 and 2 = >3, i.e., the mixture is more distasteful than an equivalent amount of 1 or 2. If the two are additive, 1 + 2 = 3. If the two are antagonistic, 1 + 2 = <3. Evolution, again, could favor the more distasteful species, which, being less grazed and more photosynthetic, would produce more offspring bearing its synergistic combination of alkaloids. Conversely, evolution would not look so favorably on the progeny if the alkaloids were antagonistic, leaving the plant more tasty than its parents. Under these oversimplified circumstances, nature would favor synergistic, over additive, over antagonistic interactions among antifeedant, antifungal, antihelminthic, antiseptic, and antiviral alkaloids (or other types of phytochemicals).

In the senior author's dozen years of compiling data on the biologically active phytochemicals in plants (Figures 6.1 and 6.2), it seems the rule rather than the

FIGURE 6.1 (Left) Dr. James A. Duke, Economic Botanist (retired), with Tommie (Al) Bass, Herbalist from Alabama.

FIGURE 6.2 Economic Botanist Dr. James A. Duke, examining specimens of yellow root with Alabama Herbalist Tommie (Al) Bass.

exception to have a suite of closely related antitumor compounds, choleretic compounds, carminative compounds, antioxidant compounds, antifeedant compounds, allelochemic compounds within a species. For example, the long-famous Madagascar periwinkle may contain more than 500 indole alkaloids, many of them antileukemic and/or antitumor. Two, vinblastine and vincristine, have been major antileukemic drugs for close to 40 years. In the mayapple (*Podophyllum peltatum*), a derivative of which was first approved for cancer treatment in 1984, there are at least four cytotoxic lignans, proven synergistic against the herpes virus. And in the yew, there are more than a dozen compounds closely related to taxol, first approved for ovarian cancer treatment in 1992. All of these billion-dollar drugs are based on compounds distasteful or downright toxic to herbivores. But toxic is relative. Only the dose determines whether it is medicine or poison. It seems logical, to us at least, to conclude that where the several compounds possess some characteristic that also makes them distasteful to grazers or repellent to fungi, etc., synergies would be passively selected and antagonisms rejected or selected against.

It is nice to read, 3 years after writing in the *Science News*,[1] that breeders have developed plants with high concentrations of allelochemicals. They find that fall armyworm, tomato fruitworm, and tomato hornworm respond differently to three allelochemicals — chlorogenic acid, rutin, and tomatine — depending on the temperature and other chemicals present. "Since the chemicals varied in their effects on a particular insect, plants would fare best if armed with all three allelochemicals."[1] Thinking is becoming the same about HIV, humans, and anti-HIV drugs. Witness their synergistic anti-AIDS cocktail (at $16,000/year). In a decade they will be seeking the natural synergies they are now reluctant to admit, as we lose the battles with evolving disease-resistant public enemies, like AIDS, tuberculosis, and yeast. But now, in modern medicine, we often just take one of the synergistic compounds that imparts a selective advantage to the plant, and use it for our own ailments, leaving behind the other compounds and their synergies. We do this, we believe, largely in our search for replicability. The whole nonhomogeneous plant, containing thousands of compounds, is not as likely to give us replicable results in a clinical trial as a single compound. For that and other reasons, modern pharmacy goes for the "silver bullet", not the "herbal shotgun". But a standardized and consistent homogeneous mixture of four active ingredients, say those four lignans in mayapple, should give us more antiherpetic activity than an equivalent amount of any one of those lignans. Would results with a consistently standardized mixture of four compounds be as replicable in clinical trials as with a single compound? Atropine, berberine, and caffeine demonstrate the rather cosmopolitan distribution of medicinal and pesticidal compounds, and they also demonstrate the potential for synergies. We think, and many agree with us, that the function of many of the secondary metabolites produced by medicinal plants is to protect the plants, not to protect us (see Chapter 2).

Atropine has viricidal activities in addition to its numerous medicinal activities. Often found in the same plant species that provides atropine is scopolamine, which is sold as a prescription transdermal drug for vertigo and sea sickness. Berberine, found in several over-the-counter (OTC) preparations is both pesticidal and medicinal. Its pesticidal properties are medicinal, e.g., amebicide, antigiardial, antileishmannic, antimalarial, bactericide, candidicide, fungicide, and viricide. Berberine may or may not be synergistic with the sanguinarine, found as an antiplaque agent in Viadent™ toothpaste. Caffeine, in addition to its antiasthmatic central nervous system (CNS)-stimulant properties, has antifeedant, herbicide, insecticide, and viricide properties. It may or may not be synergistic with the other antiasthmatic drugs, theobromine and theophylline, which co-occur with it, at least reportedly in cacao (*Theobroma cacao*).[2]

6.2 AN ALPHABETICAL EXCURSION OF EXAMPLES OF SYNERGY

6.2.1 ALOE VERA (L.) N.L.BURM. (LILIACEAE) — MEDITERRANEAN ALOE

Aloe contains several anthraquinone glycosides, e.g., aloin (barbaloin), isobarbaloin, and emodin. These anthraquinones share several activities, antiseptic, laxative, even viricidal. Some of the related sennosides have been proven synergistic as laxatives. Sydiskis et al.[3] notes that anthraquinones can inactivate enveloped viruses. Are the anthraquinones synergistic or antagonistic as antivirals? Reflecting the natural tendency of investigators to modify natural chemical compounds, often strictly for patent protection, one company recently announced (Chemical Marketing Reporter, December 30, 1985) their filing of a patent application for Carrisyn™, which "may provide all the desirable properties of Aloe... healing burns, ulcers, and other wounds of the skin and of the gastrointestinal lining." Currently Carrisyn™ is being tested for anti-AIDS activity.

6.2.2 ANTICARIOGENIC COMPOUNDS

Muroi and Kubo[4] demonstrated synergies for antibacterial activity in compounds from tea (*Camellia sinensis*): "It could be concluded that green tea extract is effective in the prevention of dental caries because of the antibacterial activity of flavor compounds together with the antiplaque activity of polyphenols.... Synergism was found in the combination of sesquiterpene hydrocarbons (δ-cadinene and β-caryophyllene) with indole; their bactericidal activities increased from 128-fold to 256-fold... the combination of 25 μg·ml^{-1} δ-cadinene and 400 μg·ml^{-1} indole reduced the number of viable {bacterial} cells at any stage of growth." More importantly, such synergies can be utilized to help prevent the development of resistance. "Usually, the rationale for using more than two antimicrobial agents is to target a broad spectrum of microorganisms and to prevent resistance mechanisms developing in microorganisms." Muroi and Kubo, investigating the old tradition that green tea prevents tooth decay, cite a recent report[5] that "proved this." Muroi and Kobo's data clearly prove synergy between indole and some terpenes found in tea (Table 6.1). Father Nature's Farmacy (FNF)[6] did have some of its compounds listed for tea. We have five plants that have quantitative data for cadinene, betel pepper, caraway, cotton, European pennyroyal, and basil. For caryophyllene, there is basil, betel pepper, biblical mint, cinnamon, citronella, clove, copaiba, cubeb, mountain mint, oregano, star anise, sage, spearmint, and thyme, which are well quantified, while we have no quantitative data for tea. For geraniol, about a dozen plants exceed tea:

TABLE 6.1
Some Bactericidal Compounds
in Tea

Compound tested	MIC (μg·ml^{-1})	MBC (μg·ml^{-1})
δ-Cadinene	800	800
β-Caryophyllene	>1600	>1600
Geraniol	400	400
β-ionone	100	200
Indole	800	1600
cis-Jasmone	800	1600
Linalool	1600	1600
Nerolidol	25	200
1-Octanol	400	400
α-Terpineol	800	1600

Modified from Muroi, H. and Kubo, I., *J. Agric. Food Chem.*, 41, 1102–1105, 1993.

carrot, citronella, palmarosa, mountain mint, thyme, wild bergamot, this latter of which (*Monarda fistulosa*) can have 20 times more than tea. For indole, we have only four with quantitation: hyacinth, jasmine, kohlrabi, and licorice. A recent study added honeysuckle, but the indole was there only at levels of 87 ppb or less. For the compound β-ionone, tea far exceeds the others, for which we have quantitative data, at 1700–2900 ppm. As a matter of fact, we only list orris and red clover; for jasmone, we quantify only jasmine (at 72 to 114 ppm, cf 0 to 20 ppm for tea).

6.2.3 ANTIEDEMIC COMPOUNDS

Recently ginkgo extracts have been promoted as a topical agent (or cosmetic) to improve peripheral circulation, hence useful as slimming and moisturizing agents due to their microvasculokinetic activity. Della Loggia et al.[7] demonstrated anti-inflammatory activity of some *Ginkgo biloba* constituents and their phospholipid complexes. Ginkgolides, bilobalide, a biflavonic fraction and some pure biflavones (especially when mixed synergistically) were comparable to indomethacin as anti-inflammatories. Ginkgolides inhibit the pro-inflammatory autacoid PAF (platelet aggregating factor). Its biflavones inhibit histamine release from mast cells and cyclic adenosine monophosphate (cAMP) phosphodiesterases. The extract also reduces production of oxygen species by activated neutrophils. Complexes with dis-tearoylphosphatidylcholine, more soluble in nonpolar solvents than the parent compounds, are still even more strongly lipophilic, resulting in increased bioavailability and activity. For example, the complex shows some five times more antiedemic activity of the ear at 25 μg/ear than the free parent. The complex of a mix of ginkgolides A and B was more potent than indomethacin while the free mixture was not quite as effective. But phospholipid alone was inactive, merely increasing the activity of ginkgolides by making them more bioavailable. Of pure bioflavones, amentoflavone was strongest with antiedemic $IC_{45} = 2$ μM/ear, followed by ginkgetin ($IC_{25} = 2$ μM/ear) and sciadopitysin ($IC_{19} = 2$ μM/ear) cf $IC_{60} = 2$ μM/ear for indomethacin. The mix of the biflavonic fraction (corresponding to ca 0.2 μM of biflavones) inhibited 73% of the edema, cf. 45% for amentoflavone, the strongest competitor. Thus the pure flavones exhibit additive or even synergic activity when mixed.[7]

6.2.4 ANTIFEEDANT COMPOUNDS

6.2.4.1 Neem (*Azadirachta indica*)

For insect control in India, Kumar and Parmar[8] prescribe oil-based formulations containing at least 300 ppm azadirachtin. Though several other constituents influence its bioactivity, salannin and azadirachtin best correlated with inhibition of Spodoptera. Looking at 42 seed sources, however, they found that azadirachtin content varied more than 2000-fold (ND to 2323 ppm), nimbin >18,000-fold (ND to 18,132 ppm), and salanin >45,000-fold (ND to 47,150 ppm), assuming that ND = 1 ppm. With such variability among the oils, it's no wonder Kumar and Parmar seemed more to support synergy than the "silver bullet". The effective doses (ED_{50}) against neonate Spodoptera larvae was 0.29 ppm for azadirachtin, >400 for nimbin, and 72 for salannin, while whole oils (none containing more than 2323 ppm azadirachtin) ED_{50} ranged from 1.8 to 3550 ppm.[8]

6.2.4.2 Jojoba (*Simmondsia chinensis*)

The toxic effects of this plant seem to be related to inhibition of food intake by some or all of the below-listed compounds. Early on, simmondsin got all the credit. But there is a whole series of compounds, as so often happens, possibly with antifeedant activity. According to VanBoven et al.,[9] simmondsin 2-ferulate, which is hydrolyzed to simmondsin in the gut, is two-thirds as active as simmondsin.

- **Known Antifeedants:**
  - Demethylsimmondsin: Antifeedant
  - Didemethylsimmondsin: Antifeedant
  - Simmondsin: Antiappetant; Antifeedant; $LD_{50} = 4000$ oral rat
  - Simmondsin-2'-ferulate: Antifeedant (two-thirds simmondsin)

- **Others:**
  - 4-Demethylsimmondsin
  - 4-Demethylsimmondsin-2'-*cis*ferulate
  - 4-Demethylsimmondsin-2'-*trans*ferulate
  - 5-Demethylsimmondsin-2'-*cis*ferulate
  - 5-Demethylsimmondsin-2'-*trans*ferulate
  - 4,5-Didemethylsimmondsin
  - Simmondsin-2'-*cis*ferulate
  - Simmondsin-2'-*trans*ferulate

- Simmondsin-3'-*cis*ferulate
- Simmondsin-3'-*trans*ferulate

6.2.5 ANTIMALARIAL COMPOUNDS

American herbalists call it Artemisia annua, "Sweet Annie;" Chinese call it "qing hao". In China artemisinin alone has effected cures in more than 2000 patients affected with *Plasmodium vivax* and *P. falciparum* (Table 6.2). Many semisynthetic derivatives of artemisinin show better solubilities and/or efficacy. There is a developing underground promotion of the herb in the U.S. for yeast infections and opportunistic infections associated with AIDS. Work by Phillipson and associates shows that several flavonoids in the crude extracts of *Artemisia annua* or its tissue cultures are apparently synergistic with artemisinin for antiplasmodial activity.

TABLE 6.2
Inhibition of *Plasmodium falciparum*

Compound	IC_{50} (nM)
Artemisinin	9.0
Artemisinin + 5 μM eupatorin	9.0
Artemisinin + 5 μM chrysosplenol-d	3.1
Artemisinin + 5 μM chrysosplenitin	2.9
Artemisinin + 5 μM cirsilineol	2.2

Modified from Liu, K. C.-S. et al., *Planta Medica*, 55, 654–655, 1989.

Furthermore, two polymethoxyflavones, casticin and artemitin, while inactive against *Plasmodium* alone, were found to selectively enhance the activity of artemisinin against *P. falciparum*. "It is interesting to note that these flavonoids co-occur with artemisinin in *A. annua* and that crude extracts of the plant may indeed offer a therapeutic advantage over the purified sesquiterpene."[11]

6.2.6 ANTIOXIDANT PANACEAS

The cumulative antioxidant index (CAI) hypothesis implies that the more antioxidant and/or the less oxidized cholesterol we carry in our bodies, the less our chances of coronary problems.[12] The CAI is calculated as follows:

$$\frac{(\text{Vitamin E}) \times (\text{Vitamin C}) \times (\beta\text{-Carotene}) \times (\text{Selenium})}{(\text{Cholesterol})}$$

The fact that the values are multiplied rather than added implies synergy rather than additive relations between these antioxidants. Rosemary is the herb of remembrance.

Can rosemary shampoos slow Alzheimeran Blues? Like many green leaves, rosemary contains β-carotene, ascorbic acid, tocopherol, and selenium. Many other antioxidants could complement the conventional vitamins. Classically, rosemary is considered a good antioxidant herb. It contains close to two dozen named antioxidants, over and beyond the CAI antioxidants. Antioxidants from rosemary, competitive with butylated hydroxyanisole (BHA) and butylated hydroxytoluene (BHT), are already a $2 million annual business in the U.S. Rosemary has received more antioxidant press than most herbs lately. Lamaison et al.[13] showed that oregano is higher than rosemary in antioxidant activity. Screening 100 mints for antioxidant activity, Lamaison and his colleagues[13] found that oregano, *Origanum vulgare* ssp., had the greatest total antioxidant activity.

The antioxidant activity of mints is due partially to rosmarinic acid, flavonoids, and other hydroxycinnamic acid derivatives. Lamaison et al. did not mention the vitamins. Fujita et al.[14] evoked data suggesting that rosmarinic acid was almost twice as good as α-tocopherol as a radical scavenger (at least to prevent peroxidation of linolenic acid). But oregano is high in rosmarinic acid (55,000 ppm) compared to rosemary (25,000 ppm). Oregano is 2.5 times more potent in total antioxidant activity. Total antioxidant activity is measured colorimetrically, with a colored free radical. Data are reported by ED_{50} in micrograms per milliliter, roughly ppm, which decreases the absorbance (color) of the free radical by 50%. The ED_{50} of oregano is roughly 16 ppm, while that of rosemary is 40. Thus, it takes 2.5 times as much rosemary to accomplish the same amount of antioxidant activity as oregano, at least under the condition of this study.[13]

There is much more than rosmarinic acid in rosemary. Chen et al.[15] compare three of more than a dozen antioxidants to rosemary. They apparently did not measure rosmarinic acid, which has several other interesting activities as well as the antioxidant activity. They mentioned tocopherol, which of the vitamins A, C, and E, usually gets the biggest press as an antioxidant, preventing various maladies. But vitamin E (tocopherol) is usually in the plant at levels of 1 to 20 ppm. We have read that rosmarinic acid is 2 to 15 times more potent than tocopherol as a radical scavenging antioxidant. Elsewhere we read that rosmarinic acid was equivalent to caffeic acid in antioxidant activity. Some mints have 50,000 ppm, rosemary has 25,000 ppm rosmarinic acid on a dry weight basis. That makes tocopherol's contribution quite trivial, compared to rosmarinic acid, at least in rosemary. We need to know how much ursolic acid, carnosol, and carnosic acid there is in the rosemary plant itself, and how much would be expected to come over in a hydroethanolic extract (tincture) or rosemary tea itself. Chen et al. present Table 1[15] which shows that there can be as much as 100,000 ppm carnosic acid in the hexane extract, 60,000 ppm in the acetone extract, and only traces in the methanolic extract. Rosmarinic acid can be close to 25,000 ppm in the plant (dry weight basis).

Cineole (200 to 10,000 ppm in rosemary) can stimulate rats even upon inhalation. Cineole is dermally absorbed 100 times more through the skin in oil-based massage than through inhalation aromatherapy and can speed up transdermal absorption of other dermally active compounds, sometimes 100-fold. Cineole also readily crosses the blood-brain barrier. That would be expected to apply also to rosemary's carvacrol, fenchone, limonene, and thymol, all of which are reported to have anticholinesterase activities. Dermal absorption is more rapid in areas rich with hair follicles, like the scalp. Would rosmarinic acid be absorbed through the scalp from a shampoo containing rosemary? Could that be dangerous?

Certainly the literature indicates that several choline- and/or acetylcholine-conserving compounds, carvacrol, carvone, cymene, cineole, fenchone, limonene, terpinene, and thymol, may be dermally absorbed and do cross the blood-brain barrier. Buchbauer[16] indicates that 100 to 150 times more of the essential-oil compounds are absorbed dermally than are obtained through aromatherapeutic inhalation. Does that mean that rosemary shampoo can help preserve brain-levels of choline and acetylcholine, fueled by bean and lentil soups? Does that mean that a daily regime of five choline-rich legume dishes plus scalp massage with rosemary/lecithin, followed by rosemary shampoo, and finally a rosemary bath, could help stave off Alzheimer's disease? Actually there are herbs considerably richer than rosemary in antioxidant and acetylcholine-conserving dermally absorbed compounds. Oil extracts of these, used in dermal massage, could then have acetylcholine conserving effects.

Thanks to synergy, rosemary may truly deserve its title as the "Herb of Remembrance". Rosemary contains at least five dermally absorbed antioxidants and at least five dermally absorbed anticholinesterase compounds, some of which readily cross the blood-brain barrier. We speculate that some of them would work like tacrine, an anticholinesterase compound recently approved by the FDA for Alzheimer's disease. It helps about 25% of patients, we have read, and disturbs about that many livers.

6.2.7 ANTIULCER ACTIVITIES

Beckstrom-Sternberg and Duke[17] indicate that ginger has 13 antiulcer compounds. This is almost double the number of antiulcer compounds found in sesame and cayenne, each containing seven. Ginger rhizomes were fractionated and assayed for antiulcer activity. One fraction, which inhibited gastric ulcers in rats, contained four compounds, α-zingiberene, β-sesquiphellandrene, β-bisabolene, and *ar*-curcumene. The total fraction exhibited 97.7% inhibition at 125 ppm. However, this concentration of the total fraction contained only trace or negligible quantities of the four compounds, resulting in a theoretical combined inhibition of only 1.47%. The total fraction was over 66 times more effective than the summed effects of the individual chemicals. This is a strong indication of synergy. Alternatively, an unknown compound may account for the remaining inhibitory activity. A necessary follow-up test for synergy would involve assaying the four pure compounds and mixtures thereof at different concentrations to determine whether other components of the fraction or the fractionation process were factors or whether synergy indeed was involved.

6.2.8 CALCIUM ANTAGONISTS

Harmala et al.[18] reported 15 calcium-antagonistic compounds from roots of *Angelica archangelica*, one, archangelicin, showed significantly higher calcium-antagonistic activity than verapamil. The calcium antagonists reported from angelica are: 2'-angeloyl-3'-isovaleryl-vaginate, archangelicin, bergapten, byakangelicin-angelate, imperatorin, isoimperatorin, isopimpinellin, 8-(2-(3-methylbutoxy)-3-hydroxy-3-methylbutoxy)-psoralen, osthole, ostruthol, oxypeucedanin, oxypeucedanin-hydrate, phellopterin, psoralen, xanthotoxin.[18]

Another plant with significant quantities of calcium-antagonistic compounds in the seeds is *Ammi majus* (bishop's weed): bergapten (400 to 3100 ppm), imperatorin (100 to 8000 ppm), isopimpinellin, oxypeucedanin (3000 ppm), oxypeucedanin-hydrate (400 ppm), xanthotoxin (2300 to 20,000 ppm).[18] So far, we have no reports of the potent archangelicin outside of *Angelica*. Nor do we know how archangelicin compares with other more widely distributed coumarins.

Note also that angelica may contain more than 1300 ppm limonene, the compound in grapefruit suspected to potentiate certain pharmaceutical calcium-blockers. Add caraway and celery seed, rich sources of limonene, to make Duke's "angelade", a mixture of apiaceous vegetables loaded with calcium blockers and hypotensive compounds. Angelade could be a superior yet safer generic calcium blocker.

6.2.9 CATHARANTHUS ROSEUS (L.) G. DON (*APOCYNACEAE*) – ROSY PERIWINKLE, MADAGASCAR PERIWINKLE

Eli Lily has been growing this most important antileukemic plant in Texas for years. There are at least nine reportedly "antitumor" compounds present in this plant, namely, leurosine, perivine, quercetin, reserpine, serpentine, β-sitosterol, ursolic acid, vinblastine, and vincristine. There are also at least nine reportedly "hypoglycemic" compounds, namely, catharanthine, leurosine, lochnerine, quercetin, β-sitosterol, tetrahydroalstonine, ursolic acid, vindoline, and vindolinine. And there are at least eight reportedly "hypotensive" compounds that include ajmalicine, choline, kaempferol, mitraphylline, reserpine, serpentine, vincamine, and vinceine.[6,19]

Geoffrey Cordell, a University of Chicago scientist, tells Duke that more than 500 alkaloids have been reported from this important medicinal plant species. If it costs $500 million to prove a new drug to be safe and efficacious, how much would it cost to test the possible combinations of these 500 alkaloids in mixtures for the potentially best synergistic anticancer, antileukemic, or hypotensive activities?

6.2.10 CITRUS

Recently Duke met with Najla Guthrie, who vigorously champions the principles of synergy in cancer prevention. She says, in effect, that the whole orange juice is better than the sum of its parts. Workers in her laboratory are performing the tests that we have urged so often when, in retrospective reading, we see that the authors' data suggest a synergy between closely related compounds in a given plant species. As long as one is testing the individual active ingredients, looking for the "magic bullet", why not mix pure ingredient A with pure ingredient B (or C...Z) and see whether the mixtures are synergistic, additive, or antagonistic. We predict you will find the magic of synergies in more combinations than you will find the evolutionarily nonadaptive antagonisms.

Two citrus flavonoids, hesperetin and naringenin, found in oranges and grapefruit, respectively, and four noncitrus flavonoids, baicalein, galangin, genistein, and quercetin, showed an IC_{50} of 18, 18, 5.9, 56.1, 140.0, and 10.4 $\mu g \cdot ml^{-1}$, respectively, against *in vitro* breast cancer cells. So et al.[20] also tried 13 different 1 + 1 mixtures of these compounds and in most cases (12 of the 13), the IC_{50} of the mixture was lower than the arithmetic mean of the two compounds' IC_{50}, indicating synergy. "All of the combinations of flavonoids, except naringenin + hesperetin, inhibited the proliferation of MDA-MB-435 human breast cancer cells *in vitro* at much lower concentrations than either of the individual compounds alone." The lack of synergism between hesperetin and naringenin may have been related to the fact that they belong to the same class of flavonoids. Combinations with quercetin, a flavonoid found in most fruits and vegetables, were most effective. This could be important if similar synergistic relations can be demonstrated for inhibition of *in vivo* tumorigenesis. Cytotoxicity was exceedingly low (LC_{50} > 500 $\mu g \cdot ml^{-1}$ in all cases).[20]

Intrigued by grapefruit juices' ability to potentiate the hypertensives, felodipine and nifedipine, they postulated it might be due to naringin, a flavonoid present as the glycoside of naringenin, structurally similar to genistein. Naringin turned out to more effectively inhibit proliferation and growth of human breast cancer cells *in vitro*.[20]

6.2.11 FUNGICIDES

Grapes contain several antifungal compounds, e.g., caffeic acid, chlorogenic acid, pterostilbene, resveratrol, and viniferin. Much as we might like to think so, these compounds were not put there for our protection. They evolved to protect the grape itself from its fungal enemies. Since evolution favors those plants that have the best defense systems against the bacteria, fungi, herbivores, insects, and viruses that plague the plant, it seems logical that synergy would be the rule rather than the exception.

If caffeic acid is viricidal at 62 ppm and ellagic acid at 200 ppm, the combination of the two would be expected to be active at 131 ppm. If there were synergy, the combination would be viricidal at less than 131 ppm. If there were antagonism, the combination would be active only at more than 131 ppm. There are also more than a dozen antioxidant and/or cancer preventive compounds in grape. Might we also assume that these work synergically? We think so, but we cannot prove it.

In 1997 one of the compounds, resveratrol, got a lot of press as a heart-healthy anticancer compound. The press[21] promoted the fruit of grape as the best source of the resveratrol. However, there may be 100 times more resveratrol in the grape leaves than in the fruits. You can possibly capitalize on the heart-healthy synergies of some of the fungicidal compounds in grape leaves by enjoying the Mediterranean "stuffed grape leaves" with a lot of celery, garlic, olive oil, onion, pimenta, and a little rice, for example, chased with no more than one glass of red wine with all its procyanidins and traces of resveratrol.

6.2.12 GARLIC (*ALLIUM SATIVUM* L.)

Both garlic and onion are reported to inhibit platelet aggregation. Ajoene has been identified as one of the active ingredients in garlic. Block et al.[22] mention at least three compounds in garlic which inhibit platelet aggregation: (1) diallyl trisulfide, (2) 2-vinyl-4H-1,3-dithiin, and (3) ajoene. Adenosine, allicin, and alliin are three more antiaggregants, at least *in vitro*. Quercetin is another antiaggregant, apparently present in traces only in garlic, at 300 ppm in chives, and from 10 ppm in white onions to 65,000 ppm in the outer skin of red onions.[23]

At least five garlic compounds inhibit cholesterol synthesis *in vitro*.[24] Results were interpreted to mean that garlic and a wild garlic relative may reduce serum cholesterol levels primarily by inhibiting cholesterol synthesis if taken in sufficient amounts. This effect may arise from a mixture of sulfur-containing thiosulfinates.

At least with rabbits, synergy seems to be involved in lipolytic activity. "The total extract (i.e., containing all constituents of garlic) was the most effective. Extracts prepared with urea and/or alcohol reacted similarly, though more weakly. On the other hand, aqueous extracts, which had been boiled, resulting in loss of the thiosulfinates, did not produce well-defined results."[25] One fat-digesting enzyme, lipase, is inhibited by agents that bind sulfhydryl compounds, and ajoene, allicin, and diallyl trisulfide all bind rapidly to sulfhydryl compounds. Allicin may "synergise" adenosine which, alone, is poorly absorbed. "Only in the presence of substances with lipophilic and hydrophilic areas in the same molecule, e.g., allicin, can adenosine be absorbed into the blood." At only 12 mg·kg^{-1} "garlic and onion oils (steam distilled), dipropyl disulfide, and diallyl sulfide were all found to have significant hypolipidemic effects (lowering fat levels) in both normal rats and hyperlipedemic rats (those with elevated fat levels), although the garlic oil consistently gave the best results."[25]

One can also add phytic acid to the fibrinolytic (antiaggregant, antithrombotic) compounds in that chemical factory called garlic. Koch and Lawson[25] mention a garlic-extract-factor ("identical with phytic acid") that increases the prothrombin time. Cycloalliin is still another compound in garlic that elevates fibrinolytic activity. They report on one garliphile whose blood took twice as long to clot when she was on garlic as when she curtailed her garlic consumption. There are many antiaggregant compounds to complement the fibrinolytic activity. Fresh garlic (100 to 150 mg·kg^{-1}; that meant 10 to 15 g for the senior author when he weighed only 100 kg) "leads to complete inhibition of thrombocyte aggregation for one to two hours after consumption." Adenosine and allicin inhibit platelet aggregation without affecting cyclooxygenase and lipoxygenase metabolites of arachidonic acid. The trisulfides inhibit platelet aggregation as well as thromboxane synthesis along with induction of new lipoxygenase metabolites. As antiaggregants, the best dry garlic powder tablets were as effective as garlic or garlic homogenates, but steam-distilled oils were only 35% as active and oil macerates only 12% as active. Ajoene, a compound unique to the oil macerates, had the highest specific activity, slightly more active than 1,2-vinyldithiin, diallyl trisulfide, or allicin and considerably more active than diallyl disulfide, allyl methyl trisulfide ($EC_{50} = 10$ μM) or 1,3-vinyldithiin. Besides diallyl trisulfide, other polysulfides, such as dimethyl trisulfide and allyl methyl trisulfide, are also active. Simple monosulfides are less active or inactive. Diallyl mono-, di-, and trisulfides have less than half the activity of ajoene. Some authors (not Lawson) believe alliin is the antiaggregant. Phenolics (see Chapter 1), structurally similar to flavonoids, are also believed to be involved. Too much garlic intake may reverse results, so moderation is best in all things. High concentrations of aggregating agents can abolish platelet inhibition caused by low-dosed garlic preparations.[25]

6.2.13 GOITROGENS

Thiocyanate, the first discovered goitrogen (generating goiter), was a less effective goitrogen at twice the concentration found in cabbage than was cabbage itself. Goitrogenicity of cabbage and other cruciferous plants can be explained as being due to combined additive or synergistic action of thiocyanate, goitrin, and allyl isothiocyanate. These compounds are enzymatically hydrolyzed from various glucosinolates. Brussels sprouts having the highest observed glucosinolate levels, namely, 1430 to 1760 ppm on a fresh weight basis and 10 times that much on a zero moisture basis if they are 90% water.[23] There are almost 100 glucosinolates known, one cabbage containing at least 12 totaling 663 ppm (fresh weight basis, estimated 6300 ppm if calculated on a dry weight basis). If the glucosinolates, there to protect these plants and their pesticidal properties, are synergistic, might not their cancer-preventive properties also be synergistic?

6.2.14 HYPERICUM

Hostettmann's[26] discussion of monoamine oxidase inhibitors (MAO-I) is interesting because several popular herbs, like St. John's wort (*Hypericum* spp.), contain significant levels of MAO-I. Disturbances in MAO levels are reported in several disorders (anxiety, depression, Huntington's chorea, Parkinson's disease, senile dementia). MAO deactivates certain neurotransmitters (catecholamines, serotonin). There are two types of MAO, namely, MAO-A and MAO-B; both are inhibited, at least *in vitro*, by *Hypericum* extracts. Inhibitors of MAO-A, in particular, look promising as antidepressants. MAO inhibition can also increase serotonin levels in the brain. Hypericin from St. John's wort inhibits both MAOs but there are other, more potent, MAOIs in *Hypericum*, namely, xanthones and flavonoids. Three xanthones from a Brazilian *Hypericum* species selectively inhibited MAO-A and MAO-B, one comparable to harman from the famed Amazonian hallucinogen, ayahuasca, which itself might be useful in depression. The authors stress further searches in the Clusiaceae, Gentianaceae, Hypericaceae, and Polygalaceae families for better selective and reversible inhibitors.[26]

In Germany, the most popular prescription drug of any type, natural or synthetic, for the treatment of mild depression is a concentrated extract of the flowers of St. John's wort, often simply called hypericum. There, just under 200,000 prescriptions per month are filled for a single brand (Jarsin), compared with about 30,000 per month for fluoxetine (Prozac). Many clinical trials show St. John's wort to be especially useful in treating mild depressive states. Studies in 3250 patients found improvement or total freedom from symptoms in about 80% of the cases treated, with only 15% not responding.[27]

There can be synergies of chemicals, there can be synergies of actions, and there can be combinations of the two. Recent developments with and studies of hypericum,

Germany's leading antidepressant, suggest that several chemical compounds may be involved and several mechanisms of actions may result from some of these chemicals as well. Not using the word synergy, Tyler[27] says the herb's multiple constituents apparently work in several different ways. These include: (1) COMT (catechol-*O*-methyl-transferase) inhibition (COMT seems to deplete "feel good" amines); (2) suppression of interleukin-6-release, affecting mood through neurohormonal pathways; and (3) MAO inhibition: monoamine oxidase is another enzyme that destroys "feel good" amines. Still not using the word synergy, Tyler says "different chemical compounds in St. John's wort work together to relieve mild depression in several different ways. The advantage of this combined action is fewer side effects for the consumer because the total response is not due to a single strong action." It is good to hear America's emeritus pharmacognocist say this. We can translate it to herbal medicine in general. The empirically proven herbs are gentler and safer because of synergistic actions of several compounds instead of strong action by a single "silver bullet". Dr. Jerry Cott, Chief of the Pharmacologic Treatment Program of the National Institute of Mental Health, tells us that, even though hypericum is a leading antidepressant, its MAO inhibitory activity is much less than we had previously believed. The MAO inhibition, and perhaps other reported activities, may occur only in the test tube. But clearly, something(s) in hypericum makes it considerably better than a placebo, perhaps also making hypericum outsell all other antidepressants combined in Germany. To top it all off, Dr. Cott sent me a draft[28] hinting at two other mechanisms: (4) modulating the serotonin receptors, and (5) increasing dopamine. This last one is particularly interesting, in that increasing dopamine levels might ease part of the addiction problems co-morbid with depression. Administering a lyophilized extract with and without 0.015% hypericin caused an increase in dopamine content 4 h after treatment and there were also suggestions of enhanced serotonin release. Hypericum extracts were suggested to exert their activity by dopaminergic activity.[27,28,29]

6.2.15 HYPERTENSION

One phthalide, 3-*n*-butyl-phthalide, is said to relax the smooth muscle linings of the blood vessels, thereby lowering blood pressure. Phthalide works directly by dilating vessels. Many current antihypertensive agents work by more roundabout mechanisms and may have troubling side effects, such as fainting, drowsiness, or impotence. Remember that drowsiness might also be a side effect of celery, because phthalides are natural sedatives as well. Perhaps this sedative activity could translate into reduced stress further translating into reduced cardiopathy. Celery is one of dozens of reputed aphrodisiacs. Sex can be good exercise and exercise is good at preventing cardiopathy. Unfortunately, celery is rather high in sodium which is often counterindicated in hypertension. But, in addition to the phthalides, celery is fairly well endowed with a few other hypotensive compounds including ascorbic acid, bergapten (sometimes phototoxic), fiber, magnesium, and rutin. So there are at least six hypotensive ingredients in celery. Duke[30] recounts a study showing that celery juice proved hypotensive in 14 of 16 males taking 40 ml (a generous shot-glass full) orally three times a day with honey or syrup.

Celery is closely related to the herb, angelica, mentioned earlier[18] as containing 15 calcium-antagonistic compounds. Celery has three of these: bergapten, 1 to 520 ppm; isopimpinellin, 4 to 122 ppm; and xanthotoxin, 6 to 183 ppm. Parsnip and parsley are even better endowed with the coumarin calcium-blockers. We're not about to suggest ingestion of coumarins in foods as calcium-antagonists; only to ask our federal health watchers, "Would angelade, i.e., juiced angelica, caraway, carrot, celery, fennel, parsley, and parsnip, be as safe and efficacious and cheap as verapamil as a calcium blocker? Might that thereby partially explain the lower incidence of cardiopathy in vegetarians?" So, celery contains hypotensive, hypocholesterolemic, and calcium-blocker phytochemicals. How about antiarrhythmic compounds? There are apigenin, apiin, magnesium, and potassium present which represent such compounds.

The senior author has replaced his antigout, hypouricemic allopurinol, with celery seed extracts and/or four stalks of celery a day for more than 8 months and experienced no gout attacks. With more than two dozen anti-inflammatory compounds (α-pinene, apigenin, ascorbic-acid, bergapten, butylidene-phthalide, caffeic acid, chlorogenic acid, cnidilide, copper, coumarin, eugenol, ferulic acid, gentisic acid, isopimpinellin, linoleic acid, luteolin, magnesium, mannitol, myristicin, protocatechuic acid, quercetin-3-galactoside, rutin, scopoletin, thymol, umbelliferone, and xanthotoxin) celery seed might prove synergetically useful in gout and other types of arthritis.

6.2.16 INFLUENZA

Weber et al.[31] studied the *in vitro* virucidal effects of garlic and some of its active components. Fresh garlic was virucidal to all viruses tested including HSV-I, HSV-II, para influenza, rhino virus type 2 (human), vaccinia, and vesicular stomatitis. Of the components, ajoene was most viricidal, then allicin, then allyl methyl thiosulfinate, then methyl allyl thiosulfinate. No comment was evident about the possibility of synergy, but since the extract killed all viruses and the individual components did not, there could be synergy. Nagai[32] had earlier shown that garlic extract protected mice against flu viruses introduced in the nose.

6.2.17 MELISSA

Varro Tyler[33] told the American Chemical Society, and we quote from his unpublished draft:

> Melissa volatile oil is used as a sedative, a spasmolytic, and an antibacterial agent. The sedative action is attributed largely to citronellal with other terpenes (see Chapter 1) such as citronellol, geraniol, caryophyllene, linalool, citral, limonene, and eugenol contributing to the effect.

Buchbauer[33a] states that mere inhalation of lemon balm oil has a "very good sedative effect.... You need not a massage." Elsewhere, Buchbauer warns that dermal absorption of these essential oil ingredients may be 100 times greater than by inhalation or "aromatherapy".

The sedative ingredients mentioned by Tyler[33] occur in many aromatic plants and may be even more concentrated in arid-land plants than in Melissa. If, as Buchbauer states, inhalation has a measurable effect, and dermal application is 100 times stronger, one might look for pronounced sedation with Melissa in the bath or in an oil-based massage. From Duke[19] we list the quantitative values (see Tables 6.3 and 6.4) for some of these reportedly sedative compounds we had calculated for Melissa.

TABLE 6.3
Alternative Sources of "Sedative" Terpenes

Terpene	Lemon balm	Lemon eucalyptus	Ginger	Juniper	Java citronella
Caryophyllene	9–238	15–780	—	12–120	84–147
Citra	—	—	0–13,500	—	8–14
Citronellal	1–2(1975)[a]	50–18,020	2–145	160	1000–2289
Citronellol	—	230–4000	2–6500	—	440–770
Eugenol	—	—	—	—	66–233
Geraniol	1–2	250–1000	2–345	—	1064–3150
Limonene	—	20–1420	17–1050	76–910	80–350
Linalool	1–10	15–180	2–1500	—	40–105

Note: All quantities are in ppm.

[a] The count is 1–975 ppm if [+]-citronellal is included.

Modified from Duke, J. A., CRC Handbook of Biologically Active Phytochemicals and Their Activities, CRC Press, Boca Raton, FL, 1992.

TABLE 6.4
Second String Sources of "Sedative" Terpenes

Terpene	Tangerine	Lemon grass	Lemon verbena	Basil	Lime
Caryophyllene	—	30–266	0–3196	30–250	2–9
Citral	1080–3480	260–182	560–7000	—	—
Citronellal	—	—	—	140	1–20
Citronellol	—	10–70	3–2,419	—	1–50
Eugenol	—	—	35–8575	—	—
Geraniol	10–36	5–420	1–1000	—	1–4
Limonene	—	42–1050	2–934	4700–7500	6500–9400
Linalool	1–10	5–35	5–8730	9–20	3–610

Note: All quantities are in ppm.

Modified from Duke, J. A., CRC Handbook of Biologically Active Phytochemicals and Their Activities, CRC Press, Boca Raton, FL, 1992.

The optimistic interpretation of the above is that we could, on a dry weight basis at least, expect no more than 975 ppm of the major "actor", citronellal, and no more than 250 ppm of the other "supporting actors" from Melissa, if our database reflects the true world. *Eucalyptus citriodora* and *Java citronella* certainly look more promising.

If lemon balm, with its 1 to 2 (to 975 ppm, if you count [+]-citronellal) ppm citronellal and no more than 250 ppm "supporting actors," is a good sedative, lemon-scented *Eucalyptus citriodora* with up to 18,000 ppm citronellal should be a better sedative, as should leaves of *Citrus limon* (up to 89,000 ppm in the essential oil), and *Cymbopogon winterianaus* (to 2250 ppm). Even the fruits of juniper, used in making gin, may contain 140 ppm citronellal, and ginger (rich in other "supporting actors" and richest in citronellol), may contain 145 ppm citronellal, all on a dry weight basis. We can see the makings of a good sedative bedtime gin and citronellal, gin and any safe lemon-scented leaf you can find, plus a little basil (high in citronellol), Biblical mint (*Mentha longifolia*, rich in caryophyllene and linalool), caraway (rich in limonene), coriander (rich in linalool), cloves (highest in eugenol and good tasting), and ginger and thyme (high in geraniol). The gin might induce sleep, if these reputed herbal sedatives did not. Our mechanical database, interpreted by humanoids, suggests that gin with eucalyptus lemonade could be a very pleasant sedative.

6.2.18 VERNONIA

In another of those zoological self-medication stories,[34] wild chimpanzees in Tanzania chew and suck on the very bitter herb Vernonia amygdalina when sick; they avoid bitter herbs when well. The plants contain at least four parasiticides, inhibiting protozoan parasites at levels as low as 4 to 11.4 $\mu g \cdot ml^{-1}$. We would wager that they are synergistic at the levels the chimpanzees seek, but we doubt that pharmaceutical firms will seek such a synergic mix. Instead they will more likely develop an unnatural synthetic analog they can patent and which will act adversely against us and our parasites, slowly, in fact, so slowly that the parasite will evolve a resistance while we will not![35]

Presuming that the chimpanzees might be suffering parasites, Jisaka et al.[34] investigated the plants activities. Several of the compounds were active against Gram-positive *Bacillus subtilis* and *Micrococcus lutea*, but none were active against Gram-negative bacteria. The structure-activity relationship correlated well with that found for the antitumor activities. However, they did not always parallel the bitterness of the compounds. Although two of the most bitter compounds, vernodalin and 4,15-dihydrovernodalin, were also most effective against the P-388 leukemia strain, the least bitter, vernodalol was also rather effective, more so than bitter compounds hydroxyvernolide, octahydrovernodalin, and hexahydrovernodalin. Moreover the bitterest compound, octahydrovernodalin, was the least effective in the P-388 system. *V. amygdalina* contained an interesting array of antiseptic and antileukemic compounds.

- **HYDROXYVERNOLIDE**: Amebicide NIG; antimalarial IC_{50} = 11.4 $\mu g \cdot ml^{-1}$ POP:105; antitumor IC_{50} (L-1210) = 1.25 BBB57:834; antitumor IC_{50} (P-388) = 0.92 BBB57:834; antiseptic 5 μg/disk BBB57:834; bitter 1.0 μg BBB57:834; cestodicide 500 mg·kg^{-1} NIG; nematicide NIG; plasmodicide IC_{50} = 11.4 $\mu g \cdot ml^{-1}$ POP:105; protisticide IC_{50} = 11.4 $\mu g \cdot ml^{-1}$ POP:105[6,19]
- **VERNODALIN**: Antifeedant JBH; antileishmanic MIC = 0.5 $\mu g \cdot ml^{-1}$ POP:105; antimalarial IC_{50} = 4 $\mu g \cdot ml^{-1}$ POP:105; antiseptic 5 μg/disk BBB57:834; antitumor IC_{50} (L-1210) = 0.26 BBB57:834; antitumor IC_{50} (P-388) = 0.14 BBB57:834; bitter 0.8 μg BBB57:834; cytotoxic (ED_{50} = 1.8) (BPCC); plasmodicide IC_{50} = 4 $\mu g \cdot ml^{-1}$ POP:105; protisticide IC_{50} = 4 $\mu g \cdot ml^{-1}$ POP:105
- **VERNODALOL**: Antifeedant JBH; antimalarial IC_{50} = 4 $\mu g \cdot ml^{-1}$ POP:105; antiseptic 50 μg/disk BBB57:834; antitumor IC_{50} (L-1210) = 0.61 BBB57:834; antitumor IC_{50} (P-388) = 0.32 BBB57:834; bitter 70 μg BBB57:834; plasmodicide IC_{50} = 4.2 $\mu g \cdot ml^{-1}$ POP:105; protisticide IC_{50} = 4.2 $\mu g \cdot ml^{-1}$ POP:105
- **VERNOLIDE**: Amebicide (=Metronidazole) NIG; antileishmanic MIC = <10 $\mu g \cdot ml^{-1}$ POP:105; antimalarial IC_{50} = 8.4 $\mu g \cdot ml^{-1}$ POP:105; antiseptic 5 μg/disk BBB57:834; antitumor IC_{50} (L-1210) = 0.11 BBB57:834; antitumor IC_{50} (P-388) = 0.11 BBB57:834; bitter 1.2 μg BBB57:834; cestodicide NIG; cytotoxic (2 ppm) (BPCC); nematicide NIG; plasmodicide IC_{50} = 8.4 $\mu g \cdot ml^{-1}$ POP:105; protisticide IC_{50} = 8.4 $\mu g \cdot ml^{-1}$ POP:105[6,19]

Since they gave no data for whole extracts, there is no way we can backtrack through their data to see if there was evidence of synergy. But, this is certainly a case where we would expect a parallel synergy between the antiseptically active compounds and the antitumorally active compounds. We see, for example, that the IC_{50}'s for vernodalin and vernolide against the P-388 tumor system were 0.11 and 0.13, respectively. If their activities were additive, you would expect the IC_{50} of the mixture to be 0.12, if synergistic, <0.12, and if antagonistic, >0.12. Clearly the authors had the capacity to test whether the pure compounds were more or less active than the mixtures, thereby supporting or refuting the theory of synergies between suites of antiseptic compounds and antitumor compounds within a single species. Even the bitter compounds might be synergistic.

Essay on Dragon's Blood

We would like to close with a recently published synergy story about one of the favorite medicines of Antonio Montero Pisco, the shamanistic forest "medicine man" who has planted the dragon's blood and some 250 other Amazonian medicinal plants at the new camp at the Amazonian Center for Environmental Education and Research (ACEER) on the Napo river in Amazonian Peru.

Dragons' blood, known to our shaman as "sangregrado", and to others as sangre de drago, is the euphorbiaceous (spurge family) weed tree, *Croton lechleri*, also very important to Shaman Pharmaceuticals Company, San Francisco, CA, who have two drugs derived from the plant in clinical trials. Shaman Pharmaceutical's Provir™ effectively treats 89% of patients with diarrhea in a phase-II trial, according to their press release of November 7, 1996. They had successfully completed a phase-II trial of 75 patients afflicted with acute diarrhea, including travelers' diarrhea and nonspecific diarrhea of unknown etiology. Overall, 89% of the 75 patients treated with their trademarked Provir™ experienced rapid return to normal bowel function. Of 71 patients available for followup, none experienced recurrence. No significant adverse reactions were reported. Of 25 traveler's diarrhea patients getting 1 to 2 $g \cdot d^{-1}$ Provir™, 72% returned to normal in 48 h instead of the usual 5 to 7 d. For reasons not discussed, the 1-g dose was more effective than the 2-g dose. Provir™ acts by inhibiting the secretion of chloride ions from the epithelial cells lining the small intestine and the subsequent accumulation of fluid in the intestine. "In contrast to conventional treatments, Provir™ is not absorbed systemically and appears to have no effect on gastrointestinal motility." This indicates a potential use for treatment of bacterial, parasitic, and viral diarrhea.

But Antonio, the ACEER "Shaman", uses whole dragon's blood topically, among other things, to treat cuts and sores. Antonio uses the whole dragon's blood rather than isolated "silver bullets" from the dragon's blood. Most pharmaceutical firms go for the isolated "silver magic bullet" rather than the synergic whole. Recently, Dutch scientists demonstrated synergy of dimethylcedrusine, pycnogenol, and taspine, parts of that whole extract called dragon's blood. They were synergistic in causing wounds to heal in exfoliated rats. The isolated compounds caused granulation of excision wounds in 4 d, while the synergistic whole extract accomplished the job in just 1 d. That is why we will use dragon's blood, should we get a cut or abrasion in tropical Peru. And if we were to develop cholera in Peru and had no doctor or pharmaceutical available, you can bet your sweet sangregrado we will be taking oral whole dragon's blood preparation. Our ACEER "shaman," Antonio, recommends the whole dragon's blood, which, at least for some indications, has proven better than the sum of its parts.

6.3 CONCLUSIONS

Humankind has been co-evolving with plants co-evolving with plant pests for more than 2 million years. In the process, humanoids ingested many, if not all, of the phytochemicals that the plants evolved to defend themselves from diseases and leaf-eating pests. Evolution would favor synergy and disfavor antagonisms in closely related phytochemical mutations which have plant-protective attributes within a plant species. Synergy in plant protective properties would enhance the survival and reproductive potential of those offspring possessing the synergies, while antagonism would impart negative survival and reproductive potential. In many cases, the medicinal activities of these phytochemicals in humans are based upon parallel physiological and/or chemical bases. In such cases, it seems logical to speculate, if not conclude, that the same evolved synergies will apply to the medicinal applications of the mixtures in humans.

Heretofore, it has proven economically more attractive to pharmaceutical firms to select one of the most active compounds within a species and to abandon the synergistic mix that, more often than not, seems to prevail. We predict that as our pathogens evolve beyond our medicines by developing resistance to our latest weapons, the pharmaceutical firms will go back to Nature's drawing board seeking new weapons. They will realize what we say here can lead them to superior medicines, namely, standardized extracts of these herbs containing several closely related synergistic phytochemicals in their evolutionary pesticidal ratios. We will then *use* rather than *ignore* this synergistic tool. Also, instead of incredibly expensive synergistic mixes of unnatural compounds, like their $16,000-a-year AIDS cocktail, they will develop better synergistic mixes of natural antiviral, antiyeast, antieboli, antibacteria, antiescherichia, anticancer, and antitumor compounds. Finally, they will have evolved from the opposite direction to the same logical point to which the crude herb industry has evolved from the other direction, i.e., to use standardized extracts of biologically active plants containing synergistic mixes of phytochemicals that evolved to protect life from predation.

In synergy, the whole is better than the sum of its parts. There will be an even greater synergy when the best of alternative and allopathic medicines are used in a new holistic medical system that uses the best of all available medicines. We want the best medicines for ourselves and our children, if we can afford them. Until the holistic "herbal shotgun" has been compared in unbiased scientific head-on trials with the solitary "silver synthetic bullet", we will not know which is best. Is the synergistic mix of compounds in *Hypericum* better (and/or cheaper, more efficacious, safer) as an antidepressant and antiviral treatment protocol than Prozac™ and Acyclovir™? Is the natural mix of sterols in saw palmetto better than pure isolated β-sitosterol or synthetic finasteride? Is the synergistic mix of parthenolides in feverfew (*Chrysanthemum parthenium*) better, cheaper, and/or safer, than Sumatriptan™ at treating migraine. Is the natural mix of lignans in mayapple (*Podophyllum peltatum*) better, cheaper, and/or safer, at killing cancers *in vivo* as it has been shown to be for arresting herpes virus *in vitro*, than isolated purified lignans, or semisynthetic derivatives thereof? Is the mixture of choline-sparing compounds in rosemary and sage better, cheaper, and/or safer, than Tacrine™ for treating and/or preventing Alzehimer's disease? We do not know the answers to any of these and a hundred other similar herbal-alternative questions. We do not know anybody who knows the answers. And until we know the answers, we cannot be sure we are getting the best medicines in our synthetiphilic society. We think synergiphilia might be safer and healthier than synthetiphilia.

REFERENCES

1. Adler, *Science News*, SN149, 389, 1996.
2. Duke, J. A., Medicinal and industrial plants from arid lands. pp. 325-343, in *Biotechnology for Aridland Plants*, Mabry, T. J., Nguyn, H. T., Dixon, R. A., and Boness, M. S., Eds., IC2 Institute, The University of Texas at Austin, 1994, 369 pp.
3. Sydiskis, R. J., Owen, D. G., Lohr, J. L., Rosler, K. H., and Blomster, R. N., Inactivation of enveloped viruses by anthraquinones extracted from plants, *Antimicrob. Agents Chemother.*, 35, 2463–2466, 1991.
4. Muroi, H. and Kubo, I., Combination effects of antibacterial compounds in green tea flavor against *Streptococcus mutans*, *J. Agric. Food Chem.*, 41, 1102–1105, 1993.
5. Onishi, M., Shimura, N., Nakamura, C., and Sato, M., A field test on the caries preventive effect of tea drinking, *J. Dental Health*, 31, 13–19, 1981.

6. Duke, J. A., *CRC Handbook of Phytochemical Constituents of GRAS Herbs, Foods, and Other Economic Plants*, CRC Press, Boca Raton, FL, 1992, 654.

7. Della Loggia, R. et al., Anti-inflammatory activity of some *Ginkgo biloba* constituents and of their phospholipid complexes, *Fitoterapia,* 67, 257–264, 1996.

8. Kumar, J. and Parmar, B. S., Physicochemical and chemical variation in neem oils and some bioactivity leads against *Spodoptera litura* F.J., *J. Agr. Food Chem.*, 44, 2137–2143, 1996.

9. VanBoven, M. et al., Determination of simmondsins and simmondsin ferulates in jojoba meal and feed by high-performance liquid chromatography, *J. Agric. Food Chem.*, 44, 2239–2243, 1996.

10. Liu, K. C.-S., Yang, S.-Y., Roberts, M. F., Elford, B. C., and Phillipson, J. D., The contribution of flavonoids to the antimalarial activity of *Artemisia annua*, *Planta Medica*, 55, 654–655, 1989.

11. Borris, R. P. and Schaeffer, J. M., Antiparasitic agents from plants, in *Phytochemical Resources for Medicine and Agriculture*, Nigg, H. N. and Seigler, D., Eds., Plenum Press, New York, 1992, 117–159.

12. Duke, J. A., Mint tease and the cumulative antioxidant index, *Trends Food Sci. Technol.*, 3, 120, 1992.

13. Lamaison, J. L., Petitjean-Freytat, C., Duband, F., and Carnat, A. P., Rosmarinic acid content and antioxidant activity in French Lamiaceae. *Fitoterapia,* LXII, 166–171, 1991.

14. Fujita, Y., Uehara, I., Morimoto, Y., Nakashima, M., Hatano, T., and Okuda, T., Studies on inhibition mechanism of autooxidation by tannins and flavonoids. II. Inhibition mechanism of caffetannins isolated from leaves of *Artemisia* species on lipoxygenase dependent lipid peroxidation, *Yakugaku Zasshi,* 108, 129–135, 1988.

15. Chen, Q. Y., Shi, H., and Ho, C. T., Effects of rosemary extracts and major constituents on lipid oxidation and soybean lipoxygenase activity, *JAOCS*, 69, 999–1002, 1992.

16. Buchbauer, G., Phytopharmaka und pharmakologie, *Deutsche Apot. Zeit.,* 130, 2407–2410, 1990.

17. Beckstrom-Sternberg, S. M. and Duke, J. A., Potential for synergistic action of phytochemicals in spices, in *Spices, Herbs, and Edible Fungi*, Charalambous, G., Ed., Elsevier, New York, 1994, 201–223.

18. Harmala, P., Vuorela, H., Hiltunen, R., Nyiredy, S., Sticher, O., Tornquist, K., and Kaltia, S., Strategy for the isolation and identification of coumarins with calcium antagonistic properties from the roots of *Angelica archangelica*, *Phytochem. Anal.*, 3, 42–48, 1992.

19. Duke, J. A., *CRC Handbook of Biologically Active Phytochemicals and Their Activities*, CRC Press, Boca Raton, Fl, 1992, 183.

20. So, F. V., Guthrie, N., Chambers, A. F., Moussa, M., and Carroll, K. E., Inhibition of human breast cancer cell proliferation and delay of mammary tumorigenesis by flavonoids and citrus juices, *Nutr. Cancer*, 26, 167–181, 1996.

21. Jang, M., Cai, L., Udeani, G., Slowing, K. V., Thomas, C. F., Beecher, C. W., Fong, H. H., Farnsworth, N. R., Kinghorn, A. D., Mehta, R. G., Moon, R. C., and Pezzuto, J. M., Cancer chemopreventive activity of reseveratrol, a natural product derived from grapes, *Science*, 275, 218–220, 1997.

22. Block, E., Ahmad, S., Jain, M. K., Crecely, R. W., Apitz-Castro, R., and Cruz, M. R., Ajoene: a potent antithrombotic agent from garlic, *J. Am. Chem. Soc.,* 106, 26, 1984.

23. Beier, R. C. and Nigg, H. N., Natural toxicants in food, in *Phytochemical Resources for Medicine and Agriculture*, Nigg, H. N. and Seigler, D., Eds., Plenum Press, New York, 1992, 247–368.

24. Sendl, A., Schliack, M., Loeser, R., Stanislaus, F., and Wagner, H., Inhibition of cholesterol synthesis in-vitro by extracts and isolated compounds prepared from garlic and wild garlic, *Atherosclerosis*, 94, 79–82, 1992.

25. Koch, H. P. and Lawson, L. D., Eds., *Garlic. The Science and Therapeutic Application of Allium sativum L. and Related Species*, Williams & Wilkins, Baltimore, 1996.

26. Hostettmann, K., Marston, A., Maillard, M., and Hamburger, M., Eds., *Phytochemistry of Plants Used in Traditional Medicine*, Proc. Phytochemical Society Europe 37. Oxford Science, New York, 1995.

27. Tyler, V. E., The honest herbalist — the secrets of Saint John's wort, *Prevention*, Feb., 74–79, 1997.

28. Upton, R., Shumake, R. L., Marquis, M. C., Graff, A., Williamson, E., Fields, S., Bunting, D., Gatherum, D. M., Walker, E. B., and Cott., J., St. John's wort Monograph, *Hypericum perforatum,* American Herbal Pharmacopoeia, unpublished, 1997.

29. Indena Pharmaceutical Co., *Hypericum perforatum* L. Inside cover. *Firoterapia,* 67, 1997.

30. Duke, J. A., *CRC Handbook on Medicinal Herbs*, CRC Press, Boca Raton, FL, 1985.

31. Weber, N. D., Furanocoumarin induction in wild parsnip — evidence for an induced defense against herbivores, *Ecology*, 71, 1926–1932, 1992.

32. Nagai, K., Experimental studies on the preventive effect of garlic extract against infection with influenza virus, *Jpn. J. Infect. Dis.*, 47, 321, 1973.

33. Tyler, V. E., Phytomedicines in western Europe: their potential impact on herbal medicine in the United States, Presented at a symposium on "Human Medicinal Agents from Plants," American Chemical Society, 203rd National Meeting, San Francisco, April 5-10, 1992 (unpublished).

33a. Buchbauer, personal communication, May 27, 1993.

34. Jisaka, M., Ohigashi, H., Takagawa, K., Huffman, M. A., and Koshimizu, K., Antitumoral and antimicrobial activities of bitter sesquiterpene lactones of *Vernonia amygdalina*: a possible medicinal plant used by wild chimpanzees, *Biosci. Biotech. Biochem.*, 57, 833–834, 1993.

35. Phillipson, J. D., Wright, C. W., Kirby, G. C., and Warhurst, D. C., Phytochemistry of some plants used in traditional medicine for the treatment of protozoal disease, in *Phytochemistry of Plants Used in Traditional Medicine*, Hostettmann, E., Marston, A., Maillard, M., and Hamburger, M., Eds., Proc. Phytochem. Soc. Europe, 37, Oxford Science Publications, New York, 1995, chap. 5.

7 Bioseparation of Compounds

Peter B. Kaufman, Leland J. Cseke, and Akira Okubo

CONTENTS

0-8493-3134-X/99/$0.00+$.50
© 1999 by CRC Press LLC

7.1 INTRODUCTION

7.1.1 TRADITIONAL METHODS

Traditional methods used for bioseparation of metabolites from plants include the use of hot water extracts to make teas or natural plant dyes. Salves and decoctions are often made from a single plant source (e.g., the monoterpenes from mints to make mint tea or the oleoresin terpenes from pitch of balsam fir used directly to treat burns) or more than one plant (as with many commercial herbal teas that utilize camomile, echinacea, mints, beebalm, lavendar, and other plants or up-regulators of the immune system such as echinacea and goldseal). The important point here is that such extracts or preparations rely on the synergistic action of several plant metabolites that are more effective than any one alone (see Chapter 6). An excellent discussion of these traditional methods is found in Penelope Ody's book, *The Complete Medicinal Herbal.*[1]

7.1.2 CONTEMPORARY METHODS

Modern methods of bioseparation utilize principles of extraction that are based on the **polarity** (relative solubility in organic solvents), solubility in water, and various alterational solubilities based on salts and pH (relative acidity or alkalinity). A good discussion of these methods is found in the CRC book by Kaufman et al., *Handbook of Molecular and Cellular Methods in Biology and Medicine.*[2]

7.2 COLLECTION, STORAGE, AND VOUCHERING OF PLANTS

7.2.1 COLLECTION OF PLANTS IN THE FIELD: DO'S AND DON'TS

When collecting plants in the field for natural product extractions, it is important to be properly prepared. Based on our experience, we suggest that you do the following:

- Wear field clothes (Figure 7.1) and cover yourself head to toe if collecting is to be in the cold of winter or when mosquitos or deer/black flies are in abundance.
- Take along a note pad and pencil to record information about the collecting site location, soil conditions, ecological habitat, date of collection, plant identity, and who collected the plant(s).

FIGURE 7.1 Casey Lu (far right) and colleagues (from Humboldt State University Department of Biological Sciences, at Arcata, CA) on a field trip at a site along the coast of the Pacific Ocean where medicinal plants were being collected. (1997 photo by Peter Kaufman.)

- Take along a pocket-size field-guide (with photos, drawings, and good, usable identification keys) to the local flora and a hand lens to help you identify the plant.
- If you are collecting live plants, take along some Zip-lock™ plastic bags of various sizes in which to put the samples after collecting plants onsite (Figures 7.2 and 7.3). Slips of recycled paper are good to have for notes on plant identity with your collected specimens that match up with your field notes about the respective collections.
- Take soil samples from each site so as to later get information on soil nutrient, soil pH, and soil type where each plant grows.
- When you collect plants for extracts, it is important to get representative samples of all parts available: roots, vegetative shoots, bark from stems (if woody plant), flowers, fruits, and seeds (if mature).
- When collecting plants in the field, **do not** take every last plant in the population, especially if the plant is rare, threatened, or endangered.
- In the process of collecting herbaceous perennial plants (plants that come from the same mother plant year after year), leave some of the original plant intact where it is growing so that it can reproduce during the current and following years. Many of these plants take years to produce even a small amout of new biomass every year.
- As native Amerinds do, *thank the plant* for providing you with material for your extracts! I think we all do this in various ways when we collect

FIGURE 7.2   Casey Lu is shown collecting a medicinal plant in the field north of Arcata, CA. (Photo by Peter Kaufman.)

plants in the home garden for food or for aesthetic purposes, or when we are collecting wild edible plants or mushrooms in the field.

- If you are collecting mushrooms or puffballs in the field, wrap the fruiting bodies in wax paper and place them in a collecting basket or other suitable container where they will not become squashed. This will help for later identification and/or making spore prints from the fruiting bodies. This is impossible with giant puffballs (*Calvatia gigantea*); these can be collected intact and placed in large paper shopping bags. Some of these mushrooms attain a diameter of 0.5 m.

7.2.2   STORAGE OF PLANTS AT LOW TEMPERATURES IN A FROZEN STATE

Once one comes in from collecting plants in the field, it is a good idea to freeze them immediately in a freezer at –20°C or a commercial freezer held at –80°C. This is to prevent any degradation of the plant material or any enzymatic changes that alter or degrade naturally occurring metabolites. Natural drying of the plant material can also be done if yield of metabolites is not critical. This is usually the case for plants used for dyeing fibers, but for extraction of medicinal compounds from plants, the use of dried plant material is **not** desirable due to degradation of naturally occurring metabolites during the drying process. Rather, it is best to rely on the use of frozen plant material. The only exception is with seeds. They are usually dried to a low moisture content to prolong seed viability. If the drying process is slow and the temperature is at ambient level, very little degradation of stored metabolites in the seeds occurs.

FIGURE 7.3   Casey Lu placing a specimen of medicinal plant in a Zip-Lock™ bag while on a field trip north of Arcata, CA. (Photo by Peter Kaufman.)

7.2.3   Vouchering of Plants Collected in the Field

7.2.3.1   Preparation of Dry Specimens

Dried plant specimens are prepared in order to have them available at any time as voucher specimens representing typical plants that were collected in the field and used for plant extracts. They are also called *herbarium specimens*. To prepare dried plant specimens, they are placed between single newspaper sheets then placed in a sandwich consisting of a dry blotter above and below the newspaper sheets. A piece of corrugated cardboard (with air spaces present) is then placed above and below each blotter. Successive sandwiches are placed atop one another then compressed between two wood-slatted frames and tied together tightly with straps. The entire assembly is then placed upright on its side over a heat source such as a radiator or a plant drier with the heat on a moderate temperature (e.g., 35 to 40°C). The specimens are allowed to dry this way for 48 h or longer. Rapid drying assures that plant pigments are well preserved; if the drying process is slow, chlorophylls will degrade and the leaves will appear yellow; flower pigments also fade badly with slow drying. If plant specimens are very high in water content, it is a good idea to replace the blotters with dry ones several times during the drying process.

Once plant specimens are dry, they can be mounted flat (with glue or cement) on heavy paper of sufficient size to accommodate the specimen and a label with information about the plant, location of collecting site, collector, date of collection, genus and species of the plant, and the family to which the plant belongs. The label may be placed in the lower right-hand corner of the sheet of heavy paper. To avoid damage to the dry, mounted specimen, the entire sheet can be covered with Saran-Wrap™. Dried sheets with plants mounted on them are usually stored in an air-tight cabinet in which one can use natural insect repellents such as dried lavendar (*Lavandula officinalis*) or neem (*Azadirachta indica*).

7.2.3.2 Keeping a Good Log Book with Plant Collection Inventory

One should always keep a log book of the plants that have been collected and then frozen. A back-up inventory on the computer is also a good idea. Why is this important? Sometimes labels with collected plant material become lost. Sometimes, one needs to quickly examine lists of collected plants without going through all of the frozen material! And sometimes, one needs to send such lists to others involved in the project.

7.2.3.3 Representative Living Plant Specimens from Field Collections

In our experience, we have found it to be a good idea to collect seeds and/or living specimens of the plants that are to be used for extraction of natural products. We do this in order to have the living plants on hand, e.g., medicinal plants, dye plants, or culinary herbs garden or in a greenhouse. These can then be used for later extractions or experimental treatments to enhance metabolite biosynthesis; this is especially important when access to the original collecting site is not possible or convenient. A good example of this is with tree of joy, *Camptotheca accuminata*, which is the source of the drug camptothecin used to treat patients with prostate cancer. Our original seed came from China. The seed used in our experiments conducted in a greenhouse came from progeny from the original Chinese seed which was grown in Louisiana. These seeds from the tree of joy progeny were kindly provided by Mr. Tracy Moore, President of XyloMed Research Inc. in Monroe, LA (Figure 7.4). The results from these studies are provided in the research essay in Chapter 3, Section 3.2 written by Atul Rustgi, Ashish Goyal, and Kathryn Timberlake.

7.3 GRINDING AND EXTRACTION PROTOCOLS

7.3.1 GENERAL EXTRACTION PROTOCOLS FOR BIOLOGICALLY IMPORTANT ORGANIC COMPOUNDS

The primary ways for extraction of organic molecules of interest to biologists and medical investigators involve breaking open the cells of the organism under investigation. Cell rupturing is accomplished in a variety of ways. The method used depends on the type of organism being considered and the type of tissue

FIGURE 7.4   (Left to right) Dr. Stanley Carpenter, LSU; Dr. Zhijun Liu, Post-Doctoral Research Fellow at LSU; Tracy Moore, President of XyloMed Research, Inc.; and Dr. John D. Tarven, LSU, while on visit to Beijing, Peoples Republic of China in November, 1996 to collect seed from different native populations of tree of joy, *Camptotheca accuminata*, the source of the drug, camptothecin, used to treat cancer patients. (Photo courtesy of Tracy Moore.)

used. For bacterial cells, one usually uses a French press so as to break open the cell walls. This involves the use of a heavy cylinder with high pressure applied to a piston that compresses the cells into a successively smaller volume within the free cylinder. As the cells leave the cylinder, the rapid drop in pressure causes the cells to lyse. Such a procedure can also be used for plant cells grown in suspension culture or for plant callus tissue. A sonicator can also be used for this purpose. In this case, repeated high frequency pulses of ultrasonic vibrations rupture the cell membranes. Animal cells and plant cells grown in culture can be ruptured with a glass tissue homogenizer. Highly lignified or silicified plant tissues within organs such as leaves, stems, roots, seeds, or fruits, are usually frozen and pulverized using liquid nitrogen in a mortar and a pestle. (Figure 7.5). Softer plant tissues can be ground in a small volume of buffer in a mortar, using white washed sand and a pestle to rupture the cells.

Once the cells have been ruptured, the actual extraction is performed using techniques that depend on the chemical properties of the compound(s) of interest. Water-soluble compounds and proteins are extracted in water or buffers. Organically soluble compounds are extracted with organic solvents. For example, since taxanes are miscible in methanol, this solvent is often used as the extraction reagent. This is by no means the only solvent that will work. Some compounds, such as cell-wall constituents, have no need to be solubilized for extraction because they can be obtained in pellet form by filtration or centrifugation followed by washing with

FIGURE 7.5   Gregg Roslund, undergraduate student at the University of Michigan, grinding a frozen plant sample with liquid nitrogen and a ceramic mortar and pestle, in preparation of analysis of medicinal products of medicinal value in this sample. (Photo by David Bay.)

buffer solutions. It is worth noting that integral membrane proteins often require the use of strong detergents, such as Triton × 100, to be extracted from the membranes.

Two steps are usually critical for good extraction. First, ruptured cells should be ground or homogenized in the extraction solvent, depending on the cell rupturing technique chosen. For example, taxanes (terpenoid compounds derived from plants) are extracted by grinding the plant tissue in organic solvent (methanol) in the same mortar and pestle that is used for the liquid nitrogen to rupture the cells. Waxes, on the other hand, can be removed from the aqueous phase coming from a French press by homogenizing and partitioning with chloroform and methanol. Second, once ground or homogenized, the extraction mixture should be allowed to stand undisturbed for 0.5 to 24 h at a temperature that will not allow degradation of the compound(s) or protein(s) of interest to occur (e.g., 4°C). This is done simply to allow time for the extraction solvent to penetrate all parts of the ruptured cells.

After these two critical steps, the resulting slurry is filtered to obtain a filtrate free of particulates or it is centrifuged in order to obtain a cell wall and/or membrane pellet fraction and a cell cytosol containing supernatant fraction. Such crude fractions can be used directly for enzyme reaction assays, or they can be subjected to further purification and clean-up procedures in order to separate and identify the compounds of interest. For example, C-18 Sep Pak columns can be used to remove chlorophylls, which interfere with subsequent analysis of taxane extractions (compare Figure 7.6A and B).

For the extraction of natural products of potential medicinal or other value in plant samples, as used in the author's laboratory, the following specific protocols have been employed:

A   **Extract Cleaning**

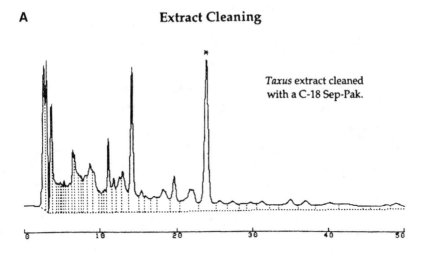

Taxus extract cleaned with a C-18 Sep-Pak.

B

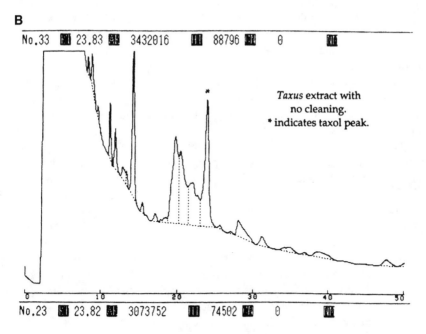

No.33 ▓ 23.83 ▓ 3432016   ▓ 88796 ▓   0   ▓

Taxus extract with no cleaning.
* indicates taxol peak.

No.23 ▓ 23.82 ▓ 3073752   ▓ 74502 ▓   0   ▓

FIGURE 7.6 HPLC taxane analysis results showing chart recordings resulting from C-18 Sep-Pak™ cleaning (A) vs. no cleaning (B) of extracts from yew (*Taxus* spp.) needles and stems.

- Protocol (1):   Hot-Water Extraction of Water-Soluble Medicinal Compounds from Plants
- Protocol (2):   Organic-Solvent Extraction of Organic Solvent-Soluble Medicinal Compounds from Plants

- Protocol (3):   Extraction of Taxanes (Diterpenoid Compounds) From Yew (*Taxus* spp.) Needles and Stems
- Protocol (4):   Extraction of Cuticular Wax from Needles of Yew (*Taxus* spp.) Plants

7.3.2   HOT-WATER AND ORGANIC-SOLVENT EXTRACTION OF WATER-SOLUBLE AND ORGANIC SOLVENT-SOLUBLE MEDICINAL COMPOUNDS FROM PLANTS

There are so many different ailments and diseases, it is a wonder we are ever healthy. Today, the development of drugs and medicines is a growing industry. U.S. companies are losing the race to be the leading developer of medicines. What are other countries doing that the U.S. is not? Most other countries have made the trend back to nature. They started looking at plants and trees for natural medicines. The U.S. has just joined this trend, but has not been able to produce as many drugs as some other countries, such as Japan. Unfortunately, it takes a lot of money to develop any new drug, and the risk of the drug failing is great.

Many of the plants that have known medicinal values have been used for centuries by different cultures. Natural medicines have been used for centuries in many parts of Asia, such as China and India. In particular, Native Americans have had a profound influence on the natural medicines of today. The Indians of North America have made the most contribution to this field.

A project conducted in Peter Kaufman's laboratory at The University of Michigan by DaRhon Conner and Nina Jain has involved the preparation of plant extracts from plants that contain natural products of know medicinal value. These extracts are prepared to be screened by Parke-Davis/Warner-Lambert Pharmaceutical Co. in Ann Arbor, MI, U.S. The plants come from a variety of sources ranging from botanical gardens to grocery stores. The plants are cut and stored in a deep freeze at –80°C. These extracts are tested in an 80-sample screen for medicinal compounds in order to determine which are effective against viruses, bacteria, mycoplasms, and fungi that are pathogenic to humans. Many of the plants are naturally toxic and harmful to humans. However, if they are administered in small amounts, the effects on humans are beneficial.

There are two different types of extracts used in this project. One is done using hot water and is called an ***aqueous extract***. The other uses organic solvents, such as methylene chloride, and is called an ***organic extract***. There are two very specific procedures that have been developed for each type of extraction. These procedures are delineated in Sections 7.3.2.1 and 7.3.2.2.

7.3.2.1   Hot Water Extraction of Water-Soluble Medicinal Compounds from Plants (Protocol 1)

First, we weigh out a 0.5-g sample of a given plant. This is then ground to a fine powder using liquid nitrogen. Next, we place the powder into a Corex centrifuge tube with hot (80°C) water and place the tube into a hot water bath for 10 min.

The tube is centrifuged at 3000 × g for 10 min. By centrifuging it, all of the particulate plant materials from the grinding get pelleted to the bottom of the tube, leaving a relatively clear liquid (the supernatant) containing the water-soluble compounds of interest. This liquid is filtered to make sure that no plant particulates remain in the filtrate. Next, the filtrate is placed in a Petri dish and frozen by placing it on dry ice. After the sample is completely frozen, it is placed in a freeze-drying apparatus and lyophilized. The sample could take anywhere from 2 to 12 h to lyophilize. Freeze-drying is done in order to remove all moisture. This yields a powdered residue and it is this powder that we send to Park-Davis/Warner-Lambert. Figure 7.7 outlines this procedure.

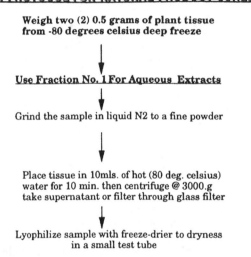

EXTRACTION PROTOCOL FOR NATURAL PRODUCT SCREENS

Weigh two (2) 0.5 grams of plant tissue from -80 degrees celsius deep freeze

Use Fraction No. 1 For Aqueous Extracts

Grind the sample in liquid N2 to a fine powder

Place tissue in 10mls. of hot (80 deg. celsius) water for 10 min. then centrifuge @ 3000.g take supernatant or filter through glass filter

Lyophilize sample with freeze-drier to dryness in a small test tube

FIGURE 7.7 Hot water extraction protocol for water-soluble plant natural product compounds that have potential medicinal value.

7.3.2.2 Organic Solvent Extraction of Organic Solvent-Soluble Medicinal Compounds from Plants (Protocol 2)

As mentioned above, this procedure uses an organic solvent to extract the compounds of interest. At least one very toxic solvent (methylene chloride) is used and therefore needs to be monitored carefully and with caution. The temperature of this extraction must also be carefully watched due to the special equipment that is used. We make use of a Soxhlet extractor which is basically a specialized glass refluxing unit that is used for organic solvent extractions. The temperature must be maintained at 80°C for 18 h in order to obtain complete extraction of a given sample. If the temperature falls below this, extraction will be slow. If the temperature goes above this, the risk of degrading the compounds of interest becomes great. Figure 7.8 outlines this procedure.

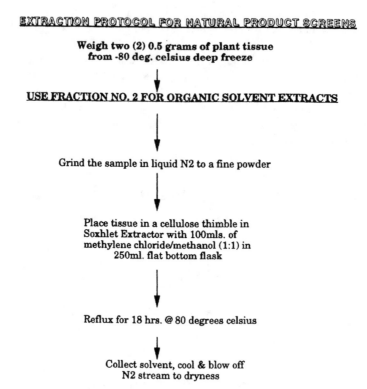

FIGURE 7.8   Organic solvent extraction protocol for organic solvent-soluble plant natural product compounds that have potential medicinal value.

7.3.2.3   Case Studies on the Purification of Crude Extracts Prior to Chromatographic Separation — Taxol and Cuticular Wax Extractions from (Yew) Plants

For bioseparation, cleanup procedures are usually necessary before samples are analyzed by high-performance liquid chromotography (HPLC) or by any of the other above-mentioned chromatographic techniques. One application that we have employed is for the analysis of taxol. Taxol is a unique taxane diterpene amide that possesses antitumor and antileukemic properties. Kilograms of this cancer chemotherapeutic agent are needed for clinical treatment of patients having breast cancer; however, taxol exists in only minute quantities — 0.01% of the inner bark and needles of Yew (*Taxus*) species. Until recently, taxol could not be synthesized, and even now, the most economical source of taxol is still from the Pacific yew (*Taxus brevifolia*). For this reason, large areas of Pacific yew forests in the Pacific northwest of the U.S. were destroyed in order to obtain this anticancer drug. The taxol extraction methods used by researchers commonly involve complicated partitioning methods in which the plant is first extracted with methanol and H_2O and then partitioned using methylene chloride to remove chlorophyll and other unwanted compounds. In the process, taxol molecules move from the aqueous methanol to the more hydrophobic methylene chloride. Unfortunately, methylene chloride is a suspected carcinogen and it seems counterproductive, in our opinion, to extract a cancer chemotherapeutic with a substance that could cause cancer! Hence, it has been our motive in studying taxol to find a cheap, efficient, and easy way to separate taxol from the thousands of other organic compounds in yew tissues. One of the primary difficulties, in this case, is the removal of chlorophylls from methanolic extracts. Chlorophylls absorb at the same wavelength as taxol and often occur in such large quantities that their resulting peaks interfere with taxane peaks. For this purpose, we use a C-18 reverse phase Sep-Pak™ column. The protocol for doing this follows. The main point here, however, is that preparatory columns are also useful to remove many unwanted compounds before the actual chromatographic analysis is performed. Figure 7.9 shows the small size of the C-18 Sep-Pak™ set-up as used in our laboratory. HPLC results using C-18 Sep-Pak™ cleaning vs. no cleaning are shown in Figures 7.6A and 7.6B, respectively.

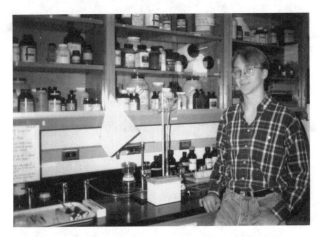

FIGURE 7.9   Leland Cseke is shown with the C-18 Sep-Pak™ column (lower corner) used in our laboratory to partially purify yew (*Taxus* spp.) plant extracts in preparation for HPLC analysis of taxane-type diterpenes found in these plants.

A new, simple and rapid method that successfully works for extraction and HPLC separation of taxol from crude extracts of *Taxus cuspidata* (Japanese Yew) needles and stems has recently been developed in our laboratory and tested by 10 groups independently with repeated success. It requires 2 h to perform steps 1 through 21 and 70 min per sample to run them in an automated Shimadzu HPLC apparatus. This long run time is used to separate the multitude of peaks that result at 228-nm spectrophotometric monitoring when *not* using the C-18 Sep-Pak™ cleanup method. The advantage of not using cleanup procedures is that the researcher saves time in preparing extracts. In this case, the researcher can go home and sleep while the Shimadzu Sil-6a autoinjector does the work. Figures 7.10A and 7.10B illustrate Shimadzu and JASCO automated HPLC apparatuses, respectively, that can be used in this procedure.

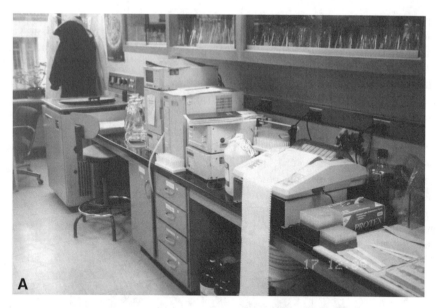

FIGURE 7.10A   Shimadzu HPLC apparatus in Peter Kaufman's laboratory at University of Michigan.

FIGURE 7.10B   JASCO HPLC apparatus with three independent pump systems, three detectors, controller, and recorder in Dr. Akira Okubo's laboratory at the University of Tokyo.

In some analytical situations, or for cases where publication is desired, it may be necessary to clean the crude extract with a C-18 Sep-Pak™ column. The clean-up procedure is given following the crude extraction protocol described in Section 7.3.2.3. [*Note*: This is by no means the only cleanup procedure for extracts. Any form of chromatography can be used to cleanup the extracts (see discussion of chromatographic methods).] The basic trick is to find the fraction eluted from the cleanup column which contains the compound of interest. In the case of taxol, this was accomplished experimentally, using purified taxol standards dissolved in 10% methanol. At this concentration of methanol, it was found that taxol binds to the C-18 packing material in the column. Then, the concentration of methanol was arbitrarily increased in several steps up to 100% methanol. A fraction of eluted mobile phase (2 to 3 ml) was collected at each concentration increase, and HPLC was performed on each fraction to determine where taxol had eluted off the column. After repeated experiments, the specific concentration of methanol that elutes taxol was identified.

In some cases, chromatography is not necessary for cleanup of the sample. Addition of adsorptive particles such as activated charcoal to the crude extracts followed by filtration may be all that is necessary to remove unwanted compounds.

7.3.2.3.1 Taxol Extraction from Fresh Taxus spp. (Yew) Tissue Resulting in Crude Extract Samples that Can be Used for HPLC (Protocol 3)

1. Obtain *Taxus* spp. plant specimens.
2. For each sample, weigh out 0.5 g of needles + 0.5 g of stems.
3. Grind to a fine powder using liquid N_2 in a mortar and pestle.
4. Let powder warm to room temperature.
5. Add 1 ml of 100% MeOH and grind vigorously.
6. Transfer this slurry to a 15 ml Corex centrifuge tube.
7. Add 2 ml of 100% MeOH and grind the remaining plant material in the mortar.
8. Add this to the tube and label.
9. Add another 1 ml of 100% MeOH to the mortar and rinse again.
10. Add this to the tube and label.
11. Place Parafilm™ over the tube and vortex or shake for 10 min.
12. Cool at −80°C for 10 min.
13. Balance the tubes by adding 100% MeOH to the lighter tube.
14. Centrifuge at 12,000 rpm for 15 min.
15. Take off the supernatant and place in a clean, acid-washed tube using a Pasteur pipette; then label.
16. Put this tube into a waterbath at 55°C.
17. Blow N_2 gas gently into tubes until the samples are completely dry.
18. Add 0.5 ml of ice-cold HPLC grade 100% MeOH and vortex for several min. with periodic placement in an ice bath. This allows one to later quantify the amount of taxol extracted because a known volume of sample has been created.
19. When the MeOH is very cool, place the 0.5 ml into a syringe and filter through a 0.2-μm filter into an HPLC bottle.
20. Add an aluminum septum to the top of the bottle and cap.
21. Parafilm™ well to prevent evaporation and label.

22. Inject sample into HPLC and run through the Curosil G column (from Phenomenex Corp.), collecting data at 228 nm. Run each sample for 70 min. at 1 ml · min⁻¹ in 36.5% acetonitrile + 63.5% 10 mM ammonium acetate at pH 4.0.

23. Use purified taxol to make several known concentrations of taxol. Run these standards on HPLC, as in step 22, to make a standard curve. Figure 7.11 shows an example of a typical standard curve for taxol.

24. Compare your data to the standard curve for taxol and calculate the percent of taxol per unit (gram) of fresh weight of tissue.

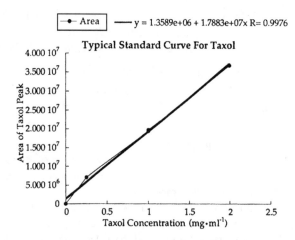

FIGURE 7.11   A typical standard curve for taxol as determined by HPLC analysis of taxol standards at a wavelength of 228 nm.

7.3.2.3.2   *Taxol Extraction from Fresh* Taxus *Spp. (Yew) Tissue Resulting in Cleaned Extracts that Can Be Used for HPLC — Use of the C-18 Sep-Pak™ Column (Protocol 4)*

1. Obtain *Taxus* spp. plant specimens.
2. Weight out 0.5 g of needles + 0.5 g of stems.
3. Grind to a fine powder using liquid N₂ in a mortar and pestle.
4. Let powder warm to room temperature.
5. Add 1 ml of 100% MeOH and grind vigorously.
6. Transfer this slurry to a 15-ml Corex centrifuge tube.
7. Add 2 ml of 100% MeOH and grind the remaining plant material in the mortar.
8. Add this to the tube and label.
9. Add another 1 ml of 100% MeOH to the mortar and rinse again.
10. Add this to the tube and label.
11. Place Parafilm™ over the tube and vortex or shake for 10 min.

12. Cool at –80°C for 10 min.
13. Balance the tubes by adding more 100% MeOH to the lighter tube.
14. Centrifuge at 12,000 rpm for 15 min.
15. Take off the supernatant and place in a clean, acid-washed tube using a Pasteur pipette; then label.
16. Dilute the volume of MeOH supernatant 10-fold with ddH$_2$O to produce a 10% MeOH sample.
17. Run all of this diluted sample through a C-18 Sep-Pak™ column (activated as per instructions from supplier). Taxol-type molecules will bind to the column at this concentration of MeOH.
18. In the case of taxol itself, wash the column first with 3 ml of 35% MeOH, then with 3 ml of 55% MeOH. This washes compounds that have less affinity for the C-18 absorbent than taxol out of the column, while the taxol remains bound.
19. Now wash the C-18 column with 2 ml of 65% MeOH and collect the sample. All taxane species of compounds (including taxol, 7-epi-taxol, 7-epi-10 deacetyl taxol, and cephalomanine) are released from the column during this elevation in MeOH concentration, leaving almost all of the problematic chlorophyll still bound to the column adsorbent phase.
20. The 2 ml of taxol-containing sample is completely dried under a stream of N$_2$ gas in a 55°C waterbath while the C-18 column can be cleaned with 100% MeOH and reused. (*Note*: No detectable amount of taxol is lost during this procedure.)
21. Add 0.5 ml of ice-cold HPLC grade 100% MeOH and vortex for several min. with periodic placement in an ice bath.
22. When the MeOH is very cool, place the 0.5 ml into a syringe and filter through a 0.2-μm filter into an HPLC bottle.
23. Add an aluminum septum to the top of the bottle and cap.
24. Parafilm™ well to prevent evaporation; then label.
25. Inject sample into HPLC and run through a Curosil G column (from Phenomenex Corp.), collect data at 228 nm. Run each sample for 30 min. at 1 ml · min^{-1} in 40% acetonitrile + 60% 10 mM ammonium acetate at pH 4.0.
26. Use purified taxol to make several known concentrations of taxol. Run these standards on HPLC as in Step 22 to make a standard curve (Figure 7.11).
27. Compare your data to the standard curve for taxol and calculate the percent of taxol per unit (gram) of fresh weight of tissue.

Note: The run time can be reduced to as little as 10 min by increasing the percentage of acetonitrile. The taxol peak is still clearly resolved in this case.

7.3.2.3.3   Extraction of Cuticular Wax from Needles of Yew
(Taxus) Plants

Waxes are lipids synthesized by plants and animals and function to keep out infectious organisms (e.g., ear wax in animals) and to prevent desiccation and serve as a barrier against fungal and bacterial pathogens in plants. In yew plants (*Taxus* spp.), waxes are synthesized in increased amounts in response to physical stresses. The protocol we have successfully used to quantitatively analyze cuticular waxes from *Taxus* needles is listed. This procedure should be adaptable for any species of plant or animal.

1. Weigh out 1.0 g of *Taxus* needles.
2. Dice needles with a razor blade and place in a mortar and pestle.
3. Add 2:1 mixture of methanol:chloroform to the mortar and grind thoroughly.
4. Transfer the resulting slurry to a centrifuge tube and centrifuge at 12,000 rpm for 15 min.
5. Remove the supernatant and place in a separate collection flask. This supernatant contains the lipids from the *Taxus* needles.
6. Place the pellet back into a mortar and pestle.
7. Add a 2:1:0.8 mixture of methanol:chloroform:water, and grind thoroughly.
8. Centrifuge this slurry at 12,000 rpm for 15 min.
9. Transfer the supernatant to the above collection flask.
10. Regrind the pellet from the centrifuge tube using the original 2:1 mixture of methanol:chloroform mixture, and repeat Steps 8 and 9.
11. While collecting the rinse liquid, vacuum filter the pellet, rinsing with 100% chloroform.
12. Transfer this rinse liquid to the collection flask.
13. Transfer all of the collected liquid to a separatory funnel and add a 1:1 mixture of chloroform:water. This will produce two phases. The lower phase will contain the dissolved lipid.
14. Empty the bottom layer of chloroform from the separatory funnel and dry it over solid sodium acetate. This absorbs water from the sample.
15. Weigh a clean rotary evaporator flask (round bottom flask) and record the weight.
16. Pipette the chloroform solution into the rotary evaporator flask.
17. Vacuum filter the sodium acetate solid, rinsing with chloroform. Add the collected rinse liquid to the rotary evaporator flask.
18. Evaporate the chloroform using a rotary evaporator until the sample is completely dry.
19. Weigh the rotary evaporator flask with the sample and subtract the recorded initial weight for the empty flask. This will give the amount of cuticular wax from the original sample of *Taxus* needles.

7.3.3 SQUARE WAVE ELECTRICAL FRACTIONATIONS

This is a relatively new procedure that allows one to fractionate plant tissues physically without any grinding. It gives a nice powder preparation that can be used for hot water or organic solvent extractions. It will virtually break-up cell walls in plant tissues and pulverize the tissue to a powder. It relies on the same principle used in ultrasound sonicating apparatus, where such vibrations can loosen particulate matter adhering to glass walls of collection vials or HPLC vials. This new protocol is being used commercially by biotech and pharmaceutical companies because it is much easier than grinding samples in liquid nitrogen and works well for both woody (lignified) and herbaceous plant tissues.

7.3.4 Use of Detergents, Alcohols, and Dimethylsulfoxide to Extract Compounds of Interest without Killing the Tissue

This is a relatively new concept being used experimentally in our laboratory to achieve maximal extraction of the metabolites of interest without actually killing the plant. It is done with seeds or green, living plants. No grinding of tissues is involved. The method relies on the partition coefficients of the chemicals used, the polarity of molecules to be extracted, and the ease with which the solvents can penetrate the tissues without killing them. Dimethylsulfoxide (DMSO) is a penetrant used in human medicine and works well in getting compounds into plants, like alcohols (methanol, ethanol, or long-chain alcohols) and detergents (which are good wetting agents and good at liberating molecules like proteins from plant membranes). In testing such compounds in various combinations, one has to run viability tests with the plants being treated to extract compounds of interest. This method is being developed at the University of Michigan in the Biomedical Engineering Program, and the Departments of Chemical Engineering and Biology by Professors Henry Wang and Peter Kaufman and Ph.D. students, Kittinan Komolpis and Artiwan Shotipruk from Thailand.

7.4 ANALYTICAL PROTOCOLS

7.4.1 Chromatographic Separation Techniques

7.4.1.1 An Overview

Chromatography is a process of separating gases, liquids, or solids in a mixture or solution by adsorption. For example, this is done by selective adsorption on particle clay, silica gel, fine powdered sugar, alumina, or on paper. As an extraction mixture flows through the adsorbent medium or phase, often in a column, each substance in the sample is separated on the basis of differences in hydrophobicity and/or ionic charges. Consequently, each substance appears in the adsorbent medium eluting from the column at a different time. This time difference is called the retention time (Rt); it represents the relative position of a compound(s) with respect to the origin (top) and the bottom of a column.

Primary methods of chromatography in use today include paper chromatography (PC), thin-layer chromatography (TLC), liquid column chromatography (LC), gas chromatography (GC) (Figure 7.12A-C), HPLC, fast protein liquid chromatography (FPLC), immobilized metal ion affinity chromatography, and antibody affinity chromatography.[3-7] Colored products or bands are obtained which can be measured with a spectrophotometer, as with an enzyme reaction. Compounds eluting at particular

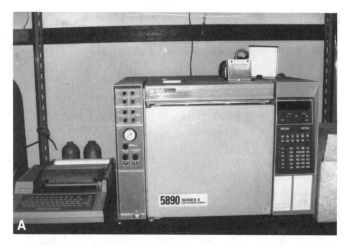

FIGURE 7.12A   Small table-model gas chromatograph (GC) in Dr. Akira Okubo's laboratory at the University of Tokyo.

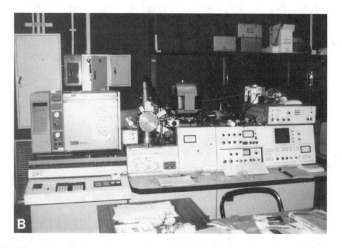

FIGURE 7.12B   Hitachi gas chromatograph (GC)/mass spectrometer (MS) in Dr. Akira Okubo's laboratory at the University of Tokyo.

times during HPLC can also be read spectrophotometrically at wavelengths equivalent to the maximum absorption peaks for such compounds. Compounds appearing at a particular Rt value during TLC can be quantified densitometrically or spectrophotometrically after scraping the compound(s) off the adsorbent phase. Highly volatile compounds (e.g., ethylene, ethane) that exist primarily in the gas phase can be subjected to gas chromatography using a flame ionization detector (FID). GC's also have other types of detectors for selective ion monitoring such as a thermal conductivity detector, or even a mass spectrometer (Figure 7.12B and C).

FIGURE 7.12C JEOL gas chromatograph (GC)/mass spectrometer (MS) in Dr. Akira Okubo's laboratory at the University of Tokyo.

7.4.1.2 Chromatographic Separation of Organic Molecules

Chromatography refers to the separation of chemical compounds by partitioning them between two phases, one of which is stationary and the other is in motion. In this process, the compounds to be separated are distributed between the stationary and mobile phases. This technique is very powerful because it allows one to separate very similar compounds, within a given extract, which may be isomers of each other. It is also faster, easier, and more economical.

There are three types of chromatography: (1) *adsorption chromatography*, (2) *partition chromatography*, and (3) *gel permeation chromatography*. TLC, gas-liquid chromatography (GLC), affinity chromatography (AC), HPLC, and gel-filtration chromatography are adaptations of these three basic types of chromatography, explained in the following sections.

7.4.1.2.1 Adsorption Chromatography

In adsorption chromatography, the compounds of interest are separated by allowing them to adsorb (bind) to the surface of a solid phase, such as DEAE cellulose (diethylaminoethyl cellulose) or charcoal. The compounds are desorbed (removed) from the solid phase by an eluting solvent such as NaCl of varying concentrations (e.g., a linear gradient of increasing NaCl concentrations from 0 to 1.0 M) or by shifting the pH of the mobile phase to lower or higher pH's. The solid phase is poured as a slurry into a column fitted with sintered glass at its base. The column typically has a length-to-width ratio of 10:1 (of the actual poured solid phase). A filter paper disk is added to the top of the buffer-covered (ca. 1 cm of buffer) solid phase, over which the compounds of interest are gently layered without disturbing the solid phase. Elution is accomplished with the help of gravity and a gradient maker. A fraction collector can be used to collect 1-, 2-, 5-, or 10-ml fractions for later analysis of enzyme activity (using enzyme assays), salt composition and concentrations (using a conductivity meter or a spectrophotometer), and protein composition and concentrations (using Bradford[7a] or Lowry et al.[7b] protein assays in conjunction with a spectrophotometer).

One specialized type of adsorption chromatography is *AC*.[8] In this case, a receptor, such as an antibody, is linked by covalent bonds to an inert solid support phase. The receptor has a high binding affinity for one of the compounds (its ligand) in the mixture of compounds from the prepared extract. Such binding is both specific and reversible. Inert solid support phases include cross-linked dextrans, cross-linked polyacrylamide, cellulose, and agarose. Due to its great selectivity, AC offers a very powerful means of achieving excellent separation and purification of biological molecules in as little as one step. Applications include purification of proteins (including antibodies and enzymes), nucleic acids, or any compound that acts as a ligand for a given bindable receptor. For example, some drugs, such as taxol, can be purified by using its monoclonal antibody as the bound receptor. After eluting unwanted compounds from the column, the bound ligand is then easily eluted from the column by shifting the mobile phase to a low, or in some cases, a high pH. This procedure can be done by immobilizing the ligand or by immobilizing the ligand's receptor to purify the ligand.

One hybrid modification of AC and ion exchange chromatography is *high performance immobilized metal-ion affinity chromatography* (HPIMAC) used to separate peptides and other organic molecules.[7] It utilizes several types of stationary phases: synthetic polymers, silica, or cross-linked agarose. Basically, chromatographic separation involves a metal-ion chelator bound to the solid support phase. This in turn binds positively charged metal ions. The metal ions frequently used are Zn^{2+}, Mg^{2+}, and Cu^{2+}. Negatively charged side groups on a given protein (or other negatively charged species such as polysaccharides, nucleotides, or nucleic acids) bind to the positively charged metal ions via ionic interactions. Nonbound compounds simply pass through the column and elute off, whereas the bound compounds stick to the solid phase until they are eluted off by use of pH shifts or a selective change in ionic strength.

Another variation of adsorption chromatography is *ion-exchange chromatography*. Here, the solid adsorbent phase has charged groups that are linked chemically to an inert solid matrix. What happens during the chromatography is that ions become electrostatically bound to the charged groups of the solid adsorbent. These ions may then be *exchanged* for ions in the mobile aqueous phase. This is accomplished by changing the ionic strength, or pH, of the eluting solvent. Two types of ion exchangers are used in ion exchange chromatography: (1) *cation exchangers*, which are exchangers with chemically bound negative charges, and (2) *anion exchangers*, which are exchangers with chemically bound positive charges. On the exchangers, the charges are balanced by counterions. For this purpose, chloride ions (Cl^-) are used for anion exchangers and positively charged metal ions are used for cation exchangers. To elute the molecule of interest from such ion exchange columns, one can use (1) changes in pH of the eluting buffer, (2) increasing ionic strength of salt (e.g., [NaCl] or [KCl] in solution, and (3) affinity selection, which depends on both charge (opposite to that of the bound macromolecule) and specific affinity for the bound macromolecule.

7.4.1.2.2   Partition Chromatography

Partition chromatography, often called liquid-liquid partition chromatography, involves two liquid mobile phases. The substances to be analyzed are separated based on their different solubilities in the two liquid phases. An inert support is used in this type of chromatography. Examples of such inert supports include sheets of paper (cellulose) as used in paper chromatography, or a thin layer of silica gel ($SiO_2 \cdot nH_2O$) or powdered alumina on a glass plate, as used in TLC.

Paper chromatography is usually carried out in a large glass tank or cabinet and involves either ascending or descending flow of the mobile phase solvents. Descending paper chromatography is faster due to gravity facilitating the flow of solvents. Large sheets of Whatman #1 or #2 filter paper (the latter is thicker) are cut into long strips (e.g., 22 × 56 cm long) for use in descending paper chromatography or a wide strip of paper (e.g., 25 cm wide) of variable height is used for ascending paper chromatography.

For descending liquid-paper chromatography, substances to be separated are applied as spots (e.g., 25 mm apart) along a horizontal pencil line placed down from the V-trough folded top of the paper. The V-trough folded paper is placed in a glass trough, held down by a glass rod, and when the tank has been equilibrated (vapor-saturated) with "running solvents" (mobile phase), the same solvent is added to the trough via a hole (later, stoppered shut) in the lid covering the chromatography tank. The lid is sealed onto the chamber with stopcock grease in order to make the chamber air-tight. After the mobile phase trails to the base of and off the paper sheet, the paper is hung to dry in a fume hood where it can then be sprayed with reagents (e.g., Ninhydrin reagent for amino acids) that give color to the separated compounds of interest in white or UV light. Some compounds of interest have their own distinctive colors, e.g., chlorophylls, and hence can be purified using this technique. In other cases, the dyes used to stain the location of the compound or protein causes irreversible covalent changes to the compound. In these cases, purification is not possible.

In ascending paper chromatography, the same basic set-up and principles apply with the exception that the mobile phase is placed at the bottom of the tank. Separation is achieved when the mobile phase travels up the paper via capillary action.

Another type of liquid-liquid partition chromatography is ***TLC*** which has several advantages over paper chromatography: (1) greater resolving power, (2) faster speed of separation, and (3) availability of a diverse array of adsorbents. The first two of these advantages are attributed to the fine particle size of the solid support adsorbent (less than 0.1 mm diameter particles) which allows more contact of this solid support with the compounds of interest as they travel up the plate. The adsorbents (e.g., silica gel, alumina, cellulose, and derivatives of cellulose) are available commercially on glass plates of various sizes. TLC plates are used in glass tanks, using ascending, or in some cases, descending chromatography. For the former, 1- to 10-µl samples of interest are spotted at 2- to 3-cm intervals across a line 15 to 20 mm from the base of the plate. The spots are allowed to dry. Then the plate is placed in the glass chromatography tank with the solvent previously placed in the bottom of the tank to a depth of 10 mm. Often it is necessary to equilibrate the vapor in the tank by placing filter paper around the sides of the tank. Next, a lid is sealed to the top of the tank with stopcock grease and the solvent is allowed to rise by capillary flow to the top of the plate. Once the mobile phase reaches the top of the plate, the plate is removed and allowed to dry. The spots are then developed with appropriate reagents for the types of compounds being separated and assessed. However, this procedure, as in paper chromatography, may result in sample destruction.

TLC can be run in two dimensions, using different solvent systems, as can paper chromatography, in order to allow for better separation of compounds. This procedure is very similar to 2-D electrophoresis. TLC is widely used to separate lipids, fructans, sugars, and hormones.

GLC is another type of partition chromatography where a high boiling point liquid is the stationary phase and an inert gas is the mobile phase. There is also an inert solid packing used in columns where these two phases are separated. Separation of compounds of interest is achieved due to differential solubility of the compounds in the mobile and stationary phases. Thus, as the carrier gas passes through the column, the compounds in the sample come off the column at different times (Rt's). A GLC apparatus basically consists of a tank of carrier gas (e.g., helium or nitrogen), an oven containing a coiled metal or glass chromatography column, a sample injection port, a detector (e.g., FID or thermal conductivity detector), and a recorder.

With FIDs, hydrogen gas is used to provide fuel for the flame. This is coupled to a flow of air to the detector to provide oxygen that allows the hydrogen to burn. A wire loop is positioned above the flame to detect compounds that pass from the column to the flame which, in turn, is connected to the recorder. FIDs are very sensitive to most organic compounds, but not to water, carbon monoxide, carbon dioxide, or the inert gases. Obviously, samples are destroyed when using this type of detector. Thus, GC is usually not used for purification of compounds. Thermal conductivity detectors, on the other hand, are less sensitive than an FID. However, they are nondestructive to the samples which allows one to completely recover a sample.

HPLC can also be grouped into this section on partition chromatography because it functions using the same principles previously discussed. It has two main advantages. First, it uses a pump to force the mobile phase through a given type of column at high pressure. The column can be made (or usually purchased) using any of the above-discussed solid absorbent phases. Hence, HPLC is commonly used to shorten the running times of any of the above types of chromatography which are usually time-restricted by gravity or capillary action. Second, because high pressure is used, much smaller adsorbent solid support particles can be used in the columns in conjunction with much smaller column volumes. Together, these two factors allow much better resolution of the compounds passing through the column due to the increased contact of the compound with the solid support adsorbent. *FPLC* works in the same manner as HPLC, but it makes use of specialized columns for use in protein purifications. The main disadvantage of HPLC or FPLC is the high cost of the specialized equipment. There is also the disadvantage of the adsorbent only being able to detect substituents on the surface of the compounds or proteins of interest, but this is a disadvantage in all forms of chromatography.

7.4.1.2.3   Gel Filtration or Permeation Chromatography

This type of chromatography is often called *gel filtration chromatography*. It involves the use of porous gel molecules of agarose, cross-linked dextran, or polymers of acrylamide, allowing the separation of compounds based on their molecular sizes/weights. One commercial series called Sephadex (Pharmacia Fine Chemicals, Inc.) is used for this purpose. These types of column packings must be hydrated before they are functional as a separation medium. The hydration process causes the pores in the Sephadex to swell to the appropriate size for the given Sephadex type. For example, G-10 Sephadex, during hydration, gains 1 ml of water per 1 g of dry gel; G-200 Sephadex gains 20 ml of water per 1 g of dry gel. Bio-gels from Bio-Rad laboratories consist of long polymers of acrylamide that are cross-linked to N,N'-ethylene-*bis*-acrylamide. These gels have a larger range of pore sizes than the Sephadex G series. Still another porous gel with an even wider pore size is agarose. It is made up of the neutral polysaccharide fraction from agar. Agarose and polyacrylamide are used to separate viruses, ribosomes, nucleic acids, and proteins. Sephadex is widely used in purification of proteins and in determining their molecular weights.

The general rationale for separation is as follows: (1) a gel having an appropriate pore size is chosen in relation to the size of the molecule of interest; (2) samples are added to the top of the gel column and are washed through using an appropriate mobile phase that is based on the solubility of the molecule of interest; (3) molecules that are too big to fit in the pores of the solid support will travel around the gel particles and hence elute first from the column; (4) molecules that fit in the pores will elute at different times according to their mobility through the gel pores.

Very fast, efficient separation of macromolecules is now possible by a technique termed *capillary zone electrophoresis*.[9] This is not a form of chromatography, but deserves recognition as an extremely powerful technique for separating compounds of interest via electrophoresis. Basically, capillary electrophoresis (Figure 7.13A and B) utilizes small-bore open capillary tubes (e.g., 200 μm internal diameter) in a system equipped with a grounded high-voltage power supply, solvent reservoir in a Plexiglas® box connected to the capillary tube, a detector, a solvent reservoir after the detector, and a power supply for current flow to ground. Once a very small sample (e.g., nanogram quantity) is loaded into the capillary tube at one end, negatively charged species, such as the negatively charged side groups of proteins, are separated by the same mechanism as ordinary electrophoresis. The major advantage here is that the capillary tube has large surface-to-volume ratio allowing rapid dissipation of the heat produced by the electric current. Consequently, much higher voltages can be used in capillary zone electrophoresis than can be used in normal electrophoresis. High voltages in normal electrophoresis tend to cause heat convection within the gel. This results in distortions and blurring of the separation bands. High voltages in capillary zone electrophoresis allow much better resolution of related species of compounds as well as much faster running times. In addition, the capillary tube's inner surface is negatively charged and, thus, attracts positively charged species of molecules. As the buffer travels through the capillary tube via the electrical current, there is an electro-osmotic flow produced which carries these positively charged species of molecules in the same direction as the negatively charged species. Hence, both negatively and positively charged species can be

FIGURE 7.13A Waters capillary electrophoresis apparatus in Dr. Akira Okubo's laboratory at the University of Tokyo.

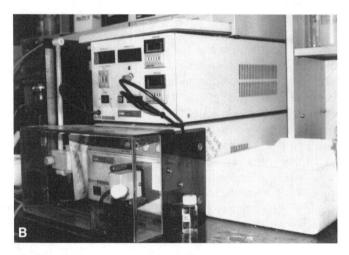

FIGURE 7.13B JASCO capillary electrophoresis apparatus in Dr. Akira Okubo's laboratory at the University of Tokyo.

separated and analyzed at the same time. Capillary zone electrophoresis utilizes several types of detectors including spectrophotometers, mass spectrometers (MS), electrochemical detectors, and radiometric detectors instead of the cumbersome stains used in ordinary electrophoresis.

7.4.1.3 Use of Mass Spectrometry to Identify Biologically Important Molecules

In order to analyze the compound(s) or protein(s) of interest, one must be able to unambiguously identify their positions or retention times (Rts) on or through the adsorbent phase used in any given type of chromatography. For example, HPLC chart recordings produced by a chart recorder monitoring a spectrophotometer during a sample run will produce a number of peaks whose identities are not necessarily known. In order to determine which peak is the compound or protein of interest, known standards are run on HPLC prior to running the collected sample. This will determine the Rt of the compound or protein of interest. The Rt is the time, in minutes, on the chart recording where a given peak of interest occurs. Unknown compounds whose Rt's are similar to those of the standards can be tentatively identified using multiple forms of chromatography, *but* MS (Figure 7.14A and B) or nuclear magnetic resonance (NMR) analysis of the collected unknown peak must be performed in order to unambiguously identify the compound of interest. A cogent application is the characterization of taxol by MS by McClure et al. (1992)[10] and NMR by Falzone et al. (1992).[11]

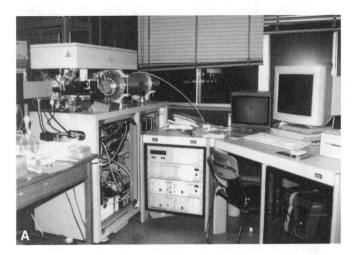

FIGURE 7.14A   Bruker TOF-mass spectrometer (MS) in Dr. Akira Okubo's laboratory at the University of Tokyo.

The amount of the compound of interest under a given peak on a chromatogram is determined by measuring the area under the peak for this compound. Several methods can be used.

1. Measure one-half the peak height times its width at the base and compare with the areas of standard peaks analyzed in the same manner to produce a standard curve (see Figure 7.11). The advantage of this technique is that it is very easy. However, only very crude estimations of the area, and hence concentration, are obtainable using this method.
2. Cut out the curve on the chart paper and weigh it in comparison to the weight of known standard peaks. This procedure is slightly more accurate than method 1 — but only if the paper is of uniform density.
3. Integrate the area under the peak using an electronic integrator/scanner (see Figure 7.10A for a view of the Shimadzu C-R4A Chromatopac

FIGURE 7.14B   Hitachi liquid chromatography (LC)-mass spectrometer (MS) in Dr. Akira Okubo's laboratory at the University of Tokyo.

electronic integrator used in our lab). This is by far the most accurate of these procedures because computer-controlled integrators can implement very complicated mathematical algorithms for integration, giving very accurate results. This procedure also has the advantage of requiring no mathematical manipulations by the researcher.

It is very difficult to say exactly what precise compound is contained in a given peak from a purified extract that shows up from HPLC. MS is very useful for clarifying such ambiguities. MS functions by bombarding a compound with high-energy electrons (or other particles), causing the loss of an electron from each molecule, yielding *molecular ions*. If the energy of the electron beam is high enough, the molecular ions will have enough excess vibrational and electronic energy to break the molecules into various positive and negative ion fragments. These ion fragments are then passed through a very strong magnetic field which, depending on the charge of the fragment, deflects the fragments at an angle into a detector. This results in a spectrum of the compound's ion fragments. Every compound has its own unique mass spectrum. A good example is the mass spectrum for taxol which has been determined by fast atom bombardment (FAB) MS.[10] In this study, three ion series are observed: (1) the M-series which are characteristic of the intact taxol molecule; (2) the T-series whose ion fragments are derived from the taxane ring, and (3) the S-series that represents the C-13 side chain of taxol (see Figure 7.15 derived from McClure et al. [1992]).[10]

GC can be combined with MS to first separate the compounds of interest in an extract according to their boiling points. Then, as each compound comes through the MS, it is broken up by the particle beam and produces a new mass spectrum. The mass spectrum for each compound is then compared to the known library of mass spectra that can be easily accessed from a computer to tell you exactly what your compound is with no ambiguity.

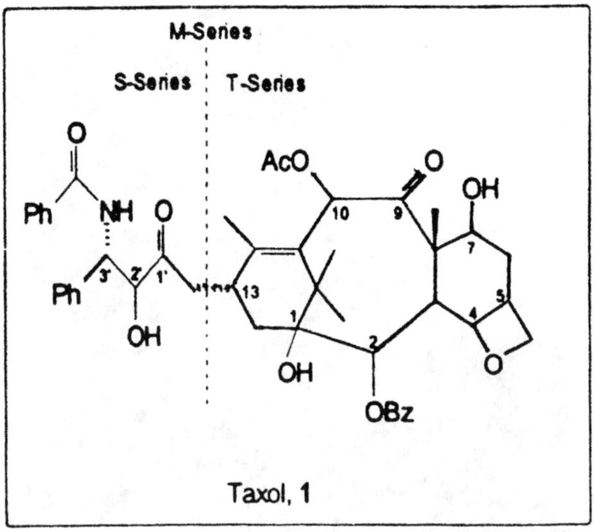

FIGURE 7.15 Illustration of taxol molecule and the three-ion series observed following fast atom bombardment (FAB) mass spectrometry of taxol. (From McClure, T. D., Schram, K. H., and Reiner, M. L. J., *J. Am. Mass Spectrom.*, 3, 672, 1992. With permission.)

7.4.2 OTHER METHODS

In addition to chromatography and MS, chemists and molecular biologists employ other ways to characterize and identify plant metabolites (Figures 7.16 and 7.17).

- *Nuclear magnetic resonance (NMR)* — This is a spectroscopic technique used to characterize the nuclear properties of elements, proton chemical shifts in substituents on molecules or protons associated with double bonds in cyclic and noncyclic hydrocarbons, and chemical shifts of isotopes (e.g., carbon, nitrogen, phosphorous, boron, silicon, and fluorine).
- *Infrared (IR) spectroscopy* — This is also a spectroscopic technique that involves measurement of the absorption spectrum of a given molecule in the infrared portion (wavelengths longer than those of visible light, occurring above the red end of the light spectrum) of the electromagnetic spectrum. Every molecule has its own characteristic IR spectrum. It refers to the absorption frequencies of single bonds to hydrogen, triple bonds, cumulated double bonds, carbonyl bonds, aromatic bands, and miscellaneous bands for such molecules.[12]
- *Melting point* — Every compound has its own characteristic melting point (the temperature, in degrees Celsius, at which a solid melts). A good place to find this kind of information is in *The Handbook of Chemistry and Physics*.[13]
- *Ultraviolet (UV) spectrum* — This is still another spectroscopic technique that refers to the absorption spectrum for a given molecule in the UV

FIGURE 7.16   Dr. Soo Chul Chang performing an electroelution of a protein for subsequent amino analysis in Peter Kaufman's laboratory at the University of Michigan.

portion of the electromagnetic spectrum (wavelengths shorter than those in the visible light portion of the spectrum and longer than those for X-rays). Every molecule has its own characteristic UV spectrum. In practical terms, it is a good idea to have an HPLC fitted with both visible and UV light sources because in HPLC analyses, we often measure the absorbance of a given compound at wavelengths that occur either in the visible or in the UV.

- *Specialized MS devices* — These include tandem MS, quadripole MS, and electrospray MS devices.[14] An electrospray MS is very useful for doing mass spectra for large molecules like azadirachtin from the neem tree (*Azadirachta indica*) and taxol from yews (*Taxus* spp.).
- *SDS/PAGE* — This refers to sodium dodecyl sulfate polyacrylamide gel electrophoresis. It is a method that allows one to separate proteins on a gel on the basis of molecular size [in 1-D (one-dimensional) gels] and

FIGURE 7.17   Hitachi amino-acid analyzer in laboratory of Dr. Akira Okubo at the University of Tokyo.

their isoelectric points [pH values where they have no net charge in 2-D (two-dimensional gels)].

- *Protein sequencing* — This technique allows one to determine the amino acid sequence of a given protein from its amino terminal end to its carboxy terminal end. Such information is helpful in determining the deduced nucleotide sequence of the specific DNA that makes this protein via transcription and translation. It also allows one to identify particular DNA binding domains, active sites (in the case of enzymes), and targeting sequences that get proteins to their ultimate destination (e.g., chloroplast, mitochondrion, cell wall, vacuole) in the cell.
- *Cloning genes by expression in Escherichia coli (the common colon bacterium)* — This technique is used by molecular biologists to amplify the amount of a given gene (DNA), especially when it and its gene product are found in low abundance in a cell, tissue, or organ. This then allows one to do biochemistry on the protein of interest which may be, for example, a rate-limiting enzyme in a metabolic pathway that leads to the synthesis of a given metabolite of interest. In order to clone the gene in *E. coli*, the DNA of interest must be transferred to *E. coli* cells so that they, in effect, become genetically transformed. Methods for transforming *E. coli* cells are discussed in detail in Wu et al. 1997.[15]

7.5   BIOASSAYS FOR ACTIVITY

7.5.1   PRESCREENING BIOASSAYS WITH YEAST AND ITS MUTANTS

Dr. Maureen McKenzie, Biochemist and Chief Executive Officer of Naniquah Corporation (see Chapter 9, Section 9.6 on Botanical Prospecting), has developed a prescreen bioassay system with yeast (*Saccharomyces cereviseae*) and yeast mutants to test HPLC-derived peaks from plant extracts for activity or lack of activity. Her bioassay system tests the effects of different dosages of an HPLC-derived compound on the growth rate of yeast cells on defined media, their morphology (cell shape, cell size, budding characteristics) and biochemistry [rates of reaction for key enzymes linked to normal and abnormal (disease-related) growth metabolism]. Such a system holds great promise for identifing particular molecules worth testing in robotic screens using enzyme-linked bioassay systems related to particular disease syndromes (see below).

7.5.2 ROBOTIC SCREENS USING ENZYME-LINKED BIOASSAY SYSTEMS

All major pharmaceutical companies employ highly specific bioassay systems to test partially or completely purified compounds for anticancer, antiviral (e.g., anti-AIDS virus), anti-heart disease, anti-Alzheimer's disease, and antineurological disease activities. Most of these bioassay systems involve enzyme-linked reaction systems that are specific to the particular disease syndrome being tested. The assays utilize 80-well plates, dose-response assays for the compound of interest, and robotics to carry out all chemical additions (enzymes, substrates, and different concentrations of the compound being tested). The system at Parke-Davis/Warner-Lambert Pharmaceutical Research Laboratories in Ann Arbor, MI utilizes three robots to do these tasks. It operates 24 h/d, 7 d/week and requires very little human input except to set up the initial programs and to collect the final data. If a compound proves to be active, and is not a "false positive", it is considered to be a positive "hit". Positive "hit" compounds may then be chemically modified to make it more effective at target sites. The company may also gain patent rights. The ultimate compound is then subjected to clinical testing, using animal models (e.g., rats, mice, chickens, monkeys). Considerable attention is currently being given to humane treatment of animals used in such testing, but animal rights activists are not convinced that this is always the case! Strict protocols for conducting these clinical tests must be performed according to FDA (Food and Drug Administration) guidelines. This may take up to 10 years before approval of the candidate drug for human use is granted by the FDA.

7.5.3 COMBINATORIAL CHEMISTRY

Combinatorial chemistry is a technique that allows vast quantities of compounds with high molecular diversity to be synthesized, screened, and analyzed in a short time, ultimately reducing the time to market. Fully automated systems include HPLC systems, LC/MS detectors, variable wavelength detectors, software that allows one to identify each peak, 3-D X-ray structure analysis apparatus, chemical synthesis robots, robotically operated biological assay systems, and bio-informatics and homology modeling systems that interface with virtual libraries. Following biological assays, preclinical drugs are the ultimate end-products of such a system. Combinatorial chemistry systems are currently much in vogue with pharmaceutical companies. Their goal initially was to use such chemistry to generate novel molecules, but now it is used as a strategic method for optimizing lead compounds by creating analogs based on a core structure that binds to the intended target.[16]

7.6   CONCLUSIONS

Bioseparation of natural products from plants can be relatively simple and inexpensive, as for example, in preparing natural/whole foods, herbal medicines and herbal teas, and natural plant dyes. Or it can be quite complicated and expensive, as for example, drug discovery from natural products, or isolation and identification of DNA, RNA, proteins, or metabolites in a given metabolic pathway. In this chapter, we have focused on the major methods employed by chemists and biologists to extract and isolate pure compounds from plants and to characterize their chemical structure. The methods involved require sophisticated and expensive apparatus for this kind of analytical work. But, there is great satisfaction to be derived in doing this kind of task because it offers one the possibility of learning how the metabolite functions in the plant, how it acts in curing diseases in humans and other animals, and whether or not such metabolites are good for us in our diets, in our craft industries, and in the fields of plant biotechnology, plant molecular biology, and plant bioengineering.

REFERENCES

1. Ody, P., Herbal remedies, in *The Complete Medicinal Herbal*, Dorling Kindersley Limited, London, 1993, 116–127.
2. Kaufman, P. B., Wu, W., Kim, D., and Cseke, L., *Handbook of Molecular and Cellular Methods in Biology and Medicine*, CRC Press, Boca Raton, FL, 1995.
3. Bruno, T. J., *Chromatography and Electrophoretic Methods*, Prentice-Hall, Englewood Cliffs, NJ, 1991.
4. Heftmann, E., Fundamentals and Techniques, in *Chromatography*, 5th ed., Part A, Journal of Chromatography Library, Vol. 51A, Elsevier Publishers, New York, 1992.
5. Heftmann, E., Applications in *Chromatography*, 5th ed., Part B, Journal of Chromatography Library, Vol. 51B, Elsevier Publishers, New York, 1992.
6. Poole, C. F. and Poole, S. K., *Chromatography Today*, Elsevier Publishers, New York, 1991.
7. Porath, J., High performance immobilized-metal-ion affinity chromatography of peptides and proteins, *J. Chromatog.*, 443, 3–11, 1988.
7a. Bradford, A. A., A rapid and sensitive method for the quantitation of microgram quantities of protein utilizing the principle of protein-dye binding, *Anal. Biochem.*, 72, 248–254, 1976.
7b. Lowry, O. H., Rosebrough, N. J., Farr, A. L., and Randall, A. J., Protein measurement with the folin phenol reagent, *J. Biochem.*, 193, 265–275, 1951.
8. Pharmacia Fine Chemicals, *Affinity Chromatography Principles and Methods*, Pharmacia Fine Chemicals, AB, Uppsala, Sweden, 1979.
9. Ewing, A. G., Wallingford, R. A., and Olefirowicz, T. M., Capillary electrophoresis, *Anal. Chem.*, 61, 271–279, 1989.

10. McClure, T. D., Schram, K. H., and Reiner, M. L. J., The mass spectrometry of taxol, *J. Am. Soc. Mass Spect.*, 3, 672–679, 1992.

11. Falzone, C. J., Benesi, A. J., and Lacompte, J. T. J., Characterization of taxol in methylene chloride by NMR spectroscopy, *Tetrahed. Lett.*, 33, 1169–1172, 1992.

12. Dean, J. A., *Lange's Handbook of Chemistry*, 14th ed., McGraw Hill, New York, 1992.

13. Weast, R. C., Ed., *Handbook of Chemistry and Physics*, 51st ed., CRC Press, Boca Raton, FL, 1971.

14. Stevenson, R., New products and technology for bioseparations at Pittcon® '97', *Am. Biotech. Lab.*, 15, 27–35, 1997.

15. Wu, W., Welsh, M. J., Kaufman, P. B., and Zhang, H. H., *Methods in Gene Biotechnology*, CRC Press, Boca Raton, FL, 1997, 1–406.

16. Glaser, V., Combinatorial chemistry evolves into strategic technology for optimizing lead compounds, *Genet. Eng. News,* 17, 1–42, 1997.

8 Case Studies

Peter B. Kaufman, James E. Hoyt, Stephanie Bergman, Barbara J. Madsen, Matthew Perry, Sara Warber, and Keewaydinoquay Peschel

CONTENTS

0-8493-3134-X/99/$0.00+$.50
© 1999 by CRC Press LLC

8.1   INTRODUCTION

In this chapter, we shall discuss classical examples of plant metabolites that are used for medicine, natural plant dyeing, fragrances and perfumes, and insect repellents. For the most part, these examples have not been discussed in any great detail in other chapters of the book. Nonetheless, they are important to us because we use them in our everyday lives. Some, such as natural plant dyes, are enjoying a resurgence of use because of their low cost and simplicity to extract for purposes of dyeing wool or other animal- or insect (silkworm)-derived fibers. The fragrance herbs we will mention have aesthetic and utilitarian uses in making herbal wreaths and dried arrangements as well as sachets to place with clothing to deter moths or to counteract undesirable smells. Perfumes derived from the flowers of plants are naturally pleasing to humans. Major industries are devoted to extracting perfumes from the flowers of plants. We also learned in Chapter 2 that they are important in luring different insects or mammals to flowers to effect pollination. And last but not least, plants provide us with many different kinds of natural pesticides that repel or kill insect predators that attack our garden crops and commercially grown crops.

8.2   CASE STUDIES WITH MEDICINAL PLANTS

8.2.1   AZADIRACHTIN FROM THE NEEM TREE (*AZADIRACHTA INDICA*)*

Neem, *Azadiractica indica* (Juss), in the family Meliaceae, or mahogany family, has a long history of use in human civilization. As with many plant species long used by humanity, its exact origins are unknown. Some have suggested southern India and Burma as the region where neem originated, while others have suggested locations ranging from Indonesia to Iran.[1] Neem grows well in dry tropical and subtropical regions, and has been introduced to hospitable regions throughout the world including Africa, Australia, the South Pacific, and the Americas. Figure 8.1A illustrates a 1-year seedling of the neem tree and a leaf from a 3-year-old tree is shown in Figure 8.1B.

Neem produces a wide variety of compounds, including several groups of tetranortriterpenoids, the azadirones, gedunins, amoorastatins, vepinins, vilasinins, c-seco meliacins, nimbolinins, salaninins, and azadirachtins. Over 100 terpenoid-based compounds have been identified. It is these compounds that provide a truly astounding variety of uses for medicine, agriculture, domestic products, and food preparation. The numerous and complex nature of these compounds has proven to be a formidable challenge to modern researchers attempting to isolate and characterize their chemistry, biosynthesis, and activity. One compound, azadirachtin or aza (Figure 8.2), has attracted special interest due to its twin effects as an insect antifeedant and as an insect growth inhibitor. Though other compounds have been found to have similar effects, aza remains the most intensively investigated component of neem. Investigations using the right and left parts of the aza molecule indicate that the antifeedant activity occurs

---

* This section was prepared by James E. Hoyt.

FIGURE 8.1A James E. Hoyt in greenhouse holding a pot containing a 1-year-old neem tree, *Azadirachta indica*. (Photo by David Bay.)

with the right part; however, the ester group attached to the C-4 carbon of the left part seems to be essential. The right part seems to be the moiety that is important for the growth inhibition action of aza. A complete understanding of the relationship between the molecule's structure and its effect on insects remains to be discovered, while the mode of action of aza and the other constituents of neem on human physiology is generally unknown at this time.

Neem has been used in medicine for thousands of years. References to the medicinal use of neem are found in the earliest Sanskrit writings and its use is described in Ayurvedic, Unani, and homeopathic medicine[1] All parts of the plant has been used in various preparations. Leaves, seeds, bark, and roots are made into teas and washes for various conditions. Seeds are also used in poultices, while twigs are used as toothbrushes. Gum from older trees is used to treat skin disorders, while flowers are used to treat intestinal worms, reduce phlegm, and suppress bile. Western medicine has confirmed the efficacy of many of these traditional uses. It has found

FIGURE 8.1B   A single neem tree leaf, about one-half its natural size. (Photo by David Bay.)

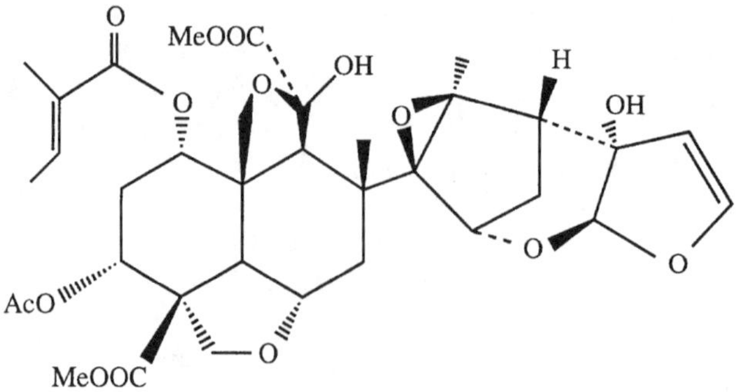

FIGURE 8.2   Structural formula of azadirachtin (aza), $C_{35}H_{44}O_{16}$.

components of neem to be effective against malaria and to be useful as an antibiotic, anti-inflammatory, antidiabetic, and antifertility agent.[2]

In agriculture, neem has been used as an insecticide and animal feed. Seeds are soaked in water which is then applied to crops as an insecticide. Crushed seed that has had its azadirachtin extracted, called "cake", is used as an animal feed. Leaves and bark are also used as an insecticidal mulch. The trees themselves are used for shade; this is no small benefit in dry tropical climates. In the home, neem leaves and twigs are used in beds, chests, and cupboards to repel insects. Neem oil is used in food storage as well as in making soap. Food containers may be treated with leaf preparations or sometimes fumigated with the smoke of burning leaves. Timber from neem wood is used in construction, where it has proven to be resistant to termites and other wood-damaging insects. Finally, it is used as firewood; its vigorous ability to resprout permits regular harvesting for this purpose.

8.2.2 MEDICINAL USES OF BALSAM FIR (*ABIES BALSAMEA*) AND ARBORVITAE/EASTERN WHITE CEDAR (*THUJA OCCIDENTALIS*)*

Presently, there are hundreds of species of plants being used for medicinal purposes. Many of these plants (or more often compounds derived from these plants) are used in Western medical practices, such as codeine and morphine from *Papaver somniferum*, but the majority are still used only in herbal healing and other forms of non-Western medicine. Yet, in the last decade there has been a resurgence in the use of medicinal plants as pharmaceutical companies search for the compounds that will be used to cure such major diseases as cancer and HIV/AIDS.

There are references to the use of medicinal plants almost as far back as recorded history. Almost every civilization has had some sort of medicine man or woman who was responsible for knowing which plants would cure which diseases. In the present day U.S. the only medicine men or women that exist are those on Indian reservations or those few who have had extensive training in herbal healing. These people treat only a minuscule portion of our population, while most people still believe that the only method of treatment for an illness comes in a pill of synthesized compounds.

We have chosen two medicinally valuable plants to discuss, the balsam fir, *A. balsamea*, and arborvitae, *T. occidentalis*. These plants were widely used by Native Americans, early settlers, and doctors into the late 1800s. Although there are no references to their uses in the present day, perhaps in the near future they will be rediscovered for their medicinal value.

8.2.2.1 Balsam Fir (*Abies balsamea*)

The balsam fir (*A. balsamea*) is in the division Coniferophyta and the family Pinaceae. The fir is also an evergreen, which signifies that its leaves are lost and replaced but not all at the same time.

---

* This section was prepared by Stephanie Bergman.

The balsam fir grows in low swampy areas and moist woods near the timberline in cooler climates. The best climate for *A. balsamea* is where the mean annual temperature does not exceed 4.4°C (40°F) and the average summer temperature is not more than 21.1°C (70°F).[3] In forests with 76.2 cm. (30 in.) mean annual rainfall, the balsam fir comprises up to 30% of the total stand, as opposed to areas with rainfall from 25.4 to 63.5 cm. (10 to 25 in.), where the species comprises less than 10% of the total forest stand.[3]

A. balsamea can be found throughout the northeastern U.S. as well as southeastern Canada. The tree has been found in the New England states, Pennsylvania, and from New York to Virginia, as well as in Minnesota, Wisconsin, and northern Michigan.[4]

The balsam fir is a slender, tapering tree that grows from 12.2 to 18.3 m (40 to 60 ft) high at maturity. Its branches are arranged in whorls of 4 to 5 with scattered branches in between. These branches diminish in length in proportion to their height from the ground, forming a pyramidal head, or crown, at the top of the tree.[5] The branches of the fir are of different sizes and grow in different directions depending on their location on the tree. The apical branches are short and ascending, while those in the midcrown are longer and horizontal. Those branches toward the base of the tree are long and pendulous.[3]

The leaves of the *A. balsamea* are needle-shaped and 1.27 to 1.90 cm. ($^1/_2$ to $^3/_4$ in.) long. They are narrow, flat, and rigid, with bright green coloring on the upper surface and silvery coloring on the lower surface due to the high prevalence of stomata there. The leaves are mostly sessile on the horizontal branches and spread in two directions so as to seem double-ranked.[5]

The bark of the balsam fir is smooth, thin, and firm. The bark, which is brownish in color, is seldom more than 1.78 cm. (0.7 in.) thick on any portion of the tree.[3] Small blisters containing resin terpenes can be found along the stem.

8.2.2.2   Arborvitae or Eastern White Cedar (*Thuja occidentalis*)

The arborvitae or eastern white cedar, like the balsam fir, is in the division Coniferophyta, yet is in the family Cupressaceae. *T. occidentalis* also is an evergreen.

The arborvitae grows in deep, cold swamps where it often forms dense stands. The arborvitae, often associated with the tamarack, can also be found on wet, rocky banks. *T. occidentalis* can be found in southeastern Canada and in the northeastern U.S. as far south as North Carolina and as far west as Illinois.[6]

The arborvitae has a conical form that grows up to 20 m (ca. 60 ft) at maturity. The tree has a rapidly tapering trunk with short horizontal branches ascending at the end. These branches form a narrow, pyramidal, compact head at the apex of the tree.[6] Although the size of the branches varies from apex to base, with the larger branches at the base and the smaller branches at the apex, the direction of the branches does not vary along the tree.

The leaves of the arborvitae are scale-like needles usually less than 0.31 cm. ($^1/_8$ in.) long. The whorled arrangement of the leaves, with alternate lateral and facial scales, create flat sprays.[7] These leaves also have a small flattened gland which contains a fragrant turpentine. The coloring on the upper surface of the leaves is of a bright green, while the lower surface is a yellowish-green.[8]

The bark of the *T. occidentalis* is thin and scaly. It has a reddish-brown coloring and is shallowly fissured into narrow connecting ridges covered with elongated scales.[7,8]

8.2.2.3   Medicinal Uses of *Abies balsamea*

The balsam fir has been used for medicinal purposes by many Native American tribes. Different parts of the tree have been used for the treatment of a great variety of diseases and ailments. The parts most frequently used are the gum (liquid balsam or resin from the bark), roots, bark, and leaves.

Many of the Native American tribes used the gum of the tree in a similar manner. The Forest Potawatomi, Menominee, and Ojibwa all applied the gum as salve to cuts and sores. Another common use of the gum of *A. balsamea* is for relieving cold symptoms or other pulmonary troubles. The Forest Potawatomi swallowed the gum fresh, while the Ojibwas applied the gum externally to relieve these symptoms. The Chippewas used the gum as an analgesic. The gum was melted on a warm stone and the fumes inhaled as a relief for headaches. The Ojibwa also used the gum, in unknown methods, for relief of chest soreness as well as a treatment for gonorrhea.[9,10]

The bark of the balsam fir was also used by many of the Native American tribes for medicinal purposes. The Menominee used a decoction (a solution prepared by boiling the inner bark in water and straining the resulting solution) to relieve chest pains.[11] This tribe also used the inner bark to prepare a poultice, as well as using it as a seasoner for medicines. The Forest Potawatomi used an infusion (the bark was steeped to extract the essence) of the inner bark for consumption (especially tuberculosis) and other internal affections (gastrointestinal and pulmonary ailments). The Ojibwa used a decoction of the inner bark as a diaphoretic (to induce sweating).[9,10]

Other parts of *A. balsamea* also were used by Native Americans in medical treatments. The roots of the balsam fir were used by the Chippewa as an antirheumatic. A decoction of the roots was sprinkled on hot stones and the herbal steam was useful in easing the pain associated with rheumatism, especially in the knees. The leaves of the tree were used by the Ojibwa in a variety of ways. One way in which they were used, the method unknown, is as a stimulant. The smoke of the leaves was inhaled as remedy for cold symptoms, while a compound containing the leaves was used as an antiseptic wash.[9,10]

People other than the Native Americans have used the *A. balsamea* for medicinal purposes. Early New England settlers (who most likely learned to use the tree from the Native Americans) used a variety of plant parts to treat various diseases and ailments. The buds of *A. balsamea* were used in epithemes to treat corns and warts. The sap was used in an unknown method as a treatment for tumors and prepared as a cataplasm (a viscous preparation intended to be warmed and applied to the body surface) in relieving the pain associated with cancer.[11,12] The settlers also made pillows from the dried leaves of the fir, whose pleasant aroma gave relief to those suffering from hay fever and colds.[13]

A medical book printed by Parke-Davis/Warner-Lambert Pharmaceutical Company in 1890 shows how *A. balsamea* was used in the medical practices of that era. The bark, prepared in capsules of 10 minims, was used as a stimulant, diuretic, and anthelmintic (destroys parasitic worms). In larger doses, the bark was used as a laxative. The bark was also recommended in treating gonorrhea, gleet, and chronic inflammation of the bladder, due to its stimulating action on the mucous tissues. The oleoresin, prepared in a fluid extract, with a dosage of 0.5 to 1 fluidrachm (1 fluidrachm is about $1/8$ fl. oz.), was used in the treatment of chronic bronchitis.[11,14]

8.2.2.4 Medicinal Uses of *Thuja occidentalis*

The arborvitae also was used in many of the medicinal practices of the Native Americans. Like the balsam fir, many parts of the tree were used by the various tribes to remedy the ailments of the tribe members. The leaves, twigs, and inner bark were the parts of the tree used most often.

The leaves were used by many of the Native American tribes in a variety of medicinal ways. The Chippewa used a compound containing the leaves as cough syrup. The Ojibwa used a decoction of the leaves as a cold remedy, while the Iroquois used the steam from a decoction of the leaves for the same purpose. The leaves of the arborvitae were used by the Menominee in a smudge (a fire made to produce dense smoke) to revive lost consciousness. The Ojibwa used an infusion of the leaves as an analgesic for headaches and a decoction of the leaves as a blood purifier. They also used the leaves in a preparation that worked as a diaphoretic in a sweatbath. Once the leaves were fried, powdered, and compressed, the Menominee also used them in a poultice to reduce swelling. The leaves of *T. occidentalis* were boiled in lard to yield a leaf oil that made an excellent salve for the Forest Potawatomi. This tribe also used the leaves in many compounds for a variety of illnesses.[9,10,13]

Other parts of the arborvitae were used by the Native Americans as well. The Menominee used an infusion of the dried inner bark that was taken during a cold to treat suppressed menses. They also used the bark as a seasoner for enhancing medicines. The Chippewa burned its twigs as a disinfectant to fumigate a house for smallpox.[9,10]

The Iroquois used a variety of parts of the arborvitae for medicinal purposes. The sprigs and leaves of *T. occidentalis* and *A. balsamea* were prepared to heal cuts, bruises, sprains, and sores. The sprigs and leaves were pounded and boiled, and the affected sores were washed with a solution as hot as one can stand. The leaves were also used as a poultice to cover the affected areas. The leaves and roots were combined to treat weakness in the hips due to an untreated broken coccyx. One handful of the leaves and two handfuls of the roots were boiled for 20 min in a gallon of water. The water was then cooled to the temperature of bath water, and dashed on the hips. Another preparation the Iroquois used was boiling the tips of the branches in a large tub of water. Patients with rheumatism put their feet in the water, as hot as they could stand, and were covered with a blanket to help relieve their pain.[15]

T. *occidentalis* was also used by people other than the Native Americans. The earliest non-Native American reference to the use of the arborvitae is in the early 1800s. The most common use for the arborvitae in this era was for the treatment of warts and venereal excrescences. These were treated by a tincture (alcohol-based solution) of the leaves and twigs.[11] A leaf tincture was also used in treating carcinomatous afflictions of the bladder, while a twig decoction was used in treating cancer.[12] A decoction of the leaves and twigs was used in treating intermittent fevers, coughs, scurvy, and rheumatism. A saturated tincture was useful as an emmenagogue (inducing menstrual discharge) when 1 teaspoonful was given three times a day.[16]

8.2.2.5 Chemical Properties of *Abies balsamea*

The oleoresin of the balsam fir, also known as Canada turpentine, is the plant constituent that was used most often by the Native Americans. Oleoresin is a solution of resins in volatile oils, which when steam distillated yield crude turpentine oil and resin.[17] When fresh, Canada turpentine is a pale-yellow liquid with a slight, greenish fluorescence, and has a viscous, honey-like consistency. When exposed to air the Canada turpentine becomes more viscous and forms a glass-like varnish.[18] This property is what makes the oleoresin such a good salve. It allows for the oleoresin to cover the sorel(s) and prohibits the entrance of water, dirt, and other agents that could cause infections. The highly viscous property of the oleoresin allows for conformation to the affected area and permits movement there without the gum pulling at or falling off the sore.[20]

Canada turpentine is composed of 23 to 24% volatile oils, 48 to 50% α- and β-canadinolic acids, 13% canadinic acid, 0.3% canadolic acid, and 11 to 12% resene. [18] Volatile oils are generally the substances that give taste and odor and are made up of various terpenes.[17] Terpenes are molecules with the general formula $(C_5H_8)_n$. These molecules are usually built from C_5 units called isopentenylpyrophosphates. Terpenes are classified as mono-, tri-, tetra-, and polyterpenes.

8.2.2.6 Chemical Properties of *Thuja occidentalis*

The arborvitae has a balsamic, terebinthinate odor and a pungently aromatic, camphoraceous and bitter taste. The odor and taste are due to the number of volatile oils.[5] The arborvitae also contains a number of essential oils, as well as sugar, wax, and resin. *T. occidentalis* also contains the bitter principle, pinipicrin, the tannic acid, pinitannuc acid and the citron-yellow, peculiar crystalline compound called thujin.[19]

A table of terpenes found in arborvitae could not be located, so instead two of the monoterpenes found in its volatile oils will be mentioned. First is fenchone (Figure 8.3) which is a cyclic monoterpene ketone. The L-form of this molecule is the one that is found in the oil of the arborvitae. Thujone is the second terpene. This is also a cyclic monoterpene ketone. Thujone is a colorless oil that produces a camphor-like smell.[17]

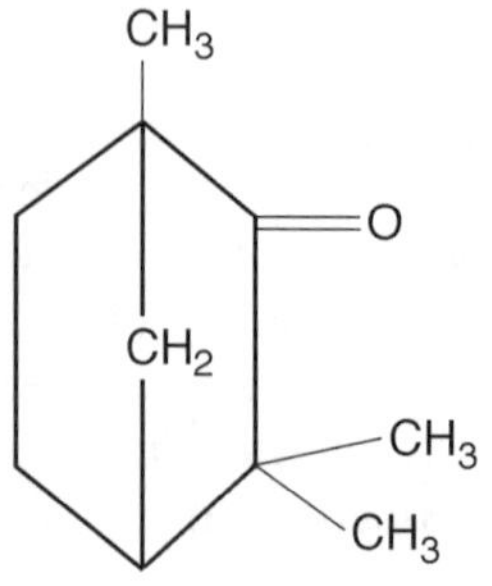

Fenchone

FIGURE 8.3 Fenchone.

8.2.2.7 Summary

This section has examined the medicinal uses of *A. balsamea* and *T. occidentalis* by Native Americans and early New England settlers, as well as Western doctors into the late 1800s. Due to the lack of references to herbal healing in the 1900s, it seems as if it became obsolete with the turn-of-the-century in the U.S. Cures for diseases in this era have mostly come from synthesized compounds, but it is very likely that these compounds are present in nature as well. The natural compound, as opposed to the synthesized one, would probably be healthier, more readily available, and cheaper. Hopefully within the next century there will be an even greater resurgence in the use of medicinal plants than there has been in the past few decades. We can only hope that there will still be herbal healers around to teach their secrets of natural healing to those in the future who realize the value of this type of healing.

8.2.3 KEEWAYDINOQUAY PESCHEL

Essay on Keewaydinoquay, Botanist, Medicine Woman, and Teacher*

Introduction

Keewaydinoquay (Woman of the Northwest Wind) is a Mashkikiquay (herbalist) of the Crane Clan, Miniss Kitigan Band, Anishinaabeg of the Three Fires Council (triumvirate sovereign nation of the Odawa, Ojibwa, and Potawatomi). A dedicated botanist, medicinalist, conservationist, and teacher, Kee's accomplishments range far in subject, but all share an origin in Native American philospophy.

Within the totality of universal harmony, all matter (and therefore all life) is originally created in balance and interconnection.

* University of Michigan former undergraduate students, Aaron Thompson and Nate McDowell, helped Sara Warber, M.D. and Keewaydinoquay prepare this essay.

The combination of these traditional beliefs and her endless intellectual curiosity has resulted in an extremely knowledgeable and powerful individual.

Background

Grandmother Kee (Figure 8.4), as nearly everyone calls her, grew up in the Leelanau and Mason counties of the northern lower peninsula of Michigan. It was in this rural pocket of the Old Ways where Kee found her true love: plants. Her mother, Minosoah-nikwe, and father, Wauboshtigwan, found great delight in plants, an interest which flowed naturally to their daughter. "Good-Cooking-Woman" had the largest floral and vegetable gardens as well as the most favored cuisine in their area, and her father, "Silver-Head", was a guardian and spokesman for the trees. He made his daughter aware of the interdependence of humans and trees and the virtues of this relationship.

FIGURE 8.4 Ojibwa botanist, medicine woman, and teacher, Keewaydinoquay Peschel, holding a plant of Polygala senega. (Photo courtesy of Keewaydinoquay Peschel.)

Apprenticeship

Native customs of the area still made provisions for an eager young person to be apprenticed to a knowledgeable elder in order that ancient teachings be preserved. Accordingly, at age nine, an ever-fascinated and inquisitive Kee was apprenticed to Nojimakway, the intrepid herbalist of Onominee, who taught Kee numerous lessons and practices of Native medicine. From Nojimakway, Kee learned the ways in which thought becomes action, "learning through doing", the techniques for transforming the healing energies of plants into forms useful to humans, and the traditional theory behind the remedies.

Western Education

During the 27 years following high school, Kee earned a Bachelor of Science and a Master's Degree in biology from Central Michigan University and completed 2 years of work toward a Ph.D. at the University of Michigan. During these years she also raised two families and taught in both secondary and higher educational institutions. A versatile writer, Kee has published many articles on both botany and Native American philosophy, including a book for the Botanical Museum at Harvard University titled *Puhpohwee for the People*. An accomplished storyteller, Kee's telling of Ojibwa legends is itself legendary.

Conservation

Kee firmly believes all things are interconnected and should exist in harmony; thus, the protection of earth and its ecosystems underlies the basic acts of her everyday life. In the Ojibwa traditions, "every action is done with consideration for the seven generations before and the seven generations after." Kee believes that the earth nourishes and supports all life and any harm humans do to it will return to affect The People. In this sense, conservation of the Earth is conservation of all interconnected life, and thus, conservation of the human kind.

Much of this passion to preserve Mother Earth came from Kee's father, who regarded trees as his guardians, "and he definitely was a guardian for the trees," says Kee. Believing that the welfare of the human kind is bound within the welfare of the tree kind, Wauboshtigwan lived a life close to the Earth and encouraged others to do the same. If he were alive in today's world, Kee believes her father would be a powerful speaker for conservation.

The years Kee spent at the University of Michigan alerted her to the numerous endangered plant species and how quickly their last available habitats are being destroyed. This is why she makes every effort to protect the endangered plants on her Great Lakes Island home. Western education and Kee's native philosophy impressed upon her the degree to which the Earth is out of balance. Therefore, in all her teachings, she strives, above all else, to instill a love of the Earth in her students' hearts.

Teaching

Now in her eighties, Kee has taught on three continents and at every level of public education from the one-room country school through the university, always inspiring and exciting her students. For example, when teaching about the botany on her home island, Kee not only gives the student the standard botanical terms, but also the plant's Ojibwa name, its translation, the medicinal value, and any associated legends. What the students cannot retain of this overwhelming introduction, they more than gain in appreciation, respect, and admiration for the plants. No longer objects waiting to be studied, plants come to life; the relationship between plants and people becomes apparent; and a stronger bond is created between student and plant life. This Ojibwa philosophy — the interconnectedness of life — permeates Kee's teachings.

Medicines

Kee's curative practices are also based around this philosophy. To her, healing is a process of restoring balance to the natural mental, physical, and spiritual health, made possible through, she says

...empowerment from Gitchi Manitou, the source of Power, possible through that catalyst which is the energy of pure intent, possible through knowledge of the virtue-filled beings of the blessed plants.

Kee explains that the energies of the plants combine with those of the people and spirits to make healing medicines. A deep respect is shown for the Earth at all times throughout medicinal preparation.

For example, it is considered crucial to maintain a relaxed and focused mind (which promotes good energy) throughout the preparation of medicinals. Before any harvesting is done, an offering of kinnickinnick (a mixture of herbs) is made to the plant species being collected, reminding The People to be grateful for this medicinal gift and asking the plants for their aid in healing. To ensure that those population areas will be available for the use of future people, collections are made only where the species are flourishing or can easily recover from the disturbance. All of these components must exist or the medicines will be unbalanced and less effective in restoring harmony.

Western and Native American Botany

Kee's background in western and Native American botany allows her a unique view-point. Western botany places its focus on the quest for knowledge and intellectual enlightenment, providing a greater factual understanding of nature and occasionally resulting in beneficial products. Conversely, Native American botany is based on the practical aspects of plants, that promotes a spiritual and holistic understanding of the uses and long-term ecology of plants. Kee finds no theoretical conflict between these two missions; yet, she has observed how they can promote a drastically different focus in their followers.

Kee's introduction to western science, via graduate school, was stimulating to her endlessly curious mind. It was also an eye-opening experience to the competitive world of science. Kee had never looked at a plant with anything other than awe and respect. So, the concept of profiting from nature was backward; yet that was what was happening all around her. She remembered...

...people who would travel to South America and spend large amounts of money simply to be the first to discover a new plant. I had no idea how important that kind of thing was to some people. She laughingly concludes, *Who is first? So what?!*

Conclusion

Keewaydinoquay's appetite for knowledge has hardly diminished over the decades, as she is still coming out to the shores and forests of her home island searching for new and endangered plants. Similarly, she will continue to follow the precepts of the Ojibwa tradition: the universe was created in balance, and thus, all things within it are inter-connected. Kee's pursuits of herbal healing, teaching, and botany exemplify her desire to help others through this tradition. When addressing the ominous challenge to protect the Earth from degradation, Kee shows the spirit of her ancestors and followers, "I want to be on the good side, even if its not the winning side."

8.3   CASE STUDIES WITH DYE PLANTS*

8.3.1   THE DYEING PROCESS (FIGURES 8.5 TO 8.10)

For thousands of years, and up until about a century ago, the color of fabrics came chiefly from the crude juices of plants. The use of the indigo plant for dyeing fabrics blue dates beyond 3000 B.C. Probably a thousand years later, people knew how to combine other substances in the dyeing process in order to make colors "fast". This later discovery is now known as ***mordanting***.

FIGURE 8.5   A skein of white, unbleached sheep's wool after washing it in warm water to remove lanolin. (Photo courtesy of Jane LaRue.)

Natural dyes come from many parts of plants: roots, leaves, bark, flowers, and fruits. The dye sources in the ancient world included such plants as madder (*Rubia tinctorum*) for red, saffron (*Crocus sativus*) and weld (*Reseda luteola*) for yellow, woad (*Isatis tinctoria*) for blue. The bark of oak (*Quercus* spp.), outer shells of fresh walnuts (*Juglans* spp.), pomegranate flowers (Punica granatum), and dyer's bugloss (*Anchusa tinctoria*) were among others used. Gall-nuts, woody swellings caused by attack of gallflies, presumably on the Lusitanian oak (*Quercus lusitanica*) later provided both a dye (brown, gray, and black) and a mordant in the form of a tannin.

---

* For dye plants not discussed in this section, References 22 through 24 are very helpful for finding information about dye plants and the dyeing process with natural dyes.

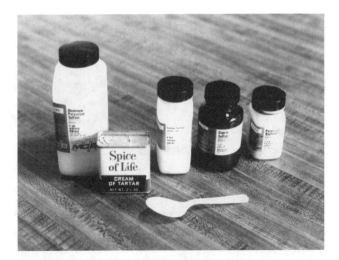

FIGURE 8.6 Mordants used in natural plant dyeing. Left to right: potassium aluminum sulfate (alum), cream of tartar, ferrous sulfate, cupric sulfate, and potassium dichromate (last-mentioned mordant is now banned by the EPA because of its toxicity to humans). (Photo courtesy of Jane LaRue.)

FIGURE 8.7 Preparing the dyebath. After plant parts are boiled in water, then allowed to remain in the cooling water for 2 to 12 h, the plant material is removed by passing it through a screen sieve so that the dyebath is free of plant debris. (Photo courtesy of Jane LaRue.)

Mordanting must be capable of combining chemically with the coloring matter (the dye) being applied, while a few plant dyes may be used directly. The majority require a mordant for permanent color. Besides tannic acid, substances that have

FIGURE 8.8 Dyeing the pre-mordanted wool in a simmering (just less than 100°C) dyebath. (Photo courtesy of Jane LaRue.)

FIGURE 8.9 Close-up of wool being held out of the dyebath with a long, wooden pot label. (Photo courtesy of Jane LaRue.)

been frequently used as mordants include the metallic salts of alum, chrome (now banned because of its toxicity to humans), iron, and copper, as well as acetic acid (as vinegar), ammonia, caustic soda, and tartaric acid. The mordant is thoroughly dissolved in soft water and the fiber is submerged in the solution. The mordant bath is usually heated to just below the boiling point of water (100°C) and the fiber is immersed in the hot mordant solution for ca. 30 min. Each skene of wool or other fiber is labelled with a pencil on masking tape to denote the mordant used. The

FIGURE 8.10 Dyed wool after washing in several changes of water, starting from dyebath temperature and ending at room temperature (ca. 25°C). (Photo courtesy of Jane LaRue.)

mordanting step is carried out just before immersing the wool or other fiber in a dyebath. Common mordants and their properties are listed in Table 8.1.

To prepare the dyebath, the plant material, whether dried or fresh, should be crushed or chopped and soaked in water overnight. Then, the mixture is boiled 0.5 to 2 h, the length of time depending on the color desired. Barks and roots require a longer boiling time than parts of herbaceous plants in order to extract the pigment(s). Most dye plants can be dried for future use, but some give brighter colors when used fresh. Table 8.2 indicates a selected group of plants used commonly for preparing natural dyes, the parts of the plant used as dye sources, and the colors of the dye one obtains from these plant sources.

8.3.2 Examples of the Use of Natural Plant Dyes

8.3.2.1 Orchil Dyes from Lichens*

A *lichen* is a combination of an alga and a fungus growing together. This composite organism has a form unlike either of its constituents. Lichens come in three basic growth forms: *crustose* (just what the name implies), *foliose* (as a thin, lobed sheet often quite firmly attached to the substrate), and *fruticose* (as a 3-D, usually branched growth attached to the substrate at only one or a few points). Lichens produce many unusual chemicals which are made by no other organisms. These chemicals (usually acids) are used by taxonomists to identify many species of lichens but people have also used some of these chemicals to make dyes. Most of these acids (and the lichens which make them) have little or no color in their natural state but must be treated chemically to make colored substances.

---

* This section prepared by Dr. Barbara J. Madsen.

TABLE 8.1
Common Mordants and Their Properties

Mordant Common name	Mordant Chemical name	As a Premordant in 1 qt water[a]	As an additive
Alum plus cream of tartar	Aluminum potassium sulfate plus potassium bitartrate	3/4 tsp alum plus 1/4 tsp cream of tartar	May be used along with wood and dyestuff
Iron	Ferrous sulfate	Primarily used as an additive to darken or "sadden" a dyebath	Use a pinch
Tin	Stannous chloride	More commonly used as an additive as it can make wool brittle	To lighten or brighten a dyebath, use a pinch well dissolved in water before adding to bath
Copper sulfate	Cuprous sulfate	Primarily used as an additive, gives wool a light blue or blue-green color	1/4 tsp in water
Vinegar	Acetic acid	1/3 cup	Frequently used to heighten the color of dyebath, especially in the red color range

[a] Amount per 1 oz medium-weight 2-ply natural wool.

Modified from Weigle, P., Natural Plant Dyeing, in *Natural Plant Dyeing: A Handbook*, Brooklyn Botanical Garden, Brooklyn, NY, 5, 1973.

TABLE 8.2
A Selected Group of Plants Used for Preparing Natural Dyes

Color of dye prepared from plant	Plants	Parts of plant used
Red	*Sanguinaria canadensis* (bloodroot)	Rhizomes
Yellow to red	*Galium verum* (Ladies' Bedstraw)	Roots
Browns, greens, yellows	*Rhododendron*	Leaves
Yellow (saffron)	*Crocus sativus* (Crocus)	Stigmas
Yellow, khaki brown	*Rhus typhina* (staghorn sumac)	Roots, berries
Yellow, gold	*Ligustrum vulgare* (privet)	Branch tips
Yellow, gold	*Solidago* (goldenrod)	Flowers
Yellow, buff, old gold, dark green	*Tagetes* (marigold)	Flowers
Gold, green-gold	*Populus* (aspen)	Leaves
Gold, orange, or red	*Allium cepa* (onion)	Bulbs scales or "skins"
Light gray	*Rubus* (blackberry)	Young shoots
Chartreuse, yellows, orange	*Dahlia*	Flowers
Brown, greys, blacks	Oak galls from *Quercus* (oak)	Galls
Purples, greens, yellows, browns	Lichens	Thallus
Yellow, brown	*Juglans nigra* (Black walnut)	Husks — tannins

The use of lichens in dyeing was mentioned by such ancient authors as Ezekiel, Theophrastus, and Pliny. The most sought after dye in the ancient world was *royal purple* (actually closer to red than to our modern idea of purple). This, the color of kings, was made from snails of the genus *Murex*. It was a beautiful, brilliant, fast color (even after thousands of years) but the snails were scarce and it took 3000 snails to make 1 g of dye extract, and the dye stank. The Phoenicians discovered how to make a similar red dye from a fruticose lichen that grew on rocks by the seashore. This dye was not fast but it had a nice fragrance (some lichens are still used in perfumes today) and was much cheaper to make.

This red lichen dye was imported to Europe about 1300 A.D. by a Florentine merchant named Frederigo Orcelli, whose name was later given to the dye substance, orchil, and slightly altered, to the genus of lichens used, *Roccella*. For 200 years, the Orcelli family held a monopoly over the collecting of the lichens from various Mediterranean islands and over the production of the red dye. Late in the 15th century, however, *Roccella* was discovered in the Canary Islands, and was later found on warm seacoasts all over the world, including Sri Lanka, Australia, Madagascar, Chili, Peru, and California. The production of orchil dye was big business — in the 1850s the Canary Islands exported 100 tons/year to England. In 1851, orchil fetched £380/ton. Orchil and other lichen dyes were important until the late 1850s, when the first aniline (coal tar) dyes were discovered by accident. The Wilton Carpet Company used some lichen dyes as late as World War II.

Orchil dye can be produced from a number of different lichen acids. The acids are broken down by ammonia in the presence of oxygen to form carbonic acid and **orcin**, which is further modified to produce **orcein**, the colored substance, which is itself a mixture of three compounds. The process is fairly simple.

- Break up or pulverize the lichen and sieve to remove impurities.
- Put in a vat with stale human urine (historically, the only source of ammonia — used until the 1800s) or ammonia and stir once a day for as long as 4 weeks.
- Add lime, potash, or wood ashes to intensify the color, or acids to modify it.
- Dry resulting paste for storage and later dissolve in water for use.

In northern European countries, many species of lichens have been used for hundreds of years to make red, brown, or yellow dyes, using a similar process. Species of *Umbilicaria*, *Parmelia*, *Pertusaria*, and *Ochrolechia* were most commonly used. At one time, every cottage in Scotland had its barrel ("litpig") of stale urine ("graith") into which the lichens ("crottles") were placed for fermentation. The dye paste (formed into cakes) from Scotland became known as "cudbear", from a corruption of the name of Dr. Cuthbert Gordon of Glasgow, who patented a process for making such lichen dyes. Harris tweeds were traditionally dyed with lichen dyes, especially before World War II, and some still are today. The cakes of dye used to be dried over peat fires so the tweads also acquired the smell of peat smoke. There has been some interest, particularly in Ireland and Scandinavia, in reviving the use of lichen dyes as a cottage industry both because the interesting variation of color makes the finished products more valuable and because (it is said) wool dyed with lichen dyes is not attacked by moths.

Some lichen dyes, including orchil, are also sensitive to acidity. In fact, litmus paper was made from an orchil dye, which changes from blue to red when exposed to acid.

The best time to collect lichens for dyeing is said to be in late August just after the season of greatest light and heat. This is when the accumulation of acids is at its maximum. Another practical note: English dyers of the 18th and early 19th centuries "learned by experience to avoid urine from beer drinkers, which is excessive in quantity, but frequently deficient in urea, ammonia, and solids, while it is abundant in water."

8.3.2.2 Saffron (Yellow) Dyes from Crocus Flowers

Saffron has been widely used as a food coloring and as a dye for cloth. It continues to be used in developing countries and by dyers worldwide. Saffron comes from the stigmas of the flowers of crocus, *Crocus sativus*. These stigmas are rich in riboflavin, which is a yellow pigment and is also a vitamin (vitamin B_2).

Mature stigmas are collected from the flowers when they develop in the spring. Over 200,000 dried stigmas, obtained from about 70,000 flowers, yield 1 lb (454 g) of saffron. The saffron has a value of $30 (U.S.) per ounce (ca. 28 g).

More information about saffron and its uses is found in *The Review of Natural Products*, 1996.[21]

8.3.2.3 Indigo (Blue Dye) from the Indigo Plant
(*Indigofera tinctoria*)

Indigo refers to the several species of *Indigofera* that are known for the natural blue colors that are obtained from the leaves and stems of this herbaceous plant. Before synthetic indigo and aniline dyes were developed, indigo plants were grown commercially in the East Indies and in South and Central America, especially during the middle ages, for the popular blue dye obtained from the plants.

The blue indigo dye is produced during the fermentation of the leaves in combination with caustic soda or sodium hydrosulfite. In this process, a paste exudes from the fermenting plant material. This is processed into cakes that in turn are finely ground to a powder. The blue color develops as the powder is exposed to the air.

The dye, indigo, is a derivative of **indican**, a glucoside found in the Indigo plants. By enzymatic hydrolysis of the glucoside by β-glucosidase derived from the leaves, **indoxyl** is formed. It is the oxidation step (exposure to air) that converts two molecules of indoxyl to a single molecule of indigo, or **indigotin** ($C_{16}H_{10}N_2O_2$).

More information about indigo and its uses is found in *The Review of Natural Products*, 1996.[21]

8.3.2.4   Brown Dyes from Black Walnut (*Juglans nigra*) Husks

The husks of black walnut, or other walnuts, are used as a brown dye source for dyeing wool. They are soaked overnight, simmered for 60 min, then strained for dyebath use. A mordant is not necessary, but ferrous sulfate can be used to give a dark brown color that is almost black. The metabolite in walnut hulls that is the brown dye is called *juglone*. It is a naturally occurring *naphthoquinone*.

8.4. CASE STUDIES WITH PLANT FRAGRANCES

8.4.1   Sources of Plant Fragrances

Plant fragrances from flowers attract various kinds of pollinators, and other fragrances that come from both flowers and vegetative parts of plants deter predators (see Chapter 2). Many of these same fragrances from plants have utilitarian uses by humans, namely, perfumes, herbal teas, sachets, herbal wreaths, and insect repellents. We grow many fragrance herbs both commercially and in the home garden. Organic gardeners grow some of the insect-repelling herbs as *companion plants* with flowers and vegetables to repel insect predators. These plants include tansy (*Tanacetum vulgare*), chives (*Allium schoenoprasum*), and marigolds (*Tagetes patula*). The fragrances, themselves, have their origin from scent-producing cells in flower petals and other flower parts as well as from glandular hairs (trichomes) found on leaves, stems, and floral bracts.

8.4.2   Uses of Plant Fragrances

8.4.2.1   Sachets

Sachets are made of dried flowers and are placed in loosely woven cloth so that the fragrance can be smelled. They are used in bathrooms and bedrooms (e.g., in bedding) to provide pleasant smells, and in linen closets to repel moths that destroy wool and other cloth garments. Dried rose (*Rosa* spp.) petals are often used in sachets to provide pleasant aromas. Dried lavendar (*Lavandula officinalis*) flowers have essential oils that not only smell nice and clean (hence, their use in soaps and perfumes), but also, destroy bacteria that cause tuberculosis and typhoid fever.[24]

8.4.2.2   Pest Repellents

Classical pest repellents from plants come from rotenone (from roots of *Derris* sp.), pyrethrum flowers (a *Chrysanthemum* sp.), lemon grass (*Cymbopogon citratus* and *C. flexuosus*), tobacco (*Nicotiana tabacum*) that contains nicotine (now banned by the EPA because it is carcinogenic to humans), and garlic and chives (*Allium sativum* and *A. schoenoprasum*, respectively). In the home garden, organic gardeners use teas made from garlic juice or juice from chive plants to control insect pests.

8.4.3 Essay on Extraction and Analysis of Essential Oils from Lavendar (*Lavandula officinalis*)*

The objectives of this section are to gain background information on the medicinal and pesticidal properties of the herb *Lavandula officinalis,* Lavender, to form working hypotheses about its uses and to test these hypotheses. The hypotheses are as follows: (1) the essential oil of lavender works both as an insecticide and as a natural herbicide in the lavender plant; (2) mixed with another oil or cream, the essential oil works as an antibacterial ointment. Background information needed to formulate these hypotheses was obtained on the essential oil of lavender as used in traditional herbal medicine, the human uses of the essential oil in the lavender plant, the plant's ability to synthesize the essential oil, documentation of the constituents of the oil, and growing the lavender plant.

The essential oil of lavender is the main product of the plant that is used in herbal medicine. In traditional herbal medicine, the lavender oil is used as an antispasmodic, a carminative, a diuretic, a sedative, a stimulant, a stomachic, and a tonic to treat such ailments as acne, colic, flatulence, giddiness, migraines, nausea, rheumatism, spasms, sprains, toothache, and vomiting.[25] Preparations of the lavender oil to treat these ailments include infusions with water and mixtures with other oils or creams.

Lavender oil is also used as the component of several prepared sedatives and such as Sedatruw, Neroflux, Beruhigungstee Satus, Nerven-Schlaf-Tee, and Chol-truw.[26] The greatest market for these products and the essential oil is in western Europe.

Lavender grows well and in abundance in western Europe, mainly France, Germany, and England. Lavender grows well in warm climates with sandy soil or soil that drains well. The plant does not need much water. However, it is necessary for lavender to have a cold season with freezing temperatures before it flowers again. Flowering usually occurs in late June to early July depending on the length of the winter. At this time, the essential oil, which is produced in the flowers, can be obtained.

The essential oil of lavender contains over 100 compounds, most of which are volatile and some of which are toxins or narcotics. Soxhlet extraction was used with hexane to obtain a mixture of hexane and essential oil from the flowers. A Shimadzu gas chromatograph was then used to document the constituents of the essential oil in lavender flowers (Figure 8.13). The documented chemicals (Table 8.3) concur with the chemicals listed in *The Encyclopedia of Common Natural Ingredients Used in Food, Drugs, and Cosmetics* (1996) by Albert Leung and Stephen Foster[27] as constituents of the essential oil in lavender.

With the amount of toxins in lavender, the following hypotheses were formulated: (1) The essential oil of lavender works as an insecticide and as a natural herbicide in the lavender plant; and (2) mixed with another oil or cream, the essential oil could be used as an antibacterial ointment.

To test the second hypothesis, pure essential oil was needed from the flowers. To obtain this, steam distillation was used. This procedure resulted in a mixture of water and the essential oil. The mixture was then frozen to separate the water from

---

* This section was prepared by Matthew Perry (Figures 8.11 and 8.12).

FIGURE 8.11   Undergraduate student Matthew Perry holding a plant of lavendar, *Lavandula officinalis*; ca. 1/10 normal size. (Photo courtesy of David Bay.)

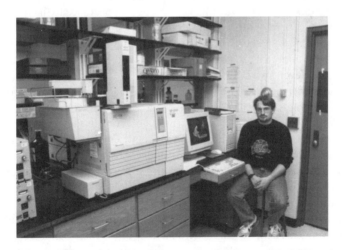

FIGURE 8.12   Undergraduate student Matthew Perry sitting next to a Shimadzu gas chromatograph/mass spectrometer used to analyze constituents in the essential oils he extracted from flowers of lavendar, *Lavandula officinalis*. (Photo courtesy of David Bay.)

the essential oil. The frozen mixture would then be melted with a hand or body lotion or cream and placed in a culture of bacteria to test the essential oil's ability to kill bacteria.

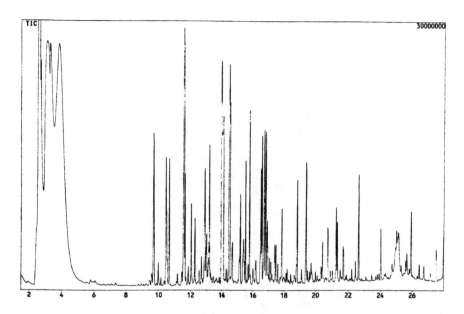

FIGURE 8.13 Gas chromatography mass spectrometer trace of essential oil constituents (see Table 8.3) in flowers of lavendar, *Lavandula officinalis*.

The result of the freezing, however, did not lead to the bacteria testing. The frozen mixture appeared to be two separate liquids, albeit both were clear. Unfortunately, a test tube of frozen water had an identical appearance to the frozen mixture of oil and water. Therefore the mixture could not be separated.

To test the first hypothesis, it would be necessary to place a number of plants in a controlled room with a large number of plant-eating insects. However, this hypothesis was never tested for lack of plants.

Since neither hypothesis was ever actually tested, no real conclusions can be made. However, the essential oil from lavender is known to be used as an antiseptic to treat eczema and cold sores.[28] Also, it was observed in the greenhouse in which the lavender was kept, that the surrounding plants were covered with aphids and the lavender was untouched. Given this qualitative observation and the information provided in *The Complete Medicinal Herbal*,[28] it can be inferred that these hypotheses are probably correct but further experimentation would be necessary.

From this point, there is much work that could be done. The first hypothesis could be fully tested. A way could be found to produce pure essential oil from the flowers of lavender. Then, the antibacterial ability of the essential oil could be tested. It would be advantageous to obtain the required number of plants before beginning any experimentation. It would also be helpful to have a knowledge of the plants and their chemistry before starting any new experiments.*

* I would like to thank Leland Cseke, University of Michigan Biology Dept., for helping to use the Shimadzu Gas Chromatograph and Horticulturist Adrienne O'Brien, U of M Matthaei Botanical Gardens, for information on how to grow the plants.

TABLE 8.3
Lavender Extract: Retention Times
and Chemical Values

Retention times	Chemical
9.667	Geranyl alcohol
10.450	Ocimene
11.567	Linalool
12.025	Terpinolene
12.525	Cyclohexane ethanol acetate
12.875	Borneol
13.958	Linalool formate
13.967	Linalool acetate
14.458	Geraniol acetate
15.150	Acetic acid
15.500	Propanoic acid
15.760	Octadien
16.500	Pyrazine
16.567	Santalene
16.708	Eremopilene
16.717	Isocaryophyllene
16.817	Dodecatriene
17.842	Copaene
18.842	Limonene
19.400	Cadinol
19.408	Naphthalene

8.5 CONCLUSIONS

This chapter provides the reader with some of the classical case studies on the uses of plants for medicinal purposes, for natural plant dyes, and for fragrances that were not covered in earlier chapters of the book. The cases cited are important, not only because they have been used by indigenous peoples in times past, but also because there is currently a resurgence in their use because they are, for the most part, less expensive, more environmentally friendly, and just as effective as many of the synthetically derived compounds used for the same purpose.

REFERENCES

1. Schmutterer, H., Ed., *The Neem Tree*, VHC, Weinheim, Germany, 1995, 1–696.
2. van der Nat, J. M., van der Sluis, W. G., de Silva, K. T. D., and Labadie, R. P., Ethnopharmacognostical survey of *Azadirachta indica* A. Juss. (Meliaceae), *J. Ethnopharm.*, 35, 1–24, 1991.
3. Heikkenen, H. J., Balsam Fir Ecology, Ph.D. Dissertation, University of Michigan, Ann Arbor, Michigan, 1957.

4. Krochmal, A. and Krochmal, C., *A Guide to Medicinal Plants of the United States*, Quadrangle Books, New York, 1973.

5. Johnson, L., *A Manual of the Medical Botany of the North America*, William Wood and Co., New York, 1884.

6. Bailey, L. H., *Cultivated Evergreens*, Macmillan, New York, 1928.

7. Vidakovik, M., *Conifers: Morphology and Variation*, Croatirion, Graficki, Zagreb, Bosnia-Herzogovina, 1991.

8. Rushford, K., *Conifers*, Facts on File, New York, 1987.

9. Moerman, D. E., *American Medical Ethnobotany: A Reference Dictionary*, Garland Press, New York, 1977.

10. Moerman, D. E., *Medicinal Plants of Native America*, Vol. 1, University of Michigan Museum of Anthropology, University of Michigan, Ann Arbor, MI, 1986.

11. Nahler, G., *Dictionary of Pharmaceutical Medicine*, Springer-Verlag, New York, 1994.

12. Hartwell, J. L., *Plants Used Against Cancer*, Quaterman, Lawrence, MA, 1982.

13. Smith, H. H., Ethnobotany of the forest Potawatomi Indians, *Bull. Pub. Mus. City of Milw.*, 7, 1933.

14. Parke-Davis and Company, *Organic Materia Medica*, Parke-Davis and Company, Detroit, 1890.

15. Herrick, J. W., *Iroquois Medical Botany*, Syracuse University Press, Syracuse, 1995.

16. Porcher, F. P., *Resources of the Southern Fields and Forests*, Walker, Evans, and Cogswell, Charleston, WV, 1869.

17. Scott, T. A., *Concise Encyclopedia Biology*, Walter de Gruyter, Hawthorne, NY, 1996.

18. Trease, G. E., *A Textbook of Pharmacognosy*, Bailliere, Tindall, and Cassell Publishers, London, 1966.

19. Grieve, M., *A Modern Herbal*, Vol. 1, Dover Publishers, New York, 1971.

20. Warber, S. and Peschel, K., *Balsam Fir and Burns*, Sara Warber and the Miniss Kitigan Drum, Ann Arbor, MI and Leland, MI, 1995.

21. Granick, B., Ed., *The Review of Natural Products*, Facts and Comparisons, Wolters Kluwer Company, St. Louis, MO, 1996.

22. Adrosko, R. J., *Natural Dyes and Home Dyeing*, Dover Publications, New York, 1971, pp. 1–154.

23. McGourty, F., Ed., Natural plant dyeing. A handbook, *Plants Gard.*, 29, 1–64, 1973.

24. Buchanan, R., *A Weaver's Garden*, Interweave Press, Loveland, CO, 1987.

25. Duke, J. A., *CRC Handbook of Medicinal Herbs*, CRC Press, Boca Raton, FL, 1985.

26. Bisset, N. G., Ed., *Herbal Drugs and Phytopharmaceuticals*, CRC Press, Boca Raton, FL, 1994.

27. Leung, A. and Foster, S., *The Encyclopedia of Common Natural Ingredients Used in Food, Drugs, and Cosmetics*, 2nd ed., John Wiley and Sons, New York, 1996.

28. Ody, P., *The Complete Medicinal Herbal*, Dorling Kindersley Publishers, London, 1993.

9 Plant Conservation

Mary Jo Bogenschutz-Godwin, James A. Duke,
Stacie Klein, and Peter B. Kaufman

CONTENT

0-8493-3134-X/99/$0.00+$.50
© 1999 by CRC Press LLC

9.1   INTRODUCTION

Far too much human-caused exploitation of fragile plant communities and ecosystems has been occurring in recent times at an accelerating pace.This is happening in tropical rain forests worldwide due to their destruction from mining, lumber, wood products, livestock grazing, and farming. In temperate regions this is due to clear-cutting forests, collection of wood from trees and shrubs for fuel, overgrazing by livestock, mining, damming river systems, and urban sprawl. In arctic regions it is the result of massive clear-cuts of boreal forests for pulpwood for paper manufacture, lumber, and wood products. The Worldwatch Institute in Washington, D.C. has been doing a great job of documenting these calamities over the past 2 decades. Their prognosis is not good for the future regarding the Earth's natural resources. Humans, with their burgeoning populations, have been engaged in overly exploitive activities that squander natural products that occur in vast ecosystems. As a result, they are living way beyond the carrying capacity in many regions of the planet.

The purpose of this chapter is to point out ways which might reverse this trend. You will see that this involves preserving natural and wilderness areas; getting involved in sustainable harvesting of plants in these ecosystems; saving rare, threatened, and endangered species of plants in "gene banks," seed banks, tissue culture banks, nurseries, botanical gardens and arboreta, and parks and shrines; and growing plants in an ecologically friendly way. If we follow these strategies, we will help sustain the supply of natural products we obtain from plants, and at the same time, help to provide a livelihood for many people who depend upon these products for their income.

9.2   PRESERVATION OF NATURAL HABITATS AND ECOSYSTEMS

9.2.1   NATIONAL PARKS

Natural resource policies aim to provide people the opportunity to enjoy and benefit from natural environments evolving by natural processes with minimal influence by human actions. The National Park Service (NPS) will ensure that lands within park boundaries are protected. Where parks contain nonfederal lands, the NPS uses cost-effective protection methods. Preservation of the character and resources of wilderness areas designated within a park, while providing for appropriate use, is the primary management responsibility. The National Parks and Conservation Association is a national nonprofit membership organization dedicated to defending, promoting, and enhancing our national parks, and educating the public about the NPS. It was established in 1919 to protect parks and monuments against private interests and commercialism and to block inappropriate development within parks. Most recently, this organization has done a magnificent job of mobilizing citizen action to prevent clear-cutting of timber and mining within and adjacent to national parks. They have also helped to protect these parks from undue human intrusion with recreational vehicles, helicopters, campers, and "vehicles" of all types (including boats, jeeps, motorcycles, mountain bikes, snowmobiles, and dune buggies). Limiting access to the national parks because of "people pressure" and consequent over-crowding has become the norm. Together, these efforts help but citizen action groups such as the National Parks and Conservation Association, the Sierra Club, the Nature Conservancy, the Wilderness Society, the Natural Resources Defense Fund, and the many other organizations who operate in the individual states, must be ever vigilant and ready for concerted action.

9.2.2 SUSTAINABLE BIOPRESERVES FOR INDIGENOUS PEOPLES

Based on a recent United Nations Conference on Environment and Development (UNCED), the U.S. has placed forest management and protection as a priority of UNCED. Further, discussions by U.S. government agencies and nongovernmental organizations have concluded that a provision needs to be included on the needs of indigenous peoples who use the forests for their livelihood, social organization, or cultural identity, and who have an economic stake in sustainable forest use.[1] Actions include promoting means for indigenous peoples and members of local communities to actively participate in decision-making processes for any proposed forest-related actions where their interests are affected.[1] Other propositions are to identify ways to enhance the value of standing forests through policy reform, more accurately reflecting the costs and benefits of alternative forestry activities, in addition to identifying economically valuable forest species, including timber and nontimber species, and the development of improved and sustainable extraction methods.[2]

Nabhan[3] has indicated that the following criteria offer the best guidelines for ensuring that indigenous peoples and other peasant communities benefit from applied ethnobotanical development, and that projects sustain rather than deplete or destroy biodiversity.

- The project should attempt to improve the objective and subjective well-being of local communities rather than seeking cheap production sites and importing inexpensive labor.
- Cultivation in fields or agroforestry management should be considered if there are threats that wild harvests will deplete the resource.
- Wildland management and sensitive harvesting practices should be introduced in cases where the resource might sustain economic levels of extraction in the habitat.
- The plant(s) chosen should offer multiple products or be adapted to diversified production systems.
- When possible, programs should build on local familiarity, use, and conservation traditions for the plant being developed.
- If possible, these programs should be based on locally available genetic resources, technologies, and social organizations to enable local people to retain control over the future of the resource.

We now turn to the topic of ethnobotany and the sustainable use of plant resources based on work of the World Wildlife Fund, UNESCO, and the Royal Botanic Gardens at Kew, United Kingdom. The People and Plants initiative is creating support for ethnobotanists from developing countries who work with local people on issues relating to conservation of plant resources and indigenous ecological knowledge. Rather than promoting the discovery and marketing of new products, emphasis is placed on subsistence use and small-scale commercialization of plants which benefit rural communities. In cases of large-scale commercialization of wild plants, emphasis is on improving harvesting methods and mechanisms that allow communities an increasing share of profits.[4]

One example is provided by the Kuna Indians of Panama. They have successfully established the world's first internationally recognized forest park created by indigenous people. The reserve provides revenues directly to the Kuna from the sale of research rights, and from ecotourists who come to learn about the rainforest. Coupled with this, it helps protect and preserve their native heritage. Scientists conducting research in the park are required to hire the Kuna to assist and accompany them during their stay. The Kuna control access to sites and require reports on all research. These terms allow the Kuna to patrol and protect outlying areas while learning from the scientists.

Head and Heismann,[5] in *Lessons of the Rainforest*, tell about the organization called Environmental Restoration in Southern Colombia. It is composed of 56 Indian communities that are organized to protect Indian lands, resources, culture, and rights in an area where the forest has been destroyed by mines and cattle ranches. The organization began a forestry program with three tree nurseries which provided seedlings to those communities that agree to plant a minimum of 1000 trees of native species. To date, one community has completed nine reforestation programs.

9.2.3 WORK OF THE NATURE CONSERVANCY

The main objective of the Nature Conservancy is to protect plants, animals, and ecological communities that represent biodiversity. To do this, they rely on conservation science to guide their work. Conservation science programs encompass biological, ecological, and technological knowledge that is used to identify and protect sensitive biodiversity, and in management methods and practices used to ensure its survival. The Natural Heritage Program and the Conservation Data Center Network programs collectively track in their databases the protected status and locations of rare and endangered species and ecological communities. Over the past 4 decades, the Nature Conservancy has protected more than 8.1 million acres (3.28 million ha) of habitat based on information about the location, range, and status of rare species. This number is even higher for total acreage protected to date — 9.3 million acres (3.77 million ha) of land in the U.S. and 40 million acres (16.19 million ha) throughout Latin America, the Carribean, and the Asia/Pacific regions. Indeed, it operates the largest system of privately owned nature preserves in the world.

In carrying out its work, the Nature Conservancy addresses ecological function and influences of people and develops better conservation planning methods and tools that will allow planning across immense biologically defined regions and the range of a particular ecological community. Stewardship of land and its resources are an important component of the work of the Conservancy. In protecting areas identified as critical for biodiversity protection, boundaries of those areas are carefully chosen to encompass important biological components and the ecological processes that sustain them. Its presence in local communities enables it to address ecosystem protection, find solutions to environmental problems, and form partnerships. An organization-wide network electronically links all the Nature Conservancy's offices to support the information systems plan which provides up-to-date information.[6]

9.2.4   CONSERVATION OF MEDICINAL PLANTS IN BELIZE, CENTRAL AMERICA*

Of the medicines of the world, 80% have been derived from plants of tropical forests. These forests are being destroyed at an alarming rate. Perhaps the greatest source of destruction comes from the poor local farmers who do not have access to any other land. They clear sections of forests to plant their crops. While these farmers need to eat, what will happen to all of the medicinal resources and research when these forests are gone? Even more pressing, what is being done to ensure the continued success of these highly useful plants, and to preserve the forests in which these species live along with many thousands of animal species?[7]

A tropical rainforest is unlike most other forests of the world in that it is climax vegetation. This means that, left alone, it is self-renewing and ecologically stable. When logging, farming, or any other outside influences occur, this can only result in degradation. This makes the concept of conservation quite simple. The best way to protect a tropical rainforest is to preserve it, banning any interference, including management. It is a rare occurrence for people nowadays to see a forest which is truly old-growth and pristine. Because there are so few of these forests left, those remaining are often believed to be in need of management and even betterment. But when these forests are influenced in these ways, they lose stability, thereby making continued management necessary. In other words, the best action we can take is to leave these pristine forests alone.[8]

But how can we utilize all of the nontimber resources (such as medicinal plants) of the tropical forest if they are set aside as preserves, thus limiting our access to them? An encouraging example of conservation of tropical forests for medicinal plants can be found in the small Central American country of Belize[9] (Figure 9.1). In 1988, the Belize Center for Environmental Studies, along with the Ix Chel Tropical Research Foundation and the Institute of Economic Botany of The New York Botanical Garden founded the Belize Ethnobotany Project (BEP).[10] Their intent was to record how all of the many different peoples of Belize (including Mayan, Gariuna, Ladino, Mennonite, Creole, and more) utilized the plants of their particular region(s).[11] With the help of traditional healers from these different Belizean peoples, many of the plants have been collected and chemically screened by the National Cancer Institute in hopes of finding medicines for cancer and HIV.[12] In addition, several of the herbs collected are prepared and sold as home remedies locally and abroad. These are marked with titles such as "Rainforest Remedies" or "Agapi".

---

* This section has been prepared by Stacie Klein, a 1977 graduate of the University of Michigan. She worked with Dr. Rita Arvigo on medicinal plants of the tropical rainforests of Belize and has prepared this essay based on her studies there.

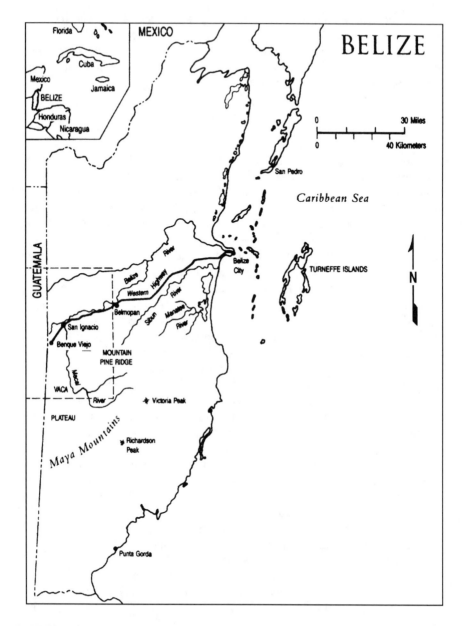

FIGURE 9.1 Map of Belize, Central America. (Modified from Arvigo, R., *Sastun*, Harper/Collins, Publishers, San Francisco, 1994.)

With correct supervision, such products have much potential for adding to the financial gain of the community. If the locals can make enough money to survive off of these products, then there will be less destruction of the forests due to their farming.

In l993, the Belizean government put aside a piece of tropical forest (6000 acres, ca. 2400 ha) to be used as a "government forest reserve, for the purpose of providing a source of native plants used locally in traditional medicines."[12] The hope was that the research, tourism, and traditional medicine collection of this area would serve to financially reward the local community, to allow for connections and idea-exchange between healers and scientists, and to educate all involved. The reserve is still in its beginning stages, but there is much positive outlook, as it is a pioneer in the field of ethnobotany. This project, stemming from the BEP (which was mostly an attempt to record data), has so far served to educate scientists, local and foreign citizens/governments, and educators. It has linked all of these people together, and they have formed a connection to the many traditional healers involved in the project.[9] The project, therefore, has been successful in emphasizing the importance of the connection between flora, fauna, medicine, and the different people of an area.

As we have seen in the example of Belize, a key method of keeping rainforests intact is to make the production of nontimber products from the forests economically and academically rewarding to the surrounding community. If the locals and natives benefit, then they will not only be less likely to clear the forest for farming but also they will be more protective of it, as it will be the source of their livelihood. Belize has thus shown that people and the forests can benefit from and live in harmony with each other. In this way, this small, developing country has set an encouraging example for the rest of the world to follow.

For more information on this topic, contact:

- Shaman Pharmaceuticals, Inc., 213 Grand Avenue, South San Francisco, CA 94080-4812.
- Michael J. Balick, Institute of Economic Botany, The New York Botanical Garden, Bronx, NY 10458-5126.
- Drs. Rosita Arvigo and Gregory Shropshire, Co-Founders, Ix Chel Tropical Research Foundation, San Ignacio, Cayo District, Belize, Central America.

9.3 PREVENTION OF DESTRUCTION OF NATURAL AND WILDERNESS AREAS

9.3.1 WORK OF THE WORLD WILDLIFE FUND

The World Wildlife Fund (WWF) has several important objectives, including (1) halting global trade in endangered animals and plants; (2) creating and preserving parks and protected areas around the world; (3) working to create strongholds for thousands of irreplaceable plant and animal species as well as protecting these and other areas from threats beyond their boundaries; (4) working with local leaders, groups, governments, and international funding institutions to coordinate conservation and improve living standards to help alleviate development pressures that may put wildlands in danger; and (5) organizing, supporting, and strengthening conservation efforts around the world.[13]

National Environmental Trust Funds, pioneered by the WWF in Bhutan and the Philippines, attract the attention of international aid agencies because they prove effective in attracting millions of dollars for conservation in addition to enlisting the participation of governments, nongovernmental organizations, local conservation organizations, and community groups. By spending the annual income from their endowments, these trust funds constitute a reliable source of long-term funding for conservation.

The Biodiversity Support Program (a USAID-funded consortium of the WWF), The Nature Conservancy, and the World Resources Institute published a work entitled, *Sustainable Harvest of Non-Timber Plant Resources in Tropical Moist Forest: An Ecological Primer*, authored by Charles N. Peters. This book is designed to help forest managers to simultaneously harvest products and to conserve forests. It provides a basis in forest ecology and addresses ways that communities can determine what and how much can be harvested over time without depleting the natural resource base on which their livelihood may depend.

The WWF uses Geographic Information Systems (GIS) technology to identify priority areas with the greatest biological wealth and the greatest degree of threat, with a focus on conservation priorities. The WWF works closely with the North American Commission for Environmental Cooperation to help ensure that its work promotes conservation initiatives, such as the North American ecoregion mapping and planning project for biodiversity management. It follows the trade agreement's effect on commodities production and health of forests, wildlife, and natural resources in North America. It also supports the Forest Stewardship Council which has developed criteria for identifying timber companies that produce environmentally sound, economically viable products. This Council consists of social, environmental, and indigenous groups from more than 24 countries, as well as representatives from the timber industry whose mission is to promote ecologically sustainable forest management. In Madagascar, the WWF brokered a debt-for-nature swap which has trained more than 350 local conservation agents and created a network of locally managed tree plantations. It is also helping to develop alternatives to cattle production and slash-and-burn agriculture in order to protect native forests.[13]

9.3.2 WORK OF THE SIERRA CLUB

The Sierra Club was founded by John Muir in 1892 in San Francisco, CA to help preserve the pristine beauty of the Sierra Nevada mountain range in California. Today, it is a national organization with chapters throughout the U.S. It continues to expand, stop abuse of wilderness lands, save endangered species, and protect the global environment. It helps to create and enlarge national parks, preserve forests, designate wilderness areas, halt dams, and prevent destruction of priceless habitats. The Sierra Club helped save Alaska's Arctic National Wildlife Refuge from oil companies, established National Park and Wilderness Preservation Systems, and safeguarded more than 132 million acres of public land.

This organization launched the Critical Ecosystems Program, which is designed to protect and restore 21 regional ecosystems in the U.S. and Canada. This program is involved in designing protection for public and private lands that are the core habitats for native species. It established task forces for each ecoregion, drawing together activists with expertise in various areas to develop strategies to save those regions. What are these strategies for the different ecoregions?

- Atlantic Coast and Great Northern Forest — preserve biodiversity by restoring and sustaining habitat for the full array of native plants and animals, establish sound forestry policy, and preserve wilderness.
- Central Appalachia, Southern Appalachian Highlands, and American Southeast — saving from development, as much as possible, the shoreline stretching 2000 miles (3200 km) from Florida to the mouth of the Rio Grand River.
- Interior Highlands, Great Lakes, Great North American Prairie — establish a system of national parks, reform Forest Service policies on grazing, oil, and gas development, and coal mining on grasslands.
- Mississippi Basin, Rocky Mountains, and Colorado Plateau — enact legislation to protect 5 million roadless acres in Utah, eliminate timber sales that threaten old-growth ponderosa pine stands, do away with subsidized timber sales in all national forests, and protect the Grand Canyon by restricting development on its boundaries.
- Southwest Deserts, Great Basin/High Desert, Sierra Nevada, Pacific Northwest, and Pacific Coast — permanently protect the remaining ancient forests on federal land.
- Alaska Rainforest, the Boreal Forest extending from Alaska to Newfoundland, Hudson Bay/James Bay Watershed, the Arctic, and Hawaii — prevent further destruction of endangered and threatened plant and animal habitats.[14]

9.4 GROWING RARE AND ENDANGERED PLANTS IN BOTANICAL GARDENS AND ARBORETA: HOW IS IT DONE?

9.4.1 INVOLVEMENT OF BOTANICAL GARDENS AND ARBORETA

According to The New York Botanical Garden, of approximately 250,000 species of flowering plants, it is estimated that some 60,000 of these may become extinct by the year 2050, and more than 19,000 species of plants are considered to be threatened or endangered around the world. More than 2000 species of plants native to the U.S. are threatened or endangered, with as many as 700 species becoming extinct in the next 10 years.[15] The New York Botanical Garden currently grows 10 species of plants on the Federal Endangered Species List. They are striving to preserve rare and endangered plants and participate with other institutions in doing this. The Garden is a Participating Institution in the Center for Plant Conservation (CPC), serving as a rescue center for six native plant species that are imminently threatened, which form part of the National Collection of Endangered Plants, and are grown and studied to be conserved.[15] The CPC is located at the Missouri Botanical Garden in St. Louis, MO. This Center is dedicated to conserving rare plants native to the U.S. in an integrated plant conservation context through a collaborative program of *ex situ* plant conservation, research, and education. It is made up of a consortium of 25 botanical gardens and arboreta.[16] A national survey by the CPC in 1988 found that over three-quarters of the endangered flora of the U.S. is in six areas; Hawaii, California, Texas, Florida, Puerto Rico and the Virgin Islands. It has designated these areas as conservation priority regions. The CPC Priority Regions Program addresses the need for conservation through programs of land conservation, management, offsite collection in seed banks, botanical gardens and other institutions, research, and site surveys.[16] The National Collection of Endangered Plants contains seeds, cuttings, and whole plants of 496 rare plant species native to the U.S. The collection is stored at 25 gardens and arboreta that form part of the CPC.

The Royal Botanic Gardens at Kew, United Kingdom support six *ex situ* and *in situ* conservation projects. The activities range from acting as the UK Scientific Authority for Plants for CITES (Convention on International Trade in Endangered Species of Wild Fauna and Flora), cooperating in the recovery and reintroduction of endangered species, and in production of management plans for sustainable development and protected areas.[17]

The Wrigley Memorial and Botanical Gardens at Catalina Island, CA is still another example. The Garden places its emphasis on California island endemic plants. Many of these plants are extremely rare, with some listed on the Endangered Species List.

9.4.2 IMPORTANCE OF ENVIRONMENTAL EDUCATION

The main purpose of environmental education is to instill an understanding and appreciation of natural resources and to develop support for preserving these resources. It promotes awareness of human impact on the environment, builds knowledge and skills needed in ascertaining environmental issues, and the ability to apply that knowledge and skills in issue remediation. There are implications here that are associated with loss of habitats, extinction of species, and their possible biomedical uses. We come to understand that indigenous inhabitants are as endangered as the forest in which they live. Tropical rainforests are considered to be nonrenewable old growth forests.

The Environmental Protection Agency (EPA) has created an environmental education office to advance and support national education efforts to develop an environmentally conscious and responsible public, and to inspire a sense of personal responsibility for the care of the environment. It awards nearly 250 grants annually worth approximately $3 million as seed money to support environmental education projects.

Among the newly formed conservation and education organizations is the not-for-profit Amazon Center for Environmental Education and Research (ACEER) Foundation with which Dr. James Duke has been associated since its inception. Since 1991, ACEER has been a dynamic force for rainforest conservation. It provides students, teachers, citizen naturalists, and researchers from around the world an opportunity to learn about the need to conserve the magnificent biodiversity and cultural richness of Amazonia (see Figures 9.2 to 9.10). ACEER

FIGURE 9.2   View from ACEER canopy walkway at the top of Amazonian rainforest north of city of Iquitos in Peru. (Photo courtesy of Dr. James A. Duke.)

FIGURE 9.3   Same as Figure 9.2, but different view. (Photo courtesy of Dr. James A. Duke.)

operates an education center in the Peruvian Amazon, north of the city of Iquitos, that is visited by more than 2000 individuals per year; the Dr. Alwyn H. Gentry Laboratory is attached to the center and is the focal point for Amazonian research at the ACEER. A major feature of ACEER's facilities is the Canopy Walkway system, the only one of its kind in South America. It allows researchers and visitors to ascend to the very top of the rainforest canopy for observation and study. In 1996, due to the efforts of ACEER Board member Dr. Jim Duke, the ACEER

FIGURE 9.4   Same as Figure 9.2, but different view. (Photo courtesy of Dr. James A. Duke.)

FIGURE 9.5   View of ACEER canopy walkway in Peruvian Amazon rainforest. (Photo courtesy of Dr. James A. Duke.)

created the ReNuPeRu Ethnobotanical Garden, a 6-ha site showcasing more than 200 economically important plants growing in their native habitat. The curator for the Garden is Don Antonio Montero Pisco, a local shaman. Ultimately, the experience gained at the Garden will be transferred to local villages to promote the sustainable economic development and use of ethnobotanicals by the peoples of Amazonia (see Figures 9.11 to 9.24). As a 501(c)[3] nonprofit organization, ACEER offers a wide range of education and research programs. In the area of education

FIGURE 9.6 Same as Figure 9.5, but different view. (Photo courtesy of Dr. James A. Duke.)

FIGURE 9.7 Same as Figure 9.5, but close-up view of ACEER canopy walkway. (Photo courtesy of Dr. James A. Duke.)

annual credit and noncredit bearing workshops on rainforest ecology, environmental education, pharmacy from the rainforest, and shamanic healing techniques and medicines are offered. The ACEER also hosts student interns, master's, doctoral and postdoctoral researchers from major universities around the world. An Adopt-a-School program fosters cultural exchanges between American and rural Amazonian schools while providing critical educational supplies for the Peruvian schools. A Peruvian Teachers Training workshop enhances environmental education curriculum

FIGURE 9.8   Individual traversing ACEER canopy walkway. (Photo courtesy of Dr. James A. Duke.)

FIGURE 9.9   Canadian Herbalist, Terry Willard, 100 ft. (ca. 30 m) high on ACEER canopy walkway. (Photo courtesy of Dr. James A. Duke.)

FIGURE 9.10 *Heliconia* plant in flower in the understory vegetation of Amazonian rainforest at ACEER. (Photo courtesy of Dr. James A. Duke.)

FIGURE 9.11 View from Machu Picchu of Andes mountain vegetation above Amazonian rainforest. (Photo courtesy of Dr. James A. Duke.)

development throughout Amazonia, while a Peruvian Scientists Training workshop instructs natural resources' scientists on how to use satellite technology and sophisticated geographic information system computer systems to study ecosystems. An ACEER research project recently mapped the spatial distribution of 15 native medicinal plant habitats. Other research has evaluated a wide range of topics including primate biodiversity, the taxonomy of bromeliads, parental behavior in a previously

FIGURE 9.12 View from Machu Picchu of Andes mountain vegetation in canyon. (Photo courtesy of Dr. James A. Duke.)

FIGURE 9.13 View of fragile forest vegetation in Peruvian Andes Mountains. (Photo courtesy of Dr. James A. Duke.)

FIGURE 9.14 View of fragile forest vegetation in Peruvian Andes Mountains. (Photo courtesy of Dr. James A. Duke.)

FIGURE 9.15 Peruvian mountain muscians in full costume performing with drums and flutes. (Photo courtesy of Dr. James A. Duke.)

undescribed species of a frog, the ecology of bats, water quality studies of lake and river systems, and more. Through the VINES (Volunteers in Environmental Service) program, volunteers from around the world may participate in ACEER education and research programs at its center in the rainforest. Another interesting educational feature is the close linkage of the not-for-profit ACEER with International Expeditions

FIGURE 9.16   Peruvian women in native garb with llamas in Andes mountain pasture above their village. (Photo courtesy of Dr. James A. Duke.)

FIGURE 9.17   Peruvian Andes mountain Amerinds in full costume providing music with sea shells. (Photo courtesy of Dr. James Duke.)

(I.E.), a closely related for-profit organization. Among many other responsibilities, I.E. conducts regular continuing-education credit courses in a series called Pharmacy from the Rain Forest. In 1997, for example, 1- to 2-week courses were given, not only in Peru (Figures 9.2 to 9.24), but also in Costa Rica, Kenya, and Tanzania. (For further information on Pharmacy Ecotour, or to join Jim Duke on

FIGURE 9.18 Peruvian Andes women and children in full costume in their mountain village. (Photo courtesy of Dr. James A. Duke.)

FIGURE 9.19 Peruvian Andes mountain village marketplace showing some of the locally grown vegetables on display for sale. (Photo courtesy of Dr. James A. Duke.)

a rainforest ecotour, call 1-800-633-4734.) The ACEER is guided by a distinguished international Board of Directors, as well as three advisory boards, one for environmental education, one for science, and the other dedicated to the ACEER's Peruvian operations. (To learn more about the ACEER Foundation, please contact: ACEER Foundation, Ten Environs Park, Helena, AL 35080; 1-800-255-8206 (phone), 205-425-1711 (fax).)

FIGURE 9.20 Peruvian Andes mountain village marketplace where many different kinds of potatoes are on display for sale. (Photo courtesy of Dr. James A. Duke.)

FIGURE 9.21 Peruvian Andes mountain village marketplace where various medicinal and culinary herbs are being sold. (Photo courtesy of Dr. James A. Duke.)

9.4.3 Importance of Cloning Rare and Endangered Plant Species for Distribution

Germplasm of vegetatively propagated plant material is cheaper to maintain in tissue culture[18], is less expensive to ship, and has the potential to yield more plants more quickly. It is one of the preferred ways to preserve rare and endangered plant species

FIGURE 9.22   Peruvian Amerind basketmaker. (Photo courtesy of Dr. James A. Duke.)

FIGURE 9.23   Peruvian village marketplace where Peruvian Amerind is eating her dinner. Items she has tied up are wrapped in banana (*Musa* sp.) leaves. (Photo courtesy of Dr. James A. Duke.)

and to distribute these species to other botanic gardens and arboreta around the world. Where conditions allow, some tissue cultured plant material can be used to reintroduce species that have become lost or extinct in the wild.

One of the preferred methods of tissue culture is shoot-tip culture (mericloning). It is becoming the preferred tissue for exchange of clonal material. Tissue cultures produced from shoot-tip cultures can produce disease-free germplasm, particularly

FIGURE 9.24 Peruvian Amerinds in tropical rainforest carving an oar from local rainforest tree wood. (Photo courtesy of Dr. James A. Duke.)

with respect to viruses. Shoot-tip explants, devoid of any vascular tissue, are typically free of any viral pathogens. This protocol was developed by George Morel in France as a way to rescue virus-infected orchid plants and rapidly propagate virus-free stock. This process is used for the micropropagation of virus-free stock of any plant species. Great success stories are seen in the shoot-tip propagation of virus-free potatoes, strawberries, cassava, pelargoniums, and orchids.

In vitro ("in glass", microorganism-free cultures) disease elimination techniques help to ensure international exchange of germplasm, particularly since viral transmission through seed is known to occur.[18] It allows for a far greater number of plants to be produced in a given time than by conventional propagation methods. The Micropropagation Unit at Kew Botanic Gardens propagates plants which are rare, endangered, or difficult to propagate conventionally. Techniques include micropropagation from vegetative material and *in vitro* germination of seeds and spores. A large number of tropical epiphytic (growing on other plants) and terrestrial (growing in the soil) orchids are grown from seed *in vitro* under sterile conditions. Of these, many are members of island floras and are in jeopardy.

9.4.4 IMPORTANCE FOR SAVING PLANTS FROM EXTINCTION IN THEIR NATIVE HABITATS

Why is it important to save rare and endangered species of plants from going extinct in their native habitats?

- At the rate at which whole ecosystems are being destroyed in the boreal forests in northern latitudes and the tropical rainforest across equatorial regions, we will see the disappearance of countless numbers of plant species, many of which have never been identified, much less studied for their potential economic utility.

- The disappearing plants may be potential sources of new medicines, foods, flavorings, natural pesticides, dyes, fibers, and wood products.
- With the extinction of plants, and the loss of the ecosystems where they exist, indigenous peoples are displaced and their cultures are irreplaceably disrupted.
- Likewise, with the extinction of plant species, many animal species that depend on the plants for food and shelter disappear. The loss of animal and insect species can also lead to the extinction of plant species where those plants rely on animal or insect pollination for reproduction.
- Just think, if even one population of plants becomes extinct, all its unique phytochemical germplasm and properties also disappear.[12]

In order to counteract this alarming loss of plant species worldwide, conservation organizations have to realize that we must, as quickly as possible, *safeguard entire natural ecosystems* from destruction by human activities. If we do this now, we not only create a sustainable environment for these plants, but also, for the animals, and indigenous peoples who reside there.

9.5   PLANT SEED BANKS FOR GERMPLASM PRESERVATION

9.5.1   PLANT INTRODUCTION STATIONS IN THE U.S.

Four regional plant-introduction stations in the U.S. are located in Pullman, WA; Ames, IA; Geneva, N.Y.; and Griffin, GA. They are responsible for the management, regeneration, characterization, evaluation, and distribution of seeds of more than one-third of the accessions of the national system (i.e., nearly 197,000 accessions of almost 4000 plant species). At Ames, IA, approximately 40,079 accessions are held; the primary crops preserved include maize, grain amaranth, oilseed brassicas (e.g., rape, canola, mustard), sweet clover, cucumber, pumpkin, summer squash, acorn squash, zucchini squash, gourds, beets, carrots, sunflower, and millets. At Geneva, N.Y., approximately 14,180 accessions are held; the primary crops preserved include tomato, birdsfoot trefoil, brassicas, and onion. At Griffin, GA, approximately 82,277 accessions are held; the primary crops preserved here include sweet potato, sorghum, peanut, pigeon pea, forage grasses, forage legumes, cowpea, mung bean, pepper, okra, melons, sesame, and eggplant. At the Pullman, WA station, approximately 60,277 accessions are held; the primary crops preserved there include common bean, onion, lupine, pea, safflower, chickpea, clovers, wild rye, lettuce, lentils, alfalfa, forage grasses, horsebean, common vetch, and milk vetch.

9.5.2   NATIONAL SEED STORAGE LABORATORY IN FORT COLLINS, CO

This laboratory houses the base collection for long-term, backup storage of the National Plant Germplasm Storage active collections. It has recently expanded and remodeled its facilities, quadrupling the storage area, and added modern research and processing laboratories. It features quality cold-storage facilities for conventional seed storage and cryopreservation (low-temperature preservation, using liquid nitrogen at –196°C) storage capacity for seeds, pollen, and vegetatively propagated germplasm. The National Seed Storage Laboratory (NSSL) can store more than one million samples. The base collection of the NSSL is not duplicated in its entirety in any other genebank. Furthermore, of the more than 268,000 accessions, about 60,628 are not duplicated at other sites.

9.5.3 INTERNATIONAL RICE RESEARCH INSTITUTE IN LOS BAÑOS, PHILIPPINES

Rice is the third best-represented crop in plant gene banks. This is most likely due to the fact that rice is a staple food crop in much of the Third World, particularly Asia. One of the main gene banks for tropical rice is at the International Rice Research Institute (IRRI). Japan and the U.S. maintain major collections of temperate rices and act as a backup for IRRI and the International Institute for Tropical Agriculture (IITA) materials. IRRI has assembled the world's largest rice collection. It represents the largest germplasm collection for any crop and is regarded as one of the best-managed gene banks. It has computerized rice collection data on samples that contain 45 morphological and agronomic characteristics for each entry. As many as 38 genetic evaluation and utilization traits are added, covering disease and pest resistance to tolerance to adverse soils and climates. Its germplasm collection is gradually regenerated and fresh seed is put in medium and long-term storage. Approximately 2000 rice varieties and much wild material still remains to be collected. The gene bank at IRRI is expected to continue growing until it reaches about 130,000 accessions.[19,19a]

9.5.4 INTERNATIONAL POTATO CENTER IN LIMA, PERU

Potato is the fourth leading world crop, exceeding all others in annual production of starch, protein, and several other important "nutrients".[20] It is susceptible to many diseases and pests and is one of the heaviest users of chemical inputs of any crop.[21] Improved potato cultivars present a great potential benefit to the economic, environmental, and nutritional future of the world potato growers and consumers.[22]

The International Potato Center (CIP) accepted the global mandate for potato genetic resources when it was founded. By 1980, more than 80% of total cultivated potato germplasm had been collected. Wild species of potato have also been systematically collected. The cultivated potato collection samples are grown annually at high altitudes and stored in conservation facilities. Duplicates of all lines are replaced by a new CIP harvest in each succeeding year.[23]

Potato cultivars are distributed worldwide from CIP. Microtubers are more tolerant of physical and environmental disturbances and delays in transit than cultures are.[22] They are now in use for distributing germplasm of potato from CIP and yam from the IITA. The CIP helped to initiate a joint database with potato gene banks around the world by sharing evaluation data and technical procedures, professional exchanges, cooperation on prioritization and organization of collecting expeditions, duplicate storage of accessions, and cooperative research.[22]

9.5.5 CRUCIFER GENETICS CENTER IN MADISON, WI

The Crucifer Genetics Center (CrGC) has been established for the purpose of developing, acquiring, maintaining, and distributing information about seed stocks of various crucifers (members of the cabbage family, Brassicaceae) as well as crucifer-specific symbionts, namely, pathogens (organisms that cause disease in crucifers). It distributes seed from various genetic stocks of rapid-cycling brassicas (short life cycle from seed to seed), some wild crucifer species, a large number of mutants of *Brassica*, *Raphanus* (radish), and *Arabidopsis* (a cress), and pathogen symbiont cultures. The CrGC has been instrumental in introducing rapid cycling brassicas into laboratory teaching experiments for students in elementary and high school and in colleges and universities for the study of plant genetics, development (flowering and fruiting), physiology (gravitropism, phototropism, and hormone action), and plant pathology. One of these plants, *Arabidopsis*, has been shown to develop from seed to seed in outer space on NASA's Space Shuttle. For humans, conservation of crucifer germplasm, as done at the CrGC, is important for humans; many of the brassicas (e.g., broccoli) are important in preventing cancer in humans.

9.5.6 COMMERCIAL SEED COMPANIES THAT SAVE AND SELL HEIRLOOM SEEDS AND SEEDS OF RARE AND ENDANGERED PLANTS

Because of the loss of crop diversity with the advent of the green revolution and the breeding of crop varieties grown as monocultures, we have lost thousands of varieties of plants because they are no longer sold. This has happened with rice, wheat, maize. With the loss of crop diversity, we have also witnessed a loss in disease and insect pest resistance, a loss of protein and essential nutrients in many of the grain crops, a loss in desirable flavor and texture in many vegetables, and an increase in the use of fertilizers, pesticides, and irrigation water. Many of the desirable cultivars of apples and roses, once widely grown, almost completely disappeared from commercial nursery catalogues.

The situation today is changing rapidly. Many of the "old-fashioned" rose cultivars or apple cultivars are now reappearing in the catalogues, primarily driven by consumer demand for more plant diversity and varieties that do not require so much in the way of fertilizer, pesticide, and water inputs. The same can be said for cucurbits (squash and melon), maize, legume crops (peas, beans, and their relatives), herbs, prairie plants, medicinal plants, woodland wild flowers, native trees and shrubs useful in landscaping and in forest restoration projects, aquatic plant species used in ponds to purify polluted water from sewage treatment plants, and species of plants which are good scavengers of heavy metal pollutants in soils. Let us cite just a few examples of sources of seeds of rare and endangered plants.

- Henry Doubleday Institute at Ryton Gardens, Coventry, U.K. has a heritage seed program whereby it distributes heirloom and rare varieties of seed plants that are generally not commercially available. The seed is not registered with the European Community, so it cannot be sold, but it can be donated. We do not know if their seeds are exportable to the U.S.
- The Seed Guild is an organization located in Lanark, U.K. that buys seed from botanical gardens throughout the world, making them available to amateur gardeners and commercial outlets. The Guild provides an opportunity to obtain unusual and rare seeds which are not generally on commercial seed lists. Their annual newsletter provides information on seed collecting expeditions and new sources of seed supply.
- Three commercial seed companies: Redwood Seed Company (P.O. Box 361, Redwood City, CA, 94064) is an alternative seed company; Sandy's Exotic Plant Seed Company (7179A Nebraska, Fairchild, WA 99011) has available rare, exotic, and unusual seeds from around the world; and Prairie Moon Nursery (Route 3, Box 163, Winona, MN 55987) sells seeds of rare ferns, cacti, forbs (herbaceous plants), grasses, sedges, rushes, trees, shrubs, vines, and prairie mixtures.

9.5.7 SEED BANKS IN BOTANICAL GARDENS ESTABLISHED FOR INTERNATIONAL SEED EXCHANGE

The Royal Botanic Gardens Kew Seed Bank, located at Wakehurst Place, U.K., was founded in 1974. It provides storage for seeds of some 4000 plant species from more than 100 countries. It is the most diverse collection anywhere in the world. It also holds a long-term collection of seeds sampled from wild populations within the U.K. and the world's arid and semi-arid lands. Their emphasis is placed on threatened plant populations and in the drylands, especially for plants of local economic value. Some 3750 plant species are conserved according to internationally accepted standards for long-term conservation. When numbers permit, seed is offered for distribution. Samples are made available through a *List of Seeds* published every other year and distributed to organizations doing research work, and subject to a commercialization agreement in the event of any commercial success, a policy of apportioning profits to the seeds' country of origin. This policy aims to abide with the spirit of the 1992 Rio Earth Summit and to keep pace with subsequent changes in national and international attitudes and legislation.[4]

The CPC, located at the Missouri Botanic Garden in St. Louis, MO, maintains a Memorandum of Understanding (MOU) with the U.S. Department of Agriculture NSSL in Ft. Collins, CO. Under this MOU, the NSSL stores seeds from rare U.S. plants in the Center's National Collection of Endangered Plants at no cost to the Center or its participating institutions. The CPC's National Collection of Endangered Plants represents perhaps the most fundamental reserve of plant germplasm for many of the rarest plants in the U.S.

9.6 BOTANICAL PROSPECTING — ETHNOBOTANICAL FIELD RESEARCH

There is a correlation between plant genetic resources and the development of new pharmaceutical products. This correlation integrates biological, ecological, chemical, medical, legal, and economic aspects. The issues can involve property, resource and access right, reciprocity, technology transfer, export, and patent and royalty rights.[23] The force behind biodiversity prospecting is the demand for new genes and chemical compounds and to research the supply of these resources in wildland diversity. Interest has increased in the pharmaceutical industry. Development and improvement of screening techniques has increased the rate for chemical testing. Ethnopharmacology is another force. This field, which involves the use of plants and animals in traditional medicine, can greatly increase the probability of finding a valuable drug. Drug exploration based on indigenous knowledge may prove to be more cost and time effective than random screenings. An example is Shaman Pharmaceuticals, a California-based company, which bases its drug exploration on plants used in traditional medicine.[24]

In the U.S., approximately 25% of prescriptions are of drugs with ingredients that are derived from plant extracts or their derivatives. The demand for genetic resources in agriculture will grow as techniques for genetic manipulation improve and research investments show a return. Between 1985 and 1990, the number of biotechnology patent applications grew by 15% annually.[25] As an example, two drugs derived from the rosy periwinkle (*Catharanthus roseus*), vincristine and vinblastine, alone earned $100 million per drug for Eli Lilly Company.[26] In addition, Paclitaxel (taxol) sold more than $600 million in 1996, and etoposide (from *Podophyllum* or mayapple) more than half that figure.

The stakes in drug development are high and the payoff is uncertain. Finding a valuable compound has a high cost since the probability of locating one with a desired action is low. It is often necessary to test as many as 10,000 substances in order to find one that may reach the drug market.[23] Developing a successful drug can require screening of some 1000 plant species. Research and development cost is generally high, an average of $231 million per drug, with nearly 12 years needed to go from source to market.[27]

International laws directly affect biodiversity prospecting. Intellectual Property Rights and Human and Indigenous Rights are measures to be used for the protection of traditional cultural manifestations (cultivated plants, medicines, and knowledge of useful properties of plants).[18] These laws guarantee rights to participate in the use, management, access, and conservation of these resources and should involve *sharing in the benefits*. The objectives of such laws should include conservation of plant and animal diversity, sustainable development of genetic resources, and the fair and equitable sharing of the resultant benefits.[23]

INBio is a private, nonprofit organization established to facilitate conservation and sustainable use of biodiversity. Other private, nonprofit intermediaries are based in developed countries. In the U.S., for example, the New York Botanical Garden, the Missouri Botanical Garden, and the University of Chicago have all contracted with private pharmaceutical companies and public research organizations to provide samples of biodiversity for pharmaceutical development. It is important that pharmaceutical companies involved in such contracts return an equitable share of their profits from any plant-derived drugs they develop from such plants to the indigenous peoples from whom these plants and the knowledge about their medical uses are obtained. Good role models are provided by Shaman Pharmaceutical Company in South San Francisco, CA and Naniquah Corporation in Girdwood, AK (Figure 9.25).

Essay on Naniquah Corporation

FIGURE 9.25   Dr. Maureen A. McKenzie, Chief Executive Officer, Naniquah Corporation in Girdwood, AK standing next to Michael Parks, President of Naniquah Corporation at site of Medicinal/Fragrance/Culinary Herbs Garden at the University of Michigan Matthaei Botanical Garden. (Photo courtesy of Peter Kaufman.)

Naniquah is an Alaskan corporation dedicated to the discovery of lead compounds for pharmaceutical development from the state's flora and ethnobotanical traditions. Encompassing 1,518,807 sq. km (586,412 sq. mi.), Alaska is one-third the size of the contiguous U.S. Great climatic variations occur in this vast area and include temperatures ranging from –26.7°C (–80°F) in winter to 37.8°C (100°F) in summer, with corresponding total darkness or daylight, and precipitation of less than 15.2 cm (6 in.) in the far north to more than 381 cm (150 in.) in the southeast. Extreme climate and rugged, complex geology make Alaska unforgiving of human occupation and sparsely inhabited (population 650,000), but exert remarkable effects on the flora that present opportunities for indentification of novel natural products.

Noteworthy is a relatively young, but complex, flora (Figures 9.26 through 9.34) characterized by many species that achieve their northernmost range extension in Alaska.[28] Influenced by warm Pacific currents, an old-growth, temperate rainforest, encompassing the Tongass and Chugach National Forests, spans the coastline from the Inside Passage in the southeast through Prince William Sound and the Gulf of Alaska to Kodiak Island. Taiga (spruce-birch) forest dominates the Interior, whereas tundra vegetation covers the cold, arid North Slope, Seward Peninsula, and wetter Western coastal plain. A unique tundra-type vegetation in the Aleutian Islands is created by cool year-round temperatures and ample rainfall. Mountain ranges throughout southern and central Alaska support alpine vegetation, whereas prehistoric glaciation patterns created refugiums, especially in the Yukon Flats, for rare plants that predate the Ice Age.[29] As an anthropological crossroad via Beringia, Alaska harbors plants from different continents in unlikely remote subarctic and arctic habitats.

FIGURE 9.26 Fireweed in Brotherhood Park in Juneau, AK, backdropped by Mendenhall Glacier and the Coast Mountains, Tongass National Forest, Alaska. (Copyright Kim Heacox/Ken Graham Agency.)

Approximately 89% of Alaska's land is government held, and over half of this is allocated to national and state parks, preserves, recreation areas, wildlife refuges, conservation areas, and military installations. Another 10% belongs to Alaska Native corporations, formed under the Alaska Native Claims Settlement Act (ANCSA) of 1971, and the remaining 1% to private interests. Despite some development from petroleum, mining, construction, tourism, fishing, and timber industries, expansive areas remain pristine. Alaska is largely inaccessible by road and, as a result, frontier lifestyles in remote settlements and Native villages coexist with modern economic and technological imperatives.

Unlike other Native Americans and indigenous peoples, Alaska Natives (Figures 9.35 and 9.36) never faced displacement to reservations. The ANCSA settlement with the Federal and State Governments transferred, by tribal demographics, $962.5 million and surface/subsurface rights for 40 million acres of land to 13 regional corporations and more than 200 village corporations.[30] Through birthright, individuals with at least one-quarter Aleut, Athabascan Indian, Tlingit, Haida, Tsimshian, Koniag or Inupiat, Yu'pik, Bering Straits (Siberian), or Chugach Eskimo heritage became shareholders in

FIGURE 9.27   Unalaska Island, AK. (Copyright Dan Parrett/Ken Graham Agency.)

FIGURE 9.28   Old growth temperate rainforest (Sitka spruce, *Picea sitchensis* and western hemlock, *Tsuga heterophylla*) in Kadashan Valley, Chichagof Island, Tongass National Forest in southeast Alaska. (Copyright Kim Heacox/Ken Graham Agency.)

FIGURE 9.29 Skunk cabbage (*Symplocarpus foetidus*) sprouts in boggy woods, Alaska. (Copyright David Job/Ken Graham Agency.)

the regional and village corporations. Two exceptions to ANCSA, a reservation and tribal government, persist today by choice of their members. The main objectives of ANCSA were to secure citizenship for Alaska Natives, with attendant legal rights and responsibilities, and independence from government welfare through economic self-determination.

Although ANCSA corporations are financially successful, Alaska Native lifestyles continue to revolve around traditional subsistence activities, such as seasonal hunting and gathering. Land, with the food and medicinal plants and wildlife it bears, is a rigorously guarded resource of the Native corporations and affiliated tribal councils. The tribal councils affirm and protect cultural and spiritual values, thousands of years old, from continuing erosion through Western lifestyle acculturation and loss of elders.

Perhaps more immediately compelling than plant conservation in Alaska is preservation of the Native peoples' traditional knowledge. Upon Western contact, Alaska Natives were considered generally healthy despite harsh living conditions.[31] The past 30 years of modernization, however, coincide with dramatic increases in diabetes,[32,33] certain cancers,[34,35] and infectious diseases[36] in the Alaska Native population. Thus, acculturation is implicated in the etiology of these conditions, and suggests disease chemopreventive roles for traditional Native foods and health practices. Of 2000 common Alaskan plants, many of the vascular species belong to the Apiaceae, Asteraceae, Betulaceae, Brassicaceae, Fabaceae, Polygonaceae, Rosaceae, and Salicaceae families[28,37] and are rich dietary sources of vitamin, antioxidant, bioflavonoid,

FIGURE 9.30   Devil's club (*Echinopanax horridum*) and black cottonwood (*Populus tri-chocarpa*) in Chugack National Forest, AK. (Copyright Dicon Joseph/Ken Graham Agency.)

sterol, and phytoestrogen compounds. Approximately 50% of the flora represents traditional medicinal plants[31] under investigation for pharmacologically active components relevant to a variety of therapeutic applications.

Naniquah has entered into two types of agreements with ANCSA corporations to achieve its drug discovery objectives through ethnobotany and random screening efforts. In compliance with Alaska law, one exchanges equity in the Company for access to land. The second provides typical royalties generated through patent rights to cooperating tribal entities for disclosures of commercially valuable natural products. Naniquah is also pursuing a collaboration with the U.S. Forest Service to assess the pharmacological value of rare plants or species threatened with extinction in Alaska.

Acknowledgments

The following individuals are gratefully acknowledged for helpful discussions in the preparation of this essay: Dr. Ken Winterberger, U.S. Forest Service, PNW Regional Research Laboratories, Anchorage, AK; Patrick M. Anderson, Esq., Juneau, AK; Carl R. Propes, Jr., MTNT, Ltd., McGrath, AK; and Ashley Schmiedeskamp, Cook Inlet Region, Inc., Anchorage, AK.

FIGURE 9.31   Richardson's saxifrage (*Boykinia richardsonii*) in Denali National Park, Alaska. (Copyright Barbara Brundege/Ken Graham Agency.)

What Does Naniquah Mean?

Naniquah (Singing Bird) was the daughter of Black Hawk, the Sac war chief born in what is now Rock Island, IL, a part of the Quad-Cities metropolitan area. Locally, Chief Black Hawk is revered for the bravery and staunch principles he demonstrated during the War of 1832, the result of a false treaty with the white man. Chief Black Hawk and his family hold sentimental association for the founders of Naniquah who believe that the Sac name for Singing Bird is appropriate for a company dedicated to the treatment of diabetes, cancer, and other devastating conditions. People afflicted with these diseases and their loved ones will indeed have cause to rejoice if Naniquah's efforts to relieve their suffering are successful. The company owes its sincere thanks to Lisa Langdon and Regina Mahieu of the Quad-Cities League of Native Americans and to Muriel Wano and her family, the remaining living descendants of Chief Black Hawk, for their counsel and blessing.

Note: Naniquah Corporation is located in Girdwood, AK. Essay was written by Maureen McKenzie, Ph.D., Chief Executive Officer.

FIGURE 9.32 Pink plumea (*Polygonum bistorta*), a member of the buckwheat family, Polygonaceae, along the Denali highway, Alaska. (Copyright Ken Graham/Ken Graham Agency.)

9.7 CONCLUSIONS

The rainforests remain a major resource of medicinal genetic resources, some of which may yield drugs to treat numerous diseases or symptoms. Approximately 15% of angiosperm species have been examined for their medicinal potential, many of these from temperate and subtropical regions. The majority of tropical plants of the rainforests, which represent about 40% of all angiosperm species, have yet to be studied. Ironically, the most diverse rainforests, those of Latin America, have been least studied, by both aborigines and modern scientists. The rainforests possess an incredible potential for supplying botanical resources and medicinal and nutritional benefits. In the process of realizing this potential, one must adequately address conservation, preservation, sustainable harvesting, socioeconomic development, and indigenous culture issues, to name a few.

FIGURE 9.33 Blooming lingonberry (*Vaccinium vitis-idaea*) and spruce (*Picea* sp.) cones in Denali National Park, Alaska. (Copyright Kim Heacox/Ken Graham Agency.)

FIGURE 9.34 Whitish gentian (*Gentiana algida*) flowers in the tundra near Thorofare Pass, Denali National Park, Alaska. (Copyright Kim Heacox/Ken Graham Agency.)

FIGURE 9.35 Barrow, Alaska. Inupiat Eskimos, Bertha Leavitt (age 80) and her grand-daughter, Nor Del, pick flowers in the Alaskan tundra. (Copyright Ken Graham/Ken Graham Agency.)

Due to the mass extinction process, as a result of rainforest destruction, it is predicted that a large number of plant species will become extinct within the next decade or so. Along with this extinction is the loss for their potential and the alteration of habitat and indigenous cultures. Currently, measures are being taken in these areas, but more needs to be done and at an accelerated rate. Education and support of development and conservation programs are stepping stones to the protection and use of these botanical resources.

> Life in a local site struck down by a passing storm springs back quickly because enough diversity exists. The assembly of life that took a billion years to evolve has eaten the storms — folded them into its genes — and created the world that created us.
>
> E. O. Wilson

FIGURE 9.36 Tlingit grandfather and grandson at Saxman, AK. (Copyright Kim Heacox/Ken Graham Agency.)

REFERENCES

1. Plotkin, M. and Famolare, L., *Sustainable Harvest and Marketing of Rain Forest Products*, Island Press, Washington, D.C., 1992.
2. Moran, K., *Ethnobiology and U.S. Policy. Sustainable Harvest and Marketing of Rain Forest Products*, Plotkin, M. and Famolare, L., Eds., Island Press, Washington, D.C., 1992.
3. Nabhan, G. P., Native plant products from the arid neotropical species: assessing benefits to cultural, environmental, and genetic diversity, in *Sustainable Harvest and Marketing of Rain Forest Products*, Island Press, Washington, D.C., 1992.

4. The Royal Botanic Gardens, Kew. People and Plants; ethnobotany and the sustainable use of plant resources, WWF-UNESCO-Royal Botanic Gardens, Kew, U.K., 1996, http://www.rbgkew.org.uk/ceb/ppi.html.

5. Head, S. and Heismann, R., *Lessons of the Rainforest*, Sierra Club Books, San Francisco, 1990.

6. The Nature Conservancy, Technology for Conservation: the information systems initiative, meeting the needs of conservation in the 21st century, 1996, http://www.tnc.org/news/techcons/index.html.

7. Smith, N., Williams, J. T., Plucknett, D., and Talbot, J., *Tropical Forests and Their Crops*, Cornell University Press, Ithaca, NY, 1992.

8. Jacobs, M., *The Tropical Rain Forest*, Springer-Verlag, Heidelberg, Germany, 1981.

9. Arvigo, R., *Sastun*, Harper/Collins Publishers, San Francisco, CA, 1994.

10. Balick, M. and Cox, P. A., *Plants, People, and Culture: The Science of Ethnobotany*, Scientific American Library, New York, 1996.

11. Arvigo, R. and Balick, M., *Rainforest Remedies*, Lotus Press, Twin Lakes, WI, 1993.

12. Balick, M., Elisabetsky, E., and Laird, S., *Medicinal Resources of the Tropical Forest*, Columbia University Press, New York, 1996.

13. World Wildlife Fund, Annual report, Shepard Poorman Communications Corp., Indianapolis, 1995.

14. Elder, J., The big picture, *Sierra*, March/April, 52–56, 1994.

15. New York Botanical Garden, Endangered and threatened plants at the New York Botanical Garden, 1995, http://www.nybg.org/Gardens/NYBG/Gardens/ndanger.html.

16. Center for Plant Conservation, Missouri Botanical Garden, 1996, http://www.mobot.org/CPC/welcome.html.

17. Royal Botanic Gardens, Kew, Conservation at the Royal Botanic Gardens, Kew, U.K., 1996, http://www.rbgkew.org.uk/conservation/index.html.

18. Akerele, O., Heywood, V., and Synge, H., *Conservation of Medicinal Plants*, Cambridge University Press, Cambridge, U.K., 1991.

19. Plucknett, D. L., Smith, N. J. H., Williams, J. T., and Anishetty, N. M., Genes in the bank, in *Gene Banks and the World's Food*, Princeton University Press, Princeton, NJ, 1987.

19a. Chang, T. T., Personal communication, 1982.

20. Niederhauser, J. S., International cooperation and the role of the potato in feeding the world, *Am. Potato J.*, 70, 385–403, 1993.

21. Martin, M., Potato production and chemical dependency, *Am. Potato J.*, Symp. Suppl., 1–4, 1988.

22. Bamberg, J. G., Huaman, Z., and Hoekstra, R., International cooperation in potato germplasm, *Crop Sci. Soc. of Am.*, 23, 177–182, 1995.

23. Reid, W. V., Laird, S. A., Gámez, R., Sittenfeld, A., Janzen, D. H., Gollin, M. A., and Juma, C., *A New Lease on Life. Biodiversity Prospecting*, a World Resources Institute Book, Library of Congress Catalog Card No. 93-60546, 1993, 1-52.

24. King, S. R., *Conservation and Tropical Medicinal Plant Research. Biodiversity Prospecting*, a World Resources Institute Book, Library of Congress Catalog Card No. 93-60546, 1992, 1-52.

25. Raines, L. J., Protecting Biotechnology's Pioneers, *Issues in Science and Technology*, Winter, 33–39, 1991-92.

26. Farnsworth, N. R., Screening plants for new medicines, in *Biodiversity*, Wilson, E. O. and Peters, F. M., Eds., National Academy Press, Washington, D.C., 1988, 83–97.

27. DiMasi, J. A., Hansen, R. W., Grabowski, H. G., and Lasagna, L., Cost of innovation in the pharmaceutical industry, *J. Health Econ.*, 10, 107–142, 1991.

28. Hulten, E., *Flora of Alaska and the Neighboring Territories. A Manual of the Vascular Plants*, Stanford University Press, Palo Alto, CA, 1968.

29. McDaniel, S., Hunting prehistory: scientists wonder if rare plants in Yukon-Charley Preserve are ice age remnants, *Anchorage Daily News*, July 21, K1-4, 1996.

30. Alaska Native Claims Settlement Act, United States Code, Title 43, Chapter 33, December 18, 1971.

31. Fortuine, R., The use of medicinal plants by Alaska Natives, *AK Med.*, Nov/Dec, 189–223, 1988.

32. Shraer, C. D., Ebbesson, S. O., Boyko, E., Nobmann, E., Adler, A., Cohen, J., Hypertension and diabetes among Siberian Yu'pik Eskimos of St. Lawrence Island, Alaska, *Pub. Health Rep.*, 111 (Suppl. 2), 51–52, 1996.

33. Murphy, N. J., Shraer, C., Thiele, M. C., Boyko, E. J., Bulkow, L. R., Doty, B. J., Lanier, A. P., Dietary change and obesity associated with glucose intolerance in Alaska Natives, *J. Am. Diet. Assoc.*, 95, 676–682, 1995.

34. Baquet, C. R., Native Americans' cancer rates in comparison with peoples of color, *Cancer*, 78 (Suppl. 7), 1538–1544, 1996.

35. Lanier, A. P., Kelly, J. J., Smith, B., Harpster, A. P., Tantilla, H., Amadon, C., Beckworth, D., Key, C., Davidson, A. M., Alaska Native cancer update: incidence rates 1989-1993, *Cancer Epidemiol. Biomarkers Prev.*, 5, 749–751, 1996.

36. Fortuine, R., Early evidence of infections among Alaska Natives, *Al. Hist.*, 2, 39–56, 1985/1986.

37. Welsh, S. L., *Anderson's Flora of Alaska and Adjacent Parts of Canada*, Brigham Young University Press, Provo, UT, 1974.

Appendix:
Information Retrieval on
Natural Products from
Plants

BASIC QUESTIONS TO ASK
by Peter Kaufman

In this book, you will find a great deal of useful and up-to-date information about a vast array of natural products from plants. How does one go about getting more information about a given natural product(s) of interest for a given plant species? We have found it useful to research information about natural products of medicinal value in plants by asking the following questions.

1. Where does this plant grow in nature and how has it been used by indigenous peoples in their traditional medicine?
2. How does one go about propagating and growing this plant in the greenhouse or field so as to be able to do experiments on it, to examine the compounds of interest, to up-regulate the biosynthesis of the compounds of interest, or to simply use it as a "show and tell" on medicinal plants to students, patients, or scientists working on the plant?
3. How does the plant synthesize the compounds(s) of interest; that is, what is the biochemical pathway leading to the synthesis of this compound(s), and what are the enzymes involved at each step in the pathway (if known)?
4. Is it possible to upregulate the biosynthesis of the compound(s) of interest by means of environmental stress treatments, by cultural practices, by herbivory, or by enhancing or turning off gene expression for particular steps in the biosynthetic pathway?
5. Can one mass-produce the compound(s) of interest in plant cell suspension cultures in greenhouse bioreactors, or even in space on board a space station?
6. Do the compound(s) that are of medicinal interest to humans have any special adaptive functions in the plant itself? Do they repel predators? Do they act as poisons? Do they prevent attack by predacious insects or pathogenic bacteria, fungi, or viruses?

0-8493-3134-X/99/$0.00+$.50
© 1999 by CRC Press LLC

7. How do the compounds of interest act at target sites in humans to prevent or arrest a particular disease? Is it a single compound doing the most effective job, or is it due to synergistic action between two or more compounds (chemically related or chemically unrelated) produced by the same plant or coming from two or more plant species?

8. If this plant is rare, endangered, or threatened in its natural habitat, what is being done to save it from extinction and to make it thrive either in the wild or in cultivation? Where are good sources of seeds or propagules of this plant commercially or from seed/plant/tissue culture ("germplasm") "banks".

SURFING THE WEB
By James E. Hoyt

The Internet Web ("the web" or "the net") provides a vast amount of data for biologists. From manuscript to posting on the web is a small step. Word processing documents, spreadsheets, and digital graphics are all easily made available to the Net.

The difficulty with the Net in its current form is two-fold. First, the data must be prepared in machine form. At first glance this may not seem to be a significant problem. Most new data are already collected as computer files, entered into graphing and spreadsheet programs and prepared on word processors. However, data collected and published before the widespread use of computers remains generally unavailable on the Net, and funding to convert these data to machine form is not readily available. The result is that the Net contains recent information, reflecting the current fashions in scientific research, but little from earlier work that may be of possible significance to a researcher's studies.

The second problem is one of finding information on the Net. The Net can be viewed as a world of data with no map. Researchers are required to produce their own "rudders" for navigating. There are some tools for searching keywords such as Yahoo and Alta Vista, but these provide only a starting point for a process that is essentially serendipitous. Librarians trained in searching the Net are extremely important. Net search tools, however, need to mature to the point where researchers can easily extract the desired data with only minimal assistance from information specialists.

The Net also provides data in truly unique and powerful forms not readily available in the past. For example, molecular modeling programs, such as RasMol, provide 3-D models of chemical compounds. A researcher can download molecule databases to her/his desktop computer and examine the molecule in various forms (spacefill, stick and ball, α-helix/β-sheet, etc.) and rotate that molecule about three axes. Viewing data and "interacting" with it in this fashion is more compelling then the 2-D images of conventional print.

SOME MEANS OF ACCESS TO NET INFORMATION

- Catalogs on-line are the oldest and best developed network resource, having been in existence since the mid 1980s. Catalogs are electronic versions of library card catalogs with the added advantage of providing rapid searching by author, subject, and keyword. Many libraries have added other services first available in paper form such as Biological Abstracts and Agricola. Some libraries are providing full-text retrieval systems for select journals, and there is a growing interlibrary network of catalogs that allows the researcher to remotely survey several libraries and place interlibrary loan requests from their home campus.

- Searching tools provide keyword searches across the net. They function like keyword searches for library catalogs, but instead of searching specific book and journal collections, they search for word matches on web pages. The results can often be interesting but not especially relevant to a researcher and, as a result, are usually the start rather then the end of a search process.
- Mail lists and news lists provide "newsletters" focused on specific areas of interest. The major difference between the two is that you must subscribe to a mail list (i.e., it has a limited distribution) while a news list is available to anyone with news reader client software (e.g., News Watcher and Nuntius). The News List includes numerous entries of interest to biologists in the "bionet" and "sci" sections; however, since they are open to all, the individual items can be of variable quality. Mail lists are generally found using search tools.

INDEX TO NAMES OF PLANTS USED IN TEXT:
COMMON AND SCIENTIFIC

INDEX TO NAMES OF CHEMICALS USED IN TEXT

A

Abietadiene cyclases, 77
Abietadiene hydroxylase, 77
Abietadiene synthase, 111, 112
Abietadienol dehydrogenase, 77
Abietadienol hydroxylase, 77
Abietic acid, 13, 76, 77, 103
Abscissic acid, 9, 12, 46, 108
Acemannan, 172
Acetaldehyde, 106
Acetic acid, 76, 256, 258, 265
Acetylcholine, 173, 178
Acetyl-CoA, 46
Acetylenes, 4
Acorone, 12
Acrylamide, 231
Acyclovir, 203
Adenine, 32, 33
Adenosine, 34, 195
Adenosine diphosphate (ADP), 38
Adenosine triphosphate (ATP), 39
s-Adenosylmethionine, 28
Agarose, 228, 231
Aglycones, 15
Agmatine, 28
AIDS cocktail, 203
Ajmalicine, 193
Ajmaline, 32
Ajoene, 194, 195, 198
Alanine, 29, 139
Albumin, 129
Alcohol aldehyde dehydrogenase, 105–106
Alcohol dehydrogenase, 105–106
Alcohols, 106
Aldehydes, 6, 7
Alkaloids, 25, 27, 30–32, 69–71, See specific
        compounds
n-Alkanes, 48
Alkenales, 83
Alkenes, 48
Allicin, 150–151, 195, 198
Alliin, 149–150, 195
Allinase, 150
Allomones, 86
Allophycocyanin, 61
Allopurinol, 198
Allyl isothiocyanate, 196
Allyl methyl thiosulfinate, 198
Allyl methyl trisulfide, 195

Allylic pyrophosphate, 109–110
Aloeemodin, 172
Aloeresin, 172
Aloesin, 172
Aloin, 172, 187
Alum, 255, 256, 258
Aluminum, 67, 84, 152
Amentoflavone, 189
Amides, 38
Amines, 27–28
Amino acids, 28–30
d-Aminobutyric acid, 29
5-Aminolevulinic acid, 61
Amoorastatins, 242
Amorphous silica gel, 41, 53, 55
Amylamine, 85
Amylases, 38, 80
α-Amylase, 38, 104, 108, 140
β-Amylase, 38, 140
Amylopectin, 27, 55–56
Amylose, 27, 55
α-Amyrin, 15
Amyrins, 14
trans-Anethole, 20, 21
Angelic acid, 9
2′-Angeloyl-3′-isovaleryl-vaginate, 193
Aniline, 259
Anthocyanidins, 22
Anthocyanins, 23, 64, 65–69, 92
Anthraquinone, 23, 172
Anthraquinone glycosides, 187
Antioomycete pathogenesis-related protein, 81
Apigenin, 198
Apiin, 198
Apiol, 20, 21
Arabinogalactan, 171
L-Arabinose, 17, 26
α-L-Arabinose, 26
β-L-Arabinose, 26
Arabinoxylans, 78
Arachidic acid, 8
Arachidonic acid, 8
Archangelicin, 192–193
Arginine, 29, 139
5-epi-Aristolochene synthase, 111, 112
Aromatics, 18–25, 85
Artemisinin, 145, 190
Artemitin, 190
Ascorbic acid, 66, 139, 191, 197, 198
Asparagine, 29

INDEX TO BOOK

DATE DUE